HAZARDOUS MATERIALS: REGULATIONS, RESPONSE, AND SITE OPERATIONS

Online Services

Delmar Online
For the latest information on Delmar Publishers new series of Fire, Rescue and Emergency Response products, point your browser to:
   http://www.firesci.com

Online Services

Delmar Online
To access a wide variety of Delmar products and services on the World Wide Web, point your browser to:
   http://www.delmar.com
   or email: info@delmar.com

A service of I(T)P®

HAZARDOUS MATERIALS: REGULATIONS, RESPONSE, AND SITE OPERATIONS

Rob Schnepp

Paul Gantt

Delmar Publishers

an International Thomson Publishing company

Albany • Bonn • Boston • Cincinnati • Detroit • London • Madrid
Melbourne • Mexico City • New York • Pacific Grove • Paris • San Francisco
Singapore • Tokyo • Toronto • Washington

NOTICE TO THE READER

Publisher does not warrant or guarantee any of the products described herein or perform any independent analysis in connection with any of the product information contained herein. Publisher does not assume, and expressly disclaims, any obligation to obtain and include information other than that provided to it by the manufacturer.

The reader is expressly warned to consider and adopt all safety precautions that might be indicated by the activities described herein and to avoid all potential hazards. By following the instructions contained herein, the reader willingly assumes all risks in connection with such instructions.

The publisher makes no representations or warranties of any kind, including but not limited to, the warranties of fitness for particular purpose or merchantability, nor are any such representations implied with respect to the material set forth herein, and the publisher takes no responsibility with respect to such material. The publisher shall not be liable for any special, consequential, or exemplary damages resulting, in whole or in part, from the readers' use of, or reliance upon, this material.

Cover/insert photos courtesy of Hans Pennink

Emergency Management Online College Courses, Disaster Planning: http://beat1.spjc.cc.fl.us/em/em.html

Delmar Staff
Publisher: Alar Elken
Acquisitions Editor: Mark Huth
Developmental Editor: Jeanne Mesick
Production Coordinator: Toni Bolognino

Art and Design Coordinator: Michele Canfield
Editorial Assistant: Dawn Daugherty
Marketing Manager: Mona Caron

COPYRIGHT © 1999
By Delmar Publishers
an International Thomson Publishing Company, Inc.

The ITP logo is a trademark under license.

Printed in the United States of America

For more information, contact:

Delmar Publishers
3 Columbia Circle, Box 15015
Albany, New York 12212-5015

International Thomson Editores
Seneca 53
Colonia Polanco
11560 Mexico D. F. Mexico

International Thomson Publishing Europe
Berkshire House
168-173 High Holborn
London, WC1V7AA
United Kingdom

International Thomson Publishing GmbH
Königswinterer Straße 418
53227 Bonn
Germany

Nelson ITP, Australia
102 Dodds Street
South Melbourne,
Victoria, 3205 Australia

International Thomson Publishing Asia
60 Albert Street
#15-01 Albert Complex
Singapore 189969

Nelson Canada
1120 Birchmont Road
Scarborough, Ontario
M1K 5G4, Canada

International Thomson Publishing Japan
Hirakawa-cho Kyowa Building, 3F
2-2-1 Hirakawa-cho, Chiyoda-ku,
Tokyo 102, Japan

International Thomson Publishing France
Tour Maine-Montparnasse
33 Avenue du Maine
75755 Paris Cedex 15, France

ITE Spain/ Paraninfo
Calle Magallanes, 25
28015-Madrid, Espana

All rights reserved. No part of this work covered by the copyright hereon may be reproduced or used in any form or by any means—graphic, electronic, or mechanical, including photocopying, recording, taping, or information storage and retrieval systems—without written permission of the publisher.

1 2 3 4 5 6 7 8 9 10 XXX 03 02 01 00 99 98

Library of Congress Cataloging-in-Publication Data
Schnepp, Rob, 1961–
    Hazardous materials : regulations, response, and site operations/
  by Rob Schnepp and Paul Gantt.
      p.   cm.
    Includes index.
    ISBN 0-8273-7999-4 (hardcover : alk, paper)
    1. Hazardous substances—Handbooks, manuals, etc.   I. Gantt, Paul, 1953–
II. Title.
T55.3.H3S36. 1998
604.7—dc21   97-44660
                                                CIP

Contents

Preface — ix
Acknowledgments — xiii
About the Author — xv

Chapter 1 Hazardous Materials Regulatory Overview — 1

INTRODUCTION/2 ■ OSHA/3 ■ HAZARD COMMUNICATION STANDARD (29 CFR 1910.120)/5 ■ HAZWOPER REGULATIONS/12 ■ SUMMARY/33 ■ REVIEW QUESTIONS/34

Chapter 2 Principles of Chemistry — 35

INTRODUCTION/36 ■ CHEMISTRY AND MATTER DEFINED/37 ■ THE PERIODIC TABLE/43 ■ COMPOUNDS/51 ■ PHYSICAL AND CHEMICAL CHANGE/53 ■ EXOTHERMIC AND ENDOTHERMIC REACTIONS/54 ■ BASIC GAS LAWS/55 ■ SUMMARY/60 ■ REVIEW QUESTIONS/61

Chapter 3 Hazard Classes — 63

INTRODUCTION/64 ■ EXPLOSIVES/65 ■ GASES/73 ■ FLAMMABLE LIQUIDS/87 ■ FLAMMABLE SOLIDS/96 ■ OXIDIZERS AND ORGANIC PEROXIDES/99 ■ POISONS, PESTICIDES, AND CARCINOGENS/105 ■ RADIOACTIVES/116

■ CORROSIVES/122 ■ SUMMARY/135 ■ REVIEW QUESTIONS/137

Chapter 4 Principles of Toxicology — 141

INTRODUCTION/142 ■ BACKGROUND/143 ■ EXPOSURE MECHANISMS/144 ■ EFFECTS OF CHEMICAL EXPOSURE/158 ■ EXPOSURE LIMITS/161 ■ TOXICOLOGY TERMINOLOGY/163 ■ SUMMARY/166 ■ REVIEW QUESTIONS/168

Chapter 5 Hazardous Materials Identification Systems — 171

INTRODUCTION/172 ■ DOT AND NFPA IDENTIFICATION SYSTEMS/173 ■ HAZARDOUS MATERIALS IDENTIFICATION SYSTEMS/185 ■ SHIPPING PAPERS AND HAZARDOUS WASTE MANIFESTS/186 ■ PENALTIES/190 ■ DOT NORTH AMERICAN EMERGENCY RESPONSE GUIDEBOOK/190 ■ DRUM PROFILES/197 ■ SUMMARY/203 ■ REVIEW QUESTIONS/SCENARIO/204

Chapter 6 Respiratory Protection — 205

INTRODUCTION/206 ■ RESPIRATORY HAZARDS/206 ■ RESPIRATORY PROTECTION FUNDAMENTALS/208 ■ RESPIRATORY PROTECTION EQUIPMENT/210 ■ TYPES OF AIR PURIFYING SYSTEMS/215 ■ AIR SUPPLYING RESPIRATORS/222 ■ RESPIRATORY PROTECTION: MAINTENANCE, STORAGE AND RECORD KEEPING OF APRS AND SCBA/230 ■ SUMMARY/232 ■ REVIEW QUESTIONS/234

Chapter 7 Personal Protective Equipment — 235

INTRODUCTION/236 ■ POSITIVE IDENTIFICATION OF PRODUCTS AND ASSOCIATED HAZARDS/237 ■ TYPE AND LEVEL OF PROTECTION/240 ■ HEALTH CONSIDERATIONS AND CHEMICAL PROTECTIVE CLOTHING/261 ■ SUMMARY/266 ■ REVIEW QUESTIONS/270 ■ COMPATIBILITY CHART EXERCISE 271

Chapter 8   Principles of Decontamination   273

INTRODUCTION/274 ■ METHODS OF DECONTAMINATION/276 ■ SITE SELECTION AND MANAGEMENT/278 ■ DECONTAMINATION PROCEDURES/283 ■ SUMMARY/291 ■ REVIEW QUESTIONS/292

Chapter 9   Incident Management and Scene Control   293

INTRODUCTION/294 ■ THE INCIDENT COMMANDER/294 ■ THE ROLE OF THE INCIDENT COMMANDER/296 ■ CHARACTERISTICS OF THE ICS/301 ■ TAKING CHARGE FROM THE BEGINNING/308 ■ ESTABLISHING SAFE WORKING AREAS—ZONES/310 ■ OVERVIEW OF JOB TITLES AND DESCRIPTIONS/313 ■ INCIDENT TERMINATION/317 ■ SUMMARY/320 ■ REVIEW QUESTIONS/321

Chapter 10   Planning   323

INTRODUCTION/324 ■ PLANNING FOR HAZARDOUS SUBSTANCES RELEASE EMERGENCIES/324 ■ EMERGENCY PLANS/325 ■ HAZARD COMMUNICATION/327 ■ OSHA PROCESS SAFETY MANAGEMENT/327 ■ ELEMENTS OF THE EMERGENCY RESPONSE PLAN/333 ■ SAMPLE INCIDENT-SPECIFIC SITE SAFETY PLAN/334 ■ SUMMARY/339 ■ REVIEW QUESTIONS/342

Chapter 11   Air and Environmental Monitoring   343

INTRODUCTION/344 ■ GAS LAWS/345 ■ DETECTION OF GASES/347 ■ THE SEQUENCE OF AIR MONITORING/349 ■ THE USE OF AIR MONITORING EQUIPMENT/349 ■ GENERAL PROCEDURES FOR AIR MONITORING/357 ■ MONITORING IN CONFINED SPACES/358 ■ MONITORING OUTSIDE/358 ■ THE EFFECTS OF LOCAL WEATHER/359 ■ SUMMARY/361 ■ REVIEW QUESTIONS/SCENARIO/362

Chapter 12 Physical Hazards — **363**

INTRODUCTION/364 ■ CONFINED SPACE/364 ■ NOISE HAZARDS/378 ■ CONTROL OF HAZARDOUS ENERGY (LOCKOUT/TAG-OUT)/378 ■ SUMMARY/382 ■ REVIEW QUESTIONS/385

Appendix A State Plan States — **387**

Appendix B Human Carcinogens — **391**

Appendix C Suggested Reading — **395**

Appendix D Emergency Response-Related Internet Directory — **397**

Acronyms — **399**

Glossary — **403**

Index — **435**

Preface

Welcome to the weird and wonderful world of hazardous materials! You are about to enter a field where the work can be scary, complicated, hot, sweaty, and potentially dangerous. Sounds great doesn't it? The truth is that if you are looking for a glamorous job, this is not it! Hazardous materials work can be very trying at times and may have some serious consequences for poor performers. If you are looking for an enormously challenging field with unlimited room for personal development then you have come to the right place. This field can also provide some very rewarding moments, but most of those are recognized quietly by those who are doing the job. Once you get out there in the suit, this point will become a whole lot more obvious. You have chosen to receive some training in a diverse and dynamic field. It is our hope that you are approaching this training with a sincere commitment and determination to understand what you are getting into. Successful completion of the course you are enrolled in does not mean you have reached the end of your hazmat journey; It only means that you have gotten your ticket punched and are allowed to *begin*. The real schooling starts your first day on the job and does not stop until you complete your final day and retire. Every day between those two landmarks should be spent learning something about your profession.

This personal commitment to going above and beyond is especially important in the hazardous materials response field. The reason is quite simple—significant hazmat incidents are relatively few and far between. Even busy metropolitan fire departments run medical and fire calls way more often than significant hazmat responses. Because we do not run "the big one" every day or even every year, it is that much more important to stay in touch and practice your trade. It is with this thinking that we offer this textbook. Our goal is to arm you with the basic information required to begin working in hazmat response or the waste site industry. The two fields are closely related and the basic concepts in either are very similar. Respiratory protection, for example, is the same whether you are at a superfund cleanup site or a rolled over tanker on the freeway. So is this the end-all, be-all textbook for hazardous materials? As much as we would like to think it is, in reality it could never be. Such a text could never be written. This textbook and others like it do not contain all the information you will ever need to work in this field. There are just too many variables, and too many idiosyncrasies, and too many different ways of doing things to put them all in one book.

Essentially, we put this book together because there appeared to be something missing from the current texts on the market. They seemed to be outdated and were difficult to read and follow. Our goal was to write a book that was user friendly, enjoyable to read, and had a little humor in it. So why did we write another book that is not going to be the final answer in hazardous materials? Because we believe you should understand the concepts and reasons why things are done, which will result in better retention. We do not necessarily subscribe to a checklist format for working in this field because each and every chemical response or cleanup scenario is somewhat different than any another. For this reason a person working in the field must truly understand what he or she is doing and be able to think creatively. It is necessary to be able to *apply* what has been learned rather than simply remember it. After years of working in the response field and more years of teaching hazmat, we feel that this book gives you the tools you really need to be successful in the field. Our belief is that if you truly apply what you learn from this book you can work in the hazmat field safely and with confidence.

After training thousands of people in this field, we have come to realize that there are two kinds of students. The first type of student muddles through the class and does just enough to get by. Once he or she has graduated, they manage to survive by doing only the minimum and do little to expand their base of knowledge. They are the ones you see on the scene constantly fumbling with equipment and never taking charge of anything. Essentially they have become a liability because they offer very little to the effort. Is this the person you want as a partner when the situation gets extreme? Sadly, there are a lot of these folks around and you may even be working with some right now.

The other type of student is very interested in the training and takes every opportunity to become involved. After graduation, they stay current in the field by reviewing learned concepts and striking out on their own for new information. These students invest in their career by learning as much as possible after being trained and strive to stay out in front of the job by staying involved. These students ultimately become better equipped to perform when they are called upon. It is imperative that you take responsibility and ownership in your career and stay current. Review and refresh your proficiency with equipment and procedures while continually adding new information to your base of knowledge.

The book itself is quite simple to use. Each chapter is full of good solid technical data along with key Safety and Note information in the margins. These special highlights represent information we feel is truly important to remember. Don't cheat and only read the highlighted material though, because you'll miss out on the other good information in the text. We have also included some stories and examples that should help reinforce the learning points. We have included high quality photos and other illustrations that are as current as we could find or produce ourselves.

The summaries at the end of each chapter are bullet summaries that refresh the concepts learned in the preceding chapter. The short sentence format puts this information in digestible bites and allows you to review the entire chapter in a short amount of time. We recommend you look over the summary before actually reading the chapter. This will help to prime the brain for the important concepts you will encounter in the reading.

This book is not intended to be the final authority on chemistry, toxicology, or any other subject found in the text. Once again, we are bringing out the important concepts of the entire job function, not attempting to make you an expert in any one particular subject. One of the strong points of the text, however, is that the technical information is simplified down to understandable concepts. We also encourage you to do the dreaded student exercises and quizzes found at the end of the chapters. As much as most people try to avoid them they really are effective in reinforcing the learning points.

Finally, we would like to invite you to make any comments or suggestions for improving the text. We are certainly aware that there are many ways to do a job and we welcome any ideas that would improve the next revision. Good luck in your training and we hope you enjoy reading the book as much as we enjoyed writing it!

Wow! Here is a guy who never had a single aspiration to become an author suddenly writing acknowledgments in a book. Talk about having some unexpected turns in the path of life! All in all it was a good experience but certainly one of the largest and most challenging things I have ever been part of. A significant point about a project this size is that it becomes way larger than any single person could ever be. This project was completed because of a team effort and there are some important people that I would like to thank.

First of all, without the love and support of the best family on the planet I would have accomplished nothing. My poor wife and kids have been there throughout my crazy fire department career, paramedic school, having me fly over all over the country to teach, and now this book. They all have a tremendous amount of patience and my wife Dalynn just smiles and tells me not to worry about anything. My kids, Kate and Andrew, are a constant source of inspiration to me and both have an incredible sense of humor. They always seemed to keep me smiling when I secretly wanted to jump off a bridge.

I would also like to say thanks to good ol' Mom and Dad. They had to put up with me as a kid and read those same comments from my teachers year after year, "Rob just doesn't seem to apply himself in school. . .".

Finally, to all my friends and family who were so kind in asking "How's that book coming along?" I can finally answer "Done!"

R. S.

Without a doubt, my part in making this book a reality would not have been possible without the support of my family, including my very supportive wife and children. Another group that I would like to acknowledge are those who have worked with me throughout my career in this field. To those students and colleagues with whom I interact on a regular basis I thank you and wish you all the best.

P. G.

Acknowledgments

Certain instructors in every field feel that in one way or another they have "written the book" in their respective discipline. We also felt strongly about our field and had every intention of writing a book some day. What we didn't realize when we began this process was just how difficult a task it would be and the amount of time that would be needed to actually bring it to fruition.

Much like handling a hazardous materials incident, writing a book is largely a team effort that requires considerable dedication, hard work, and a willingness to be flexible. As we wrote the book we realized that the field was changing on an almost daily basis and that our approach to training changed as well. We also realized that we learn from our students and those other professionals with whom we work. We continue the learning as we stay active in the hazardous materials field. For this reason, we wish to thank the following people for their work in helping to make this project a success.

The team of instructors and professionals at Safety Compliance Management not only contributed to our continued learning of new ideas and ways to make learning hazardous materials fun and effective, but also contributed directly to the ideas and information in the book. This team includes Denise Barndt, the one who most clearly understands the benefits of planning ahead and looking down the road at the big picture; Mark Rea, one of the most prolific hazardous materials trainers around, whose unique approach to training includes the use of historical examples to underscore critical points; Michael Francis, the medical expert for the firm, contributed by helping to better explain the importance of reducing and minimizing the health effects of chemical exposures; and lastly, Cynthia Janiger and Laura Gantt provided the countless hours of administrative support so necessary in hunting down the pictures, files, and other documents that help to make your learning experience a success.

We also want to thank Jeanne Mesick. This woman is part saint and part bounty hunter. She could find us and shake us down for information no matter where we tried to hide. Without her assistance (threats) and patience this textbook would no doubt be nonexistent.

Finally, it would be inappropriate not to acknowledge the huge contribution that has been made by the thousands of students with whom we have interacted. In almost every class we learn something new from them. Yes, hazardous materials is an ever-changing field. No one will ever know it all, and we must keep learning. When we as instructors learn from our students and have fun in the process, we know that we have been successful in our efforts. We wish you success in your learning experience.

Special thanks to Chief Doug Barnes of the Menands Fire Department, members of the West Albany Fire Department Hazardous Materials Team and all those who assisted us with the photographs for the color insert. Additionally, the authors and Delmar Publishers would like to thank the following reviewers, individuals, and companies who contributed to this textbook:

Chris Hawley, Hazardous Materials Specialist
Baltimore County Fire Department
Towson, MD

Walter Patterson, Fire Safety and Environmental Manager
Edinboro University
Edinboro, PA

Larry Ceretto, Assistant Dean, Fire Science
Milwaukee Area Technical College, South Campus
Oak Creek, WI

Mike Connors
Naperville Fire Department
Naperville, IL

Mike Bucy
Emergency Media Services
Valparaiso, IN

William Tyler
EIC Environmental Health and Safety
Everett, WA

Captain John Pangborn, Chief Training Officer
Jersey City Fire Department
Jersey City, NJ

Ron Wakeham, Adjunct Instructor
National Fire Academy
Emmitsburg, MD

Larry Collins
Eastern Kentucky University
Richmond, KY

James Madden
Lake Superior State University
Sault Ste. Marie, MI

James Mueller, Adjunct Instructor
Jersey City Fire Department
Jersey City, NJ

Clinton Smoke
Northern Virginia Community College
Annandale, VA

Air Products Corporation
Allentown, PA

North Safety Products
Charleston, SC

Mine Safety Appliances
Pittsburgh, PA

John McClintic
Firefighter/Paramedic
Alameda County Fire Department
California Bay Area

Chief Bill McCammon
Alameda County Fire Department
California Bay Area

About the Author

Paul Gantt has served as the President of Safety Compliance Management, Inc. (SCM) for over eight years. Paul regularly develops and delivers a variety of occupational and environmental services ranging from training programs to consulting and expert witness. Paul is a Class I Registered Environmental Assessor with the California EPA. His areas of expertise include OSHA regulations and compliance, hazardous materials handling/response, and disaster preparedness. He also served 15 years in the fire service during which time he held the positions of Firefighter/Paramedic, Fire Captain, Fire Marshal, Battalion Chief, Deputy Chief, and Fire Chief.

Chapter 1

Hazardous Materials Regulatory Overview

Learning Objectives

Upon completion of this chapter, you should be able to:

- Identify the relationship between the following regulations and the HAZWOPER regulation:
  - Resource Conservation and Recovery Act (RCRA).
  - Comprehensive Environmental Response, Compensation, and Liability Act (CERCLA).
  - Superfund Amendment and Reauthorization Act (SARA).
- Identify the types of issues that lead to the promulgation of the HAZWOPER regulation.
- Identify the major components of 29 CFR, Part 1910.120—Hazardous Waste Operations and Emergency Response (HAZWOPER).
- Describe the differences between the requirements for operations at hazardous waste cleanup sites, treatment storage and disposal facilities (TSDFs), and emergency response activities.
- Identify other applicable regulations related to employee health and safety at hazardous waste cleanup sites.
- Describe the applicable training programs required for workers involved in hazardous waste cleanup or emergency response activities.

INTRODUCTION

Before any study of hazardous waste training programs can occur, we must understand the background of what we will call the regulatory drivers. Regulatory drivers are those core regulations that provide the reason that we do certain things. It would be nice to believe that everyone, whether in business or government does things because they are the "right" things to do; however, this is not the case. All too often the only reason that certain things occur is that there is a regulatory mandate, or, as we have previously referred to it, a regulatory driver, that causes things to happen.

In the case of hazardous waste handling and response, the main regulatory driver is the HAZWOPER regulation that was passed by the federal Occupational Safety and Health Administration (OSHA) in its final form in March 1989. HAZWOPER became effective in March, 1990 and was placed into the Code of Federal Regulations, book number 29, Part 1910.120. A comprehensive law enacted in response to conditions that were taking place in the mid- to late 1980s, HAZWOPER actually stands for hazardous waste operations and emergency response and is commonly referred to as 29 CFR, Part 1910.120.

If we are to understand the *why* of the HAZWOPER law, we should look closely into what was going on at the time that it was proposed. This would involve a careful study of the early and mid-1980s when there was an increasing public awareness of *toxic* materials in the environment. The environmental consciousness of the nation was emerging. This increased level of concern over the environment and its *toxic* hazards, coupled with the increased federalism that was taking place in the country, led to the development of a series of regulations to limit or control the production of so-called bad materials, as well as to limit their spread into the environment. The following is a partial listing of the regulations that were developed during this period.

- Clean Air Act
- Clean Water Act
- Consumer Product Safety Act
- Federal Food, Drug and Cosmetic Act
- Flammable Fabrics Act
- Federal Hazardous Substance Act
- Federal Insecticide, Fungicide, and Rodenticide Act
- Federal Water Pollution Control Act
- Hazardous Materials Transportation Act
- Occupational Safety and Health Act
- Poison Prevention and Packaging Act
- Resource Conservation and Recovery Act
- Superfund Amendments and Reauthorization Act

■ NOTE
HAZWOPER became effective in March 1990 and was placed into the Code of Federal Regulations, book number 29, part 1910.120.

- Safe Drinking Water Act
- Toxic Substance Control Act

In this era of environmental awareness a regulation known as RCRA was enacted to promote development of federal and state programs for otherwise unregulated land disposal of waste materials. It regulates anyone engaged in the creation, transportation, treatment, and disposal of hazardous waste and mandates that specific handling and disposal procedures be followed when working with hazardous wastes. It was one of the precursors to the HAZWOPER regulation.

The Comprehensive Environmental Response, Compensation, and Liability Act of 1980

A second major piece of regulation worthy of some discussion is the Comprehensive Environmental Response, Compensation, and Liability Act of 1980. More commonly known as CERCLA or *Superfund,* this law provided for the cleanup and proper reclamation of the pre-RCRA abandoned hazardous waste sites. Additionally it required that parties immediately remove the hazardous substances released into the environment. Because CERCLA mandated Superfund cleanup of specific sites, it too was one of the reasons for the development of the HAZWOPER regulation.

The Superfund Amendments and Reauthorization Act of 1986

> **SARA**
> Superfund Amendments and Reauthorization Act of 1986

As a result of the issues associated with the passage of CERCLA, the most significant piece of regulation relative to the HAZWOPER regulation was the Superfund Amendments and Reauthorization Act of 1986. This act, known as **SARA,** was the actual forerunner to the HAZWOPER regulation in that it mandated OSHA to develop worker protection regulations for persons who work with hazardous wastes. Those who enacted this law felt so strongly about the need to have such regulations that they also closed one of the loopholes with the OSHA regulations by also requiring the Environmental Protection Agency (EPA) to pass an identical regulation for workers not otherwise covered by the OSHA regulations.

OSHA

In order to fully understand the HAZWOPER law, we must understand the agency responsible for its enforcement—OSHA. OSHA was enacted with the passing of the Occupational Safety and Health Act in 1970. Although some regulations had been enacted prior to this, the act was the first real attempt by the federal government to significantly impact worker safety.

The OSHA law requires that "each employer shall furnish to each of his employees employment and a place of employment which are free from recognized hazards that are causing or are likely to cause death or serious physical harm to his employees." This condition is accomplished through a series of continually evolving mechanisms, the most obvious of which is the ability of OSHA to develop regulations that cover various aspects of employment. At the base of the regulations is the requirement that all employers must do the following:

- Provide competent inspection of each work site
- Prohibit the use of unsafe equipment by employees
- Require trained, experienced equipment operators to use the various equipment
- Instruct employees in hazard recognition and avoidance
- Instruct employees in pertinent regulations and procedures

In making these mandates, OSHA also states that the legal responsibility for ensuring a safe and healthy workplace cannot be shifted through insurance or other forms of substitution. There is no option but to comply. If compliance is not achieved, the OSHA regulations allow for the enforcement of these mandates through such actions as safety audits and inspections, warnings, fines and penalties, and imprisonment for flagrant, serious, or repeated violations.

Another important point that needs to be discussed when attempting to understand the powers and authority of OSHA is that federal OSHA does not have jurisdiction in all areas of the country. Although federal OSHA has the overall responsibility for occupational safety and health within the federal Department of Labor, it does not have primary responsibility for worker safety in some states that have their own plans for protecting employees within that state. Such states with their own plans are known as **state plan states**.

state plan states
the concept of the state plan states is that the various levels of government work together to enact regulations that are appropriate for every area of the country

The Constitution allows the various levels of government to develop regulations that take into account more local issues and problems. This is the concept of local control or home rule. It takes place from the federal level, to the state, down to the regions within a state such as counties, and eventually to the local level.

In order to better understand this concept, consider this. Although the federal government has responsibility for the entire country, it cannot pass regulations that are most appropriate for the entire country, given the diversity that exists. Take for example speed laws. The federal laws are broadest and represent the maximum level. Each state, county, and local government is free to adopt more restrictive speed laws. Note that the laws passed at each of the lower levels must be at least as restrictive as the federal mandates. Laws should not contradict others from higher levels of government.

As with speed laws, if the states choose to have laws more restrictive than the federal mandate, they may do so. If an individual city wants to adopt even lower speeds, it may do so. In this case, at no time would passing a lower speed limit contradict the federal mandate.

Such is the case with many of the states that are considered state plan states relative to OSHA. In these cases, the states have the option of adopting the federal regulations after they are passed, or they might choose to review them and then issue more stringent requirements based on the conditions in that state.

Chapter 1 Hazardous Materials Regulatory Overview

■ **NOTE**
At no time can a state issue or enforce less restrictive regulations than the federal version because this would result in treatment of citizens in certain states that would be less than the federally mandated laws.

■ **NOTE**
Approximately half of the states have adopted their own OSHA requirements in which the state, not the federal government, regulates worker health and safety within that state.

■ **NOTE**
State plan states can pass more restrictive laws than those passed by federal OSHA.

At no time can a state issue or enforce less restrictive regulations than the federal version because this would result in treatment of citizens in certain states that would be less than the federally mandated laws. Approximately half of the states have adopted their own OSHA requirements in which the state, not the federal government, regulates worker health and safety within that state. See Appendix A for a list of state plan states.

Another important point regarding state plan states is that the state is not only responsible for enacting regulations, but also for enforcing them. It may be wise for each of us to review whether the state we are working or living in is a state plan state because it will be that agency, and not federal OSHA that will ultimately have jurisdiction. State plan states can pass more restrictive laws than those passed by federal OSHA.

In the case of HAZWOPER laws, most states have adopted the federal law in a form that is very close to the original federal version. Even in California, which we know to be a state plan state, as well as one that has a very high level of environmental consciousness, the HAZWOPER law is more or less in line with the federal version. Therefore, it is safe to use the federal version of the regulations, found in 29 CFR, Part 1910.120, as the basis for our study.

A final issue that needs to be understood regarding the OSHA's jurisdiction is that federal OSHA regulations do not necessarily apply to all aspects of public employment. Even in non state plan states, federal OSHA regulations are not always applicable to employees of the federal or state governments in those states. We discussed this issue in our earlier discussion of SARA. For this reason EPA, which has jurisdiction, was also instructed to pass an identical regulation to cover those employees who might otherwise not be covered.

While this is the case, when it came to enacting the HAZWOPER regulation, the writers of SARA Title 3, the law that required OSHA to develop the HAZWOPER regulation in the first place, recognized the importance of a workplace safety law covering hazardous waste handling and response. In this instance, not only was OSHA told to develop the HAZWOPER regulations, but it also mandated that the U.S. EPA also develop similar regulations to cover those public employees who might otherwise not be covered by the OSHA regulation.

HAZARD COMMUNICATION STANDARD (29 CFR 1910.120)

Prior to enacting the HAZWOPER regulations, what other regulations were in effect that related to the issues of worker safety? This question is often asked and is well founded given the high number of regulations that were enacted in the 1970s and 1980s.

If we review the previous list of laws passed during this flurry of activity, we find very few have any mention of worker safety. Many dealt with public protection, as in the case of the Clean Air Act, the Clean Water Act, and others. Only in the Occupational Safety and Health Act, do we find any efforts focused toward protecting the worker from environmental hazards.

> **NOTE**
> This series of regulations, more commonly known as the "employee right to know" laws have been around since the 1970s and have undergone significant expansion and modification.

In reviewing the OSHA regulations, we find that one of the first worker safety laws that related to the hazards of chemicals and other materials was known as the Hazard Communication Standard. This series of regulations, more commonly known as the "employee right to know" laws have been around since the 1970s and have undergone significant expansion and modification. The rationale for these laws is that employees who work with materials in the workplace, have a *right to know* about the hazards of the materials with which they work. Since many of these hazards are invisible to them, or very complicated at best, employers are required to do a number of things to ensure that the employees are provided with information relative to the hazards. Some of these requirements include:

1. Identify and list hazardous chemicals used in the workplace. Before any definitive programs can be enacted in the workplace to protect the workers, the hazards must first be identified. This identification must take place at the onset of the program and continue on a regular basis to ensure that the identification of the materials used in the facility are maintained up-to-date.

2. Obtain material safety data sheets on all hazardous chemicals. Once the hazards have been identified, it is necessary to review each of them for the specific problems that might be associated with them. This is accomplished through the use of a **material safety data sheet** (MSDS). Each manufacturer or importer of the chemical is required by law to provide a copy of the MSDS to those who use the product. The information contained on an MSDS is standardized, although the format for the material varies significantly between manufacturers. Figure 1-1 shows a sample MSDS. Not all MSDSs are created equal and in many cases they will look different depending on the agency that created it. Take some time and review several different MSDS sheets to fully understand the differences between them.

material safety data sheet (MSDS)
written information on a specific compound that expresses such items as physical hazards, signs and symptoms of exposure, toxicology information, and other pertinent data

3. Ensure all hazardous chemicals are labeled properly. Proper labeling of the material is critical if the employee is to recognize the material as hazardous and take the appropriate action. The labeling requirements include a variety of warnings and systems such as that used by the Department of Transportation.

4. Develop and implement a written hazard communication program. Written programs are the foundation of many OSHA regulations. Written programs force each employer to commit to various actions required by OSHA and have the advantage of providing a mechanism whereby individual employees, or labor organizations, can monitor whether the activities listed are being carried out. After the items have been identified and MSDSs have been obtained, the employer is required to develop a written plan to disseminate the information and provide for the maintenance of the program.

5. Train workers to recognize the hazards associated with chemicals used in the workplace. The foundation of the training program is that workers not

Chapter 1   Hazardous Materials Regulatory Overview

GASES AND EQUIPMENT GROUP
MATERIAL SAFETY DATA SHEET

SECTION 1: PRODUCT IDENTIFICATION

PRODUCT NAME: Argon, compressed
CHEMICAL NAME: Argon   **FORMULA:** Ar
SYNONYMS: Argon gas, Gaseous argon, GAR
MANUFACTURER: Air Products and Chemicals, Inc.
7201 Hamilton Boulevard
Allentown, PA 18195 - 1501
PRODUCT INFORMATION: 1 - 800 - 752 - 1597
MSDS NUMBER: 1004   **EFFECTIVE DATE:** August 1997   **REVISION:** 5

SECTION 2: COMPOSITION - INFORMATION ON INGREDIENTS

Argon is sold as pure product > 99%.
CAS NUMBER: 7440-37-1
EXPOSURE LIMITS:
   OSHA: Not established.
   ACGIH: Simple asphyxiant
   NIOSH: Not established.

SECTION 3: HAZARDS IDENTIFICATION

EMERGENCY OVERVIEW

Argon is a nontoxic, odorless, colorless, nonflammable gas stored in cylinders at high pressure. It can cause rapid suffocation when concentrations are sufficient to reduce oxygen levels below 19.5%. It is heavier than air and may concentrate in low areas. Self Contained Breathing Apparatus (SCBA) may be required.

EMERGENCY TELEPHONE NUMBERS
800 - 523 - 9374   in Continental U.S., Canada and Puerto Rico
610 - 481 - 7711   outside U.S.

POTENTIAL HEALTH EFFECTS INFORMATION
   INHALATION: Simple asphyxiant. Argon is nontoxic, but may cause suffocation by displacing the oxygen in air. Lack of sufficient oxygen can cause serious injury or death.
   EYE CONTACT: No adverse effect.
   SKIN CONTACT: No adverse effect.
   CARCINOGENIC POTENTIAL: Argon is not listed as a carcinogen or potential carcinogen by NTP, IARC, or OSHA Subpart Z.
EXPOSURE INFORMATION
   ROUTE OF ENTRY: Inhalation

MSDS # 1004   Argon   1 of 5

Figure 1-1   *A sample portion of a typical material safety data sheet.* Courtesy of Air Products and Chemicals, Inc. Presented for illustrative purposes only. Please contact your supplier for a current revision.

TARGET ORGANS: None
EFFECT: Asphyxiation (suffocation)
MEDICAL CONDITIONS AGGRAVATED BY OVEREXPOSURE: None
SYMPTOMS: Exposure to an oxygen deficient atmosphere (<19.5%) may cause dizziness, drowsiness, nausea, vomiting, excess salivation, diminished mental alertness, loss of consciousness and death. Exposure to atmospheres containing 8-10% or less oxygen will bring about unconsciousness without warning and so quickly that the individuals cannot help themselves.

SECTION 4: FIRST AID

INHALATION: Persons suffering from lack of oxygen should be moved to fresh air. If victim is not breathing, administer artificial respiration. If breathing is difficult, administer oxygen. Obtain prompt medical attention.
EYE CONTACT: Not applicable.
SKIN CONTACT: Not applicable.

SECTION 5: FIRE AND EXPLOSION

FLASH POINT	AUTOIGNITION TEMP	FLAMMABLE LIMIT
N/A	Nonflammable	Nonflammable

EXTINGUISHING MEDIA: Argon is nonflammable and does not support combustion. Use extinguishing media appropriate for the surrounding fire.
HAZARDOUS COMBUSTION PRODUCTS: None
SPECIAL FIRE FIGHTING INSTRUCTIONS: Argon is a simple asphyxiant. If possible, remove argon cylinders from fire area or cool with water. Self contained breathing apparatus may be required for rescue workers.
UNUSUAL FIRE AND EXPLOSION HAZARDS: Upon exposure to intense heat or flame cylinder will vent rapidly and or rupture violently. Most cylinders are designed to vent contents when exposed to elevated temperatures. Pressure in a container can build up due to heat and it may rupture if pressure relief devices should fail to function.

SECTION 6: ACCIDENTAL RELEASE MEASURES

Evacuate all personnel from affected area. Increase ventilation to release area and monitor oxygen level. Use appropriate protective equipment (SCBA). If leak is from container or its valve, call the Air Products emergency telephone number. If leak is in user's system close cylinder valve and vent pressure before attempting repairs.

SECTION 7: HANDLING AND STORAGE

STORAGE: Cylinders should be stored upright in a well-ventilated, secure area, protected from the weather. Storage area temperatures should not exceed 125° F (52° C) and area should be free of combustible materials. Storage should be away from heavily traveled areas and emergency exits. Avoid areas where salt or other corrosive materials are present. Valve protection caps and valve outlet seals should remain on cylinders not connected for use. Separate full from empty cylinders. Avoid excessive inventory and storage time. Use a first-in first-out system. Keep good inventory records.
HANDLING: Do not drag, roll, or slide cylinder. Use a suitable handtruck designed for cylinder movement. Never attempt to lift a cylinder by its cap. Secure cylinders at all times while in use. Use a pressure reducing regulator or separate control valve to safely discharge gas from cylinder. Use a check valve to prevent reverse flow into cylinder. Do not overheat cylinder to increase pressure or discharge rate. If user experiences any difficulty operating cylinder valve, discontinue use and contact supplier. Never insert an object (e.g., wrench, screwdriver, pry bar, etc.) into valve cap openings. Doing so may damage valve causing a leak to occur. Use an adjustable strap-wrench to remove over-tight or rusted caps.

Argon is compatible with all common materials of construction. Pressure requirements must be considered when selecting materials and designing systems.
SPECIAL REQUIREMENTS: Always store and handle compressed gases in accordance with Compressed Gas Association, Inc. (ph.703-412-0900) pamphlet CGA P-1, *Safe Handling of Compressed Gases in Containers*. Local regulations may require specific equipment for storage or use.

Figure 1-1
Continued

CAUTION: Compressed gas cylinders shall not be refilled except by qualified producers of compressed gases. Shipment of a compressed gas cylinder which has not been filled by the owner or with the owner's written consent is a violation of federal law (49 CFR 173.301).

SECTION 8: PERSONAL PROTECTION / EXPOSURE CONTROL

ENGINEERING CONTROLS: Provide good ventilation and/or local exhaust to prevent accumulation of high concentrations of gas. Oxygen levels in work area should be monitored to ensure they do not fall below 19.5%.
RESPIRATORY PROTECTION
  GENERAL USE: None required.
  EMERGENCY: Use SCBA or positive pressure air line with mask and escape pack in areas where oxygen concentration is < 19.5%. Air purifying respirators will not provide protection.
OTHER PROTECTIVE EQUIPMENT: Safety shoes and leather work gloves are recommended when handling cylinders.

SECTION 9: PHYSICAL AND CHEMICAL PROPERTIES

APPEARANCE: Colorless gas.
MOLECULAR WEIGHT: 39.95
SPECIFIC GRAVITY (Air =1): At 70°F (21.1°C) and 1 Atm: 1.38
FREEZING POINT/MELTING POINT: -308.9°F (-189.4°C)
GAS DENSITY: At 70°F (21.1°C) and 1 Atm: 0.103 lbs/ft^3 (1.65 kg/m^3)
SOLUBILITY IN WATER: Vol/Vol at 32 °F (0°C): 0.056

ODOR: Odorless.
BOILING POINT: -302.2°F (-185.9°C)
SPECIFIC VOLUME: 9.7 ft^3/lb. (0.606 m^3/kg
VAPOR PRESSURE: Not applicable @ 70°F

SECTION 10: STABILITY AND REACTIVITY

CHEMICAL STABILITY: Stable
CONDITIONS TO AVOID: None
INCOMPATIBILITY: None
HAZARDOUS DECOMPOSITION PRODUCTS: None
HAZARDOUS POLYMERIZATION: Will not occur.

SECTION 11: TOXICOLOGICAL INFORMATION

Argon is a simple asphyxiant.

SECTION 12: ECOLOGICAL INFORMATION

The atmosphere contains approximately 1% argon. No adverse ecological effects are expected. Argon does not contain any Class I or Class II ozone depleting chemicals. Argon is not listed as a marine pollutant by DOT (49 CFR 171).

SECTION 13: DISPOSAL

UNUSED PRODUCT / EMPTY CONTAINER: Return cylinder and unused product to supplier with the cylinder valve tightly closed and the valve caps in place. Do not attempt to dispose of residual or unused quantities.
DISPOSAL: For emergency disposal, secure the cylinder and slowly discharge gas to the atmosphere in a well ventilated area or outdoors.

Figure 1-1 *Continued*

SECTION 14: TRANSPORT INFORMATION

DOT HAZARD CLASS: 2.2
DOT SHIPPING NAME: Argon, compressed
REPORTABLE QUANTITY (RQ): None
DOT SHIPPING LABEL: Nonflammable Gas
IDENTIFICATION NUMBER: UN 1006

SPECIAL SHIPPING INFORMATION: Cylinders should be transported in a secure upright position in a well ventilated truck. Never transport in passenger compartment of a vehicle.

SECTION 15: REGULATORY INFORMATION

U.S. FEDERAL REGULATIONS:

ENVIRONMENTAL PROTECTION AGENCY (EPA)

CERCLA: Comprehensive Environmental Response, Compensation, and Liability Act of 1980 requires notification to the National Response Center of a release of quantities of hazardous substances equal to or greater than the reportable quantities (RQ s) in 40 CFR 302.4.

CERCLA Reportable Quantity: None.

SARA TITLE III: Superfund Amendment and Reauthorization Act of 1986

SECTION 302/304: Requires emergency planning on threshold planning quantities (TPQ) and release reporting based on reportable quantities (RQ) of EPA's extremely hazardous substances (40 CFR 355).

Argon is not listed as an extremely hazardous substance.

Threshold Planning Quantity (TPQ): None

SECTIONS 311/312: Require submission of material safety data sheets (MSDSs) and chemical inventory reporting with identification of EPA defined hazard classes. The hazard classes for this product are:

IMMEDIATE HEALTH: No PRESSURE: Yes
DELAYED HEALTH: No   REACTIVITY: No
                      FLAMMABLE: No

SECTION 313: Requires submission of annual reports of release of toxic chemicals that appear in 40 CFR 372.

Argon does not require reporting under Section 313

40 CFR Part 68 - Risk Management for Chemical Accident Release Prevention: Requires the development and implementation of risk management programs at facilities that manufacture, use, store, or otherwise handle regulated substances in quantities that exceed specified thresholds.

Argon is not listed as a regulated substance.

TSCA - TOXIC SUBSTANCES CONTROL ACT: Argon is listed on the TSCA inventory.

OSHA - OCCUPATIONAL SAFETY AND HEALTH ADMINISTRATION

29 CFR 1910.119: Process Safety Management of Highly Hazardous Chemicals. Requires facilities to develop a process safety management program based on Threshold Quantities (TQ) of highly hazardous chemicals.

Argon is not listed in Appendix A as a highly hazardous chemical.

STATE REGULATIONS

CALIFORNIA:

Proposition 65: This product does NOT contain any listed substances which the State of California requires warning under this statute.

SCAQMD Rule: VOC = N/A

MSDS # 1004   Argon

Figure 1-1
Continued

SECTION 16: OTHER INFORMATION

HAZARD RATINGS:

NFPA RATINGS:		HMIS RATINGS:	
HEALTH:	0	HEALTH:	0
FLAMMABILITY:	0	FLAMMABILITY:	0
REACTIVITY:	0	REACTIVITY:	0
SPECIAL:	SA*		

*Compressed Gas Association recommendation to designate simple asphyxiant.

N/A=Not Applicable.

MSDS # 1004 Argon

Figure 1-1
Continued

only have a right to know, but also have a right to understand. This understanding is best conveyed through a comprehensive training program to teach employees about the types of information found on the MSDS, the location of where the MSDS file is maintained, and selected information relative to the terms and information found of the MSDS. Although everyone agrees on the importance of such training programs, there is little or no agreement on the amount or type of training that is necessary to comply. This item is cited as one of the main problems with the Hazard Communication Standard in that while it specifies that workers should be trained, like most other OSHA regulations, it does little to specify the amount of time needed to become trained.

This confusion may have led to the need for OSHA to enact the very restrictive HAZWOPER regulations. The hazard communication laws were in effect prior to the enactment of the HAZWOPER laws, but almost everyone involved in the hazardous waste industry believed that the basic provision of the hazard communication regulations did not provide adequate protection for those involved in hazardous waste operations. The rationale in this belief was the fact that there were no MSDS for the materials involved in most of the cleanup operations.

For example, consider the situation of employees involved in cleaning up a Superfund site prior to 1989. The basic training requirements of the hazard communication standards required that an MSDS be provided to give workers the information necessary to safely work with the materials present. Imagine just how unrealistic this would be in a site where the materials were largely unknown, or composed of a mixture of various waste materials.

HAZWOPER REGULATIONS

Now that we have a background about OSHA and the types of issues that were taking place at the time of the development of the HAZWOPER regulation, let us shift our attention onto the specifics of what is contained in it. Because it is a very complex regulation, it took considerable time to be developed.

The process started in the early 1980s. The Interim Final Rule, which contained most of the provisions that we now have in effect, was issued on December 19, 1986. More discussion occurred, and the Proposed Final Rule was issued on August 10, 1987. The Final Rule actually went into effect on March 6, 1990.

In our study of the specific points contained within the regulation, it is probably best to look at it in its entirety, section by section. To do this we must understand the layout that is used in the codification process. Unlike the format used by most of us if we were to develop an outline, the federal regulations use the following system of hierarchy:

Lower case letters in parenthesis such as (a), (b), (c) are used to denote the main divisions or paragraphs. Subdivisions of the paragraphs or main divisions are denoted in the following order:

- Arabic numbers such as (1), (2), (3) are used to denote the next major divisions.
- Lower case Roman numerals such as (i), (ii), (iii), (iv) are used to denote the next major subdivisions.
- Upper case letters such as (A), (B), (C) are used to denote the next major subdivisions.

Paragraph (a)(1): Scope

> **■ NOTE**
> One of the most important and frequently misunderstood issues involving the HAZWOPER regulations is whom the law is intended to cover.

One of the most important and frequently misunderstood issues involving the HAZWOPER regulations is whom the law is intended to cover. All too often, people apply these laws in situations that were never the intent of the regulatory driver. In fact, many OSHA inspectors and staff persons give out the wrong information relative to what is referred to as the scope of the regulation.

The first main subpoint of paragraph (a), subsection (1) defines the scope of the regulation. It simply answers the question of who is suppose to follow this particular law. All laws do not apply to everyone; take for example laws covering trucks and other large vehicles. No doubt you have seen signs that state that the speed limit for such vehicles is one speed while the speed limit for cars is different. The law regulating truck speeds does not apply to someone operating a car or motorcycle.

Such is the case with the HAZWOPER law which starts off in section (a) with scope. In it, five subgroups listed, are including:

1. Cleanup operations required by a governmental body, whether federal, state, local, or other involving hazardous substances that are conducted at uncontrolled hazardous waste sites (including, but not limited to, the EPA's National Priority Site List (NPL), state priority site lists, sites recommended for the EPA NPL, and initial investigations of government identified sites which are conducted before the presence or absences of hazardous substances has been ascertained;
2. Corrective actions involving cleanup operations at sites covered by the Resource Conservation and Recovery Act of 1976 (RCRA) as amended;
3. Voluntary cleanup operations at sites recognized by federal, state, local, or other government bodies as uncontrolled hazardous waste sites;
4. Operations involving hazardous wastes that are conducted at treatment, storage, and disposal (TSD) facilities regulated by 40 CFR parts 264 and 265 pursuant to RCRA; or by agencies under agreement with U.S. EPA to implement RCRA regulations;
5. Emergency response operations for releases of, or substantial threats of releases of, hazardous substances without regard to the location of the hazard.

Although the actual scope of the regulation is somewhat limited to those areas in the foregoing list, there has been a movement to apply this regulation to other areas as a recognized standard. Many industries have adopted the training requirements of the HAZWOPER regulations as the minimum requirement for those who work within their specific industry, despite the fact that the regulations do not mandate such training, since that particular operation is not part of the scope of the regulation.

Take for example the petrochemical industry. At a typical petroleum refinery, although it could be argued that there might be some hazardous waste issues, the site does not qualify as one of the items listed above. It is certainly not a Superfund cleanup site, nor is there an uncontrolled release of hazardous wastes, yet many petroleum refineries mandate that workers and contractors certify to the 40-hour HAZWOPER level as a general site worker.

The reason cited for this is relatively easy to understand. What other recognized level of training is there for those who might work with, or around, hazardous chemicals? Well, there is always the hazard communication standard, but as we know, it is fairly vague and does not specify any number of hours required for certification. Given this, almost by default, many refinery workers and contractors are required to take the general site worker classes as a means of training to a recognized standard relative to the safe handling of hazardous materials and wastes.

Paragraph (a)(2): Applications

The second subdivision within paragraph (a), denoted as (2), defines the application of the regulation. In this case, the term application is used to further describe the specific areas of the regulation relative to whom it is intended to apply.

If we study this portion of the regulation, we find that sections (a) through section (o) applies to those operations specified in (a)(1)(i) through (iii). That is, the majority of the regulation applies to those operations that are hazardous waste cleanup operations as noted above.

Further, it states that section (p) is the only section that applies to those operations listed in section (a)(1)(iv). This section denotes those operations conducted at licensed treatment storage and disposal facilities (TSDF).

The last section, section (q), is the section applicable to those operations discussed under section (a)(1)(v). Therefore, if you are involved in emergency response operations, section (q) is the section that applies to you.

One last item to review as we discuss application is that in many cases, OSHA regulations overlap with certain operations. The HAZWOPER regulation references many other OSHA regulations throughout. In all cases, we need to remember that OSHA requires that the most restrictive regulation applies if there are two regulations that seem to fit the given situation. Remember, you need to have a good understanding of not just the HAZWOPER law as codified in 29 CFR, Part 1910.120, but also other regulations that are referenced in this regulation.

■ **NOTE**
If you are involved in emergency response operations, section (q) is the section that applies to you.

■ **NOTE**
OSHA requires that the most restrictive regulation applies if there are two regulations that seem to fit the given situation.

Paragraph (a)(3): Definitions

One of the very nice things that OSHA does in this regulation is to use part of the very first paragraph to define certain terms that are used in it. These are contained in paragraph (a), subsection (3). Unlike other sections, this subsection does not contain any further subdivisions, but rather, the terms are listed alphabetically.

Paragraph (b): Safety and Health Program

Paragraph (b) which contains some important subdivisions, describes a very important aspect of hazardous waste site cleanup operations: All employers who are covered by this part of the regulation must develop and implement a written safety and health program for their employees who are involved in hazardous waste operations. Remember that this section only applies to those operations covered under the Scope as defined in (a)(1)(i), (ii), and (iii). These are hazardous waste cleanup operations, not TSDFs or emergency response operations.

In summary, this section requires that a written health and safety program be implemented. The program must have the following minimum components:

- An outline of the organizational structure
- Comprehensive work plan for activities
- Outline of the training program
- Outline of the medical surveillance program
- Standard operating procedures
- Listing of the necessary interface between general program and site-specific activities
- Program to inform contractors and subcontractors of the hazards
- Requirements for a pre-entry briefing
- Review procedures for the health and safety program
- Requirements for a site-specific safety and health plan that contains the following:
  - Risk and hazard analysis of the site
  - Specific training requirements for workers at the site
  - Personal protective equipment requirements for operations at the site
  - Medical surveillance requirements for those operating at the site
  - Air monitoring requirements for the site
  - Site control plan
  - Decontamination procedures
  - Emergency response plan

- Confined space entry procedures
- Spill containment program and procedures

Paragraph (c): Site Characterization and Analysis

This paragraph outlines what should be obvious—the need to evaluate the hazards present prior to any significant activity taking place. Within the paragraph we find that minimum information regarding the site is required to be obtained.

Among other things, the paragraph requires that a preliminary evaluation of the site's characteristics be made prior to entry by a trained person in order to identify any potential site hazards and to aid in the selection of appropriate employee protection methods. Included in this evaluation would be all suspected conditions immediately dangerous to life or health or which may cause serious harm.

During the course of the initial evaluation of the site hazards, this paragraph requires that personnel who are part of the evaluation be provided with appropriate levels of **personal protective equipment** (PPE). At a minimum, the personnel who are part of the initial entry are required to have respiratory protection if a respiratory hazard is suspected, and that the level of respiratory hazard will be at least to Level B (See Chapter 7 for a description of Level B protection).

Paragraph (d): Site Control

One very important safety issue that relates to activities at hazardous waste cleanup sites is that of access to the site, or more correctly, restricting access to the site. The regulation mandates that prior to any activity taking place, the site must be fully mapped out to determine the extent of the hazards, work zones must be established, and workers must be instructed in the use of the buddy system for activities taking place at the site.

Paragraph (e): Training

One of the most important sections of the HAZWOPER regulation is paragraph (e), which covers the minimum training requirements. Because this section is probably what causes most of us to undertake this course of study in the first place, it might be wise to spend some time reviewing in detail the requirements outlined in it.

As we have previously discussed, prior to the issuance of the HAZWOPER regulations, there were few or no requirements specific to worker training for persons who worked at hazardous waste sites, which resulted in a high potential for injury or exposure to those mostly untrained employees. Because Superfund money was used in many cases to fund the cleanup of such sites, the federal government felt compelled to develop minimum guidelines for activities at these waste cleanup sites. A summary of the specific training requirements for those

---

personal protective equipment (PPE)
correct clothing and respiratory equipment needed to perform a job involving hazardous materials and to protect the worker

■ **NOTE**
One very important safety issue that relates to activities at hazardous waste cleanup sites is that of access to the site, or more correctly, restricting access to the site.

persons who work at hazardous waste sites are found in the following section. Three training levels are specified: general site worker, occasional site worker, and supervisor.

General and Occasional Site Workers The General Site Worker classification is the group of workers for whom the regulations were primarily intended. Examples of those who qualify as general site workers are noted in the regulations where it states that this includes persons "such as equipment operators, general laborers, and supervisory personnel." Obviously, these are not the only ones who work at such waste cleanup sites and who might need to be trained in safety practices at such sites. The actual list of those classified as a general site worker is actually understood almost by default. By this we mean that because this regulation was intended for all employees who work at waste sites and who are involved in cleanup, it would follow that all workers would be required to comply with this level unless one of the other training levels listed would apply. We then need to see which groups of workers do not require this level of certification in order to see who is considered a general site worker. This group is known as Occasional Site Worker, which is discussed later in this section. For now, we focus on the role of the general site worker and on the requirements for certification to this level.

The term *general site worker* denotes those who are generally at the site where the cleanup operations take place and whose job responsibilities place them in possible contact with hazardous waste above the **permissible exposure limits** (PEL) established by OSHA. With this level of training, they are authorized to use the various levels of personal protective equipment and engage in activities that are considered to be fairly hazardous.

Because of this potential exposure, training for certification as a general site worker is mandated to be at least 40 hours, with an additional 24 hours of on-the-job supervised field training/experience. The need for the on-the-job experience is obvious when one reviews the list of items that must be covered during the training. Many of the items in the following bulleted list required to be covered in a training program cannot be discussed during the classroom training programs because they are site specific. Items such as "names of personnel and alternates responsible for site safety and health," or the "safety, health, and other hazards present on the site" could only be addressed during the site-specific training programs.

- Names of personnel and alternates responsible for site safety and health
- Safety, health, and other hazards present on the site
- Use of personal protective equipment
- Work practices by which the employee can minimize risks from hazards
- Safe use of engineering controls and equipment on the site
- Medical surveillance requirements
- Recognition of signs and symptoms that might indicate exposure to hazards

permissible exposure limits (PEL)
an OSHA term used to denote an exposure limit

- Elements of the Site Health and Safety Plan including:
  - Decontamination procedures
  - Site Emergency Response Plan
  - Confined space entry procedures
  - Site-specific spill containment procedures

Occasional site workers are those whose job responsibilities place them at the site on only an occasional basis and whose potential exposure is below the established PEL. Such workers might include well drillers or engineers who only occasionally visit the site for a specific purpose and whose exposure levels are very low.

Another group of workers who fall into the occasional site worker classification are those who work at a hazardous waste site on a regular or continuing basis, but whose exposure will never exceed the PEL. The regulations mandate that the sites be "fully characterized" and determined not to present any potential for exposure to the workers. Additionally, the workers would not be required to wear respiratory protection, because the need for respiratory protection implies that an exposure level above the PEL is present.

Occasional site worker training programs are mandated to be at least 24 hours with an additional 8 hours of on-the-job, supervised field training/experience.

Although their job responsibilities vary significantly, both classifications of site workers are mandated to have a training program that contains the same elements. Because the general site workers course contains an extra 16 hours of instruction, plus 16 additional hours of on-the-job supervised field training/experience, the topics are generally covered in greater depth. Topics such as the use of personal protective equipment and decontamination practices are emphasized because general site workers may be required to be involved in such activities.

Once trained and certified, the regulations mandate that the training program be refreshed and updated annually. Both the general and occasional site workers each must receive annual refresher training of at least 8 hours per year. Once this refresher training program is complete, the certification is good for another year, during which time the employee is required to receive another 8-hour refresher class in order to remain certified.

Supervisor Training As we saw, the training for general and occasional site workers requires that each receive a certain amount of supervised field experience under the "direct supervision of a trained, experienced supervisor." Supervisory qualifications are also mandated in the regulations. In this case, it states that supervisors who are certified under this regulation are required to undergo the same initial training program as mandated for workers. Whether the supervisor takes either the 24-hour or the 40-hour initial course depends on the level of involvement of the supervisor. Obviously, those supervisors who will be engaged in supervising general site workers (employees who need the 40-hour course) would need to have taken the 40-hour training. Supervisors of occasional site workers would be able to take only the 24-hour program initially. As with the other site workers, supervisors are also required to have the supervised field experience prior to being certified and also are required to take the annual refresher training to maintain their certification as a site worker.

Following completion and certification of the initial training, supervisors are required to undergo an additional training program in order to be certified as a supervisor. This program is mandated to be at least 8 hours. The regulations are not as specific with the elements of the supervisor training program but state that it should cover "such topics as . . . the employer's safety and health program and the associated employee training program, PPE program, spill containment program, and health hazard monitoring procedures and techniques." Obviously there is considerable leeway in this and such training programs can run the gamut of being very good to absolutely useless. Figure 1-2 depicts students receiving instruction in cleanup practices during a training evolution.

One last note regarding the training requirements as outlined in paragraph (e) is that while the regulation is specific as to the types of personnel who are mandated to take the site worker training programs, it does not preclude other types of employees from taking this training program. For example, in many facilities, the 40-hour training class has become the standard for training those employees who work within that industry. Such is the case in the high technology

Figure 1-2 *Students learning to use loose absorbent material to pick up a gasoline spill.*

industry where many firms require parts of their workforce to receive the 40-hour training. The students receive certification as general site workers even though they are not involved in activities at hazardous waste cleanup sites.

The rationale for this is that these workers are exposed to similar hazards on the job at the high technology firms due to their exposure to similar types of hazardous materials found in waste cleanup sites. Many other industries have followed suit and require 40-hour certification for employees and contractors whose job responsibilities place them in close contact with hazardous materials and hazardous wastes.

Paragraph (f): Medical Surveillance

This paragraph outlines the requirements for medical examinations and medical surveillance for certain groups of workers covered by the regulations. In summary, the paragraph mandates that the following workers receive annual medical examinations:

- Those who are exposed to levels of the chemicals above the permissible exposure limits for more than 30 days per year
- Those who wear respirators for more than 30 days per year
- Members of hazmat teams

Additionally, certain other groups of employees are required to have medical examinations provided to them on an as-needed basis. This includes anyone who is injured as a result of exposure to chemicals. In such cases, they would be seen by a physician who would determine the frequency of follow-up examinations if any.

A final note on this section is that while medical examinations are required, there is no mandated type of examination. All the regulation states is that the examination be done on an annual basis and that it include a patient medical and work history. Obviously this gives considerable leeway for a wide range of physical examinations. I am sure that we all can recall some of our own doctors who are very thorough, and others who do a cursory examination and take little time. It does reference a number of standards that can be consulted, but no standards are required.

Paragraph (g): Engineering Controls, Work Practices, and PPE for Employee Protection

Paragraph (g) discusses items that could reasonably construed as common sense. This portion states that steps should be taken to reduce employee exposure to hazards at the site through the use of engineering controls, personal protective equipment, and work practices. Items such as the use of specific types of personal protective equipment are discussed.

Paragraph (h): Monitoring

Paragraph (h) covering monitoring actually addresses two major issues: establishing a monitoring plan for the environment or site, and monitoring the exposure of the personnel working at the site. The items contained in these monitoring programs include requirements for determining the presence of a flammable atmosphere or whether the levels of an airborne contaminate reach the Immediately Dangerous to Life or Health (IDLH) levels.

Paragraph (i): Informational Program

Another of the more basic sections, and frankly one that seems to go without saying is paragraph (i) covering the requirement for employers to provide employees with information regarding the hazards at the site. The paragraph states that the program shall be in writing (remember that OSHA likes written programs), and that it define the actual level of hazards present at the site.

Paragraph (j): Handling Drums and Containers

Because many of the activities related to work with hazardous waste cleanup involves the use of drums and containers, the writers of the HAZWOPER regulations decided to make specific requirements for the handling of such items. In this paragraph all "unlabelled drums shall be considered to contain hazardous materials and handled accordingly," and "drums and containers that cannot be moved without rupture, leakage, or spillage shall be emptied" prior to moving. Such obvious statements point out the fact that prior to this, there had been little discussion of such worker safety issues. Figure 1-3 illustrates students practicing product transfer from one container to another.

Paragraph (k): Decontamination

The issues of decontamination are paramount in activities involving hazardous waste cleanup sites. As discussed in Chapter 8, the process of decontamination is necessary to limit the exposure of those working at the site to the hazards present on the site and to reduce the spread of the materials beyond the boundaries of the site. Implementation of a decontamination procedure is required before any employee or equipment may leave an area of potential hazardous exposure. Further, the paragraph requires that standard operating procedures be established to minimize exposure through contact with exposed equipment, other employees, or used clothing and that showers and change rooms be provided where needed. Figure 1-4 shows students preparing to decontaminate a fellow worker in level A protection.

Figure 1-3 *Students learning the proper methods of transferring flammable liquids.*

Figure 1-4 *Decon scenario.* Courtesy of Mine Safety Appliances.

Paragraph (l): Emergency Response at Uncontrolled Hazardous Waste Sites

Paragraph (l) covers the topic of an emergency occurring at a hazardous waste cleanup site. In it we find the requirements to develop a written emergency response plan to handle possible on-site emergencies prior to beginning hazardous waste operations. Such plans must address the following topics:

- Personnel roles
- Lines of authority
- Training and communications
- Emergency recognition and prevention
- Safe places of refuge
- Site security
- Evacuation routes and procedures
- Emergency medical treatment
- Emergency alerting

Paragraph (m): Illumination

Imagine if you will a law that requires that you provide adequate lighting to work in an area that has been determined to be hazardous, and one where the potential for exposure and hazards is extremely high. Paragraph (m) outlines that minimum lighting be provided in various areas of the site.

Paragraph (n): Sanitation for Temporary Workplace

Once again, we find some specific requirements outlined, only this time concerning sanitation. In paragraph (n) we find items regarding toilets, sleeping areas, and washing facilities. An example of the types of items covered is found in (n)(3)(v) where it states that "doors entering toilet facilities shall be provided with entrance locks controlled from inside the facility."

Paragraph (o): New Technology Programs

Another of the small paragraphs covered by the regulation is the one involving new technologies. In this we find the requirement for employers to look for new ways to accomplish the types of hazardous tasks in a less hazardous manner.

Paragraph (p): Certain Operations Conducted under the Resource Conservation and Recovery Act of 1976 (RCRA)

In paragraph (p) we find all of the issues dealing with treatment, storage, and disposal facilities covered. Training programs required for workers, decontamination programs, new technology programs, and medical surveillance requirements related to activities at TSDFs are covered.

If you recall the discussion regarding the scope of the regulation, you will remember that paragraphs (a) through (o) deal with activities at hazardous waste cleanup sites. Paragraph (p) then is a mini version of all of the previous paragraphs as they relate to the TSDF activities. Remember that none of the previous paragraphs cover such activities. Paragraph (p) is the major section of the regulation that applies to activities at TSDFs. All of the requirements within the HAZWOPER regulations for these sites is contained in paragraph (p).

An area worthy of note is the training requirements for those employees at TSDFs. They are similar to those for occasional site workers, but are actually not the same. In section (p)(8)(iii), the minimum training requirement is 24 hours. However, unlike the occasional site worker requirement, there is no mention of a mandated number of hours of on-site training that is required. Additionally, the type of training topics is not specifically addressed in as much detail as was noted in paragraph (e) for the site cleanup workers.

Like the site cleanup workers, employees at TSDFs are also required to have an 8-hour annual refresher class. An obvious question at this point is to ascertain whether the 8-hour refresher class for the site workers would satisfy the requirements for certification as a TSDF worker.

Paragraph (q): Emergency Response to Hazardous Substance Releases

Again, if we review the scope of the HAZWOPER regulations we remember that only paragraph (q) covers those activities related to emergency response. Paragraph (q) includes a detailed description of how to handle emergency responses and the need to use the **Incident Command System** (ICS). OSHA, in outlining these procedures, actually makes law many of the practices used by the fire service in handling emergencies.

One of the more important areas of paragraph (q) relates to the mandated amount of training that emergency response personnel must receive. Reviewing this topic in detail, we find some specific information regarding the levels of training, but none regarding which level applies to whom. Specifically, the regulations cite that the training "shall be based on the duties and function to be performed by each responder of an emergency response organization." This leaves the responsibility for determining the level of training up to the employer. Employers are charged with the responsibility of determining what action they wish their employees to undertake in the event of a release of hazardous materials and then to train and certify them to that level.

■ NOTE
Paragraph (p) is the major section of the regulation that applies to activities at TSDFs. All of the requirements within the HAZWOPER regulations for these sites is contained in paragraph (p).

incident command system (ICS)
management tool used by responders to coordinate response to an incident

Chapter 1 Hazardous Materials Regulatory Overview

> **■ NOTE**
> Employers are charged with the responsibility of determining what action they wish their employees to undertake in the event of a release of hazardous materials and then to train and certify them to that level.

Once we determine what we expect the employees to do, the regulations mandate that we provide training to one of five responder levels as listed:

Level 1. First Responder Awareness level (FRA)
Level 2. First Responder Operational level (FRO)
Level 3. Hazardous Materials Technician
Level 4. Hazardous Materials Specialist
Level 5. Hazardous Materials Scene Commander

In looking at the specific requirements for each of these, we note that levels 1 through 4 are placed in order of progressively more knowledge and job responsibilities. For example, to be certified and function at level 3 dictates that you have been previously trained and have the knowledge of levels 1 and 2. In other words, the levels build as you go up.

The exception to this rule comes into effect with level 5. Personnel certified to level 5 need not have completed all of the levels below them. In this case, they are mandated to be trained up through level 2 and then receive additional training to certify them to level 5, only through level 2.

Following is a discussion of each of these levels as outlined in the regulations. It is important that you understand which of these levels you will be trained and certified to, because this will dictate your level of involvement in a hazardous materials release.

First Responder Awareness Level The first level of emergency response training is the First Responder Awareness level (FRA). The term is actually misleading when the role of these personnel is evaluated. They technically are not "first responders" as the term implies, because their primary role is to not respond to an emergency scene, but rather to identify the incident as an emergency and leave the area. So their actions do not involve a response in the traditional sense, but rather a reaction to the incident.

The regulations specify that this level is designed for those individuals who are "likely to witness or discover a hazardous substance release and who have been trained to initiate the emergency response sequence by notifying the proper authorities of the release." They would take no further action beyond notifying the authorities of the release. This level of certification is severely limited in that the individuals at this level are nothing more than people who are *aware* of the spill.

In studying the regulations, we find that there are no required number of hours to be eligible for certification to this level. If you recall, this is consistent with the manner in which OSHA deals with almost all of their regulations.

Instead, as with other regulations, OSHA lists a number of areas to which the student must "objectively demonstrate competency." In the case of First Responder Awareness level, these include the following:

1. An understanding of what hazardous substances are and the risks associated with them in an incident.

2. An understanding of the potential outcomes associated with an emergency created when a hazardous material is released.
3. The ability to recognize the presence of a hazardous substance in an emergency.
4. The ability to identify the hazardous substances, if possible.
5. An understanding of the role of the First Responder Awareness individual in the employer's emergency response plan (including site security and control), and the U.S. Department of Transportation's Emergency Response Guidebook (now called the North American Emergency Response Guidebook.)
6. The ability to realize the need for additional resources and to make appropriate notifications to the communications center.

This is a fairly significant list of items to cover in a training program. Although it is true that no specified number of hours are listed for certification to this level, most training programs designed to provide this certification range from 4 to 8 hours in their attempt to meet the listed competencies. Remember however, that because there are no listed hours for this level of certification, you might find some trainers or employers who certify their staff on the basis of a very minimal program comprised mostly of video tapes.

The next obvious question regarding this level of training is just who might benefit from this level of certification. Some examples are obvious and include employees who work in areas where hazardous materials are used, stored, or transported. These could include forklift drivers in a warehouse, shipping and receiving personnel at a site where chemicals are used, or almost any worker in a typical facility where chemicals are handled routinely. Another group of personnel who often receive this training are more traditional emergency response personnel including police officers, emergency medical services personnel (paramedics and emergency medical technicians), and public works or other governmental employees whose job involves response to chemical spills.

First Responder Operational Level The second level of emergency response training is the First Responder Operational level (FRO). The FRO level of training is designed for those workers who actually respond to an incident from an area outside of the spill. Once at the scene, personnel at this level are expected to initiate some type of action to minimize the effects of a spill or release of a hazardous material. As the regulations state, these workers "respond to releases or potential releases of hazardous substances as part of the initial response to the site for the purpose of protecting nearby persons, property, or the environment from the effects of the release. They are trained to respond in a defensive fashion without actually trying to stop the release. Their function is to contain the release from a safe distance, keep it from spreading, and prevent exposures."

Some examples of the types of actions that someone might initiate could include turning off the flow of a gas or liquid from a remote area, evacuating an area potentially threatened by a chemical release, or putting down absorbents to keep a spill from entering the environment. In the course of their activities, they may have to use some types of personal protective equipment, however in this area they get close to overstepping their boundaries because they are not expected to ever enter an area where they might contact the spilled material. In this case, the personal protective equipment is only to serve as a backup in the event that the material does go beyond where it is expected to be. The First Responder Operations level personnel are to use distance as their primary source of protection. Personal protective equipment is secondary protection for them. It could be argued that if they actually are relying on the PPE for protection, they are too close and might be operating outside of the regulations because they might be in an area where there is exposure.

> ■ **NOTE**
> First Responder Operations level personnel are to use distance as their primary source of protection. Personal protective equipment is secondary protection for them.

Unlike the Awareness level, a set number of hours are part of the minimum training requirement for certification to this level. The regulations specify that the training programs be at least 8 hours and that the students "objectively demonstrate competency" in the following areas in addition to those listed for the Awareness level:

1. Knowledge of the basic hazard and risk assessment techniques.
2. How to select and use proper personal protective equipment provided to the First Responder Operational level.
3. An understanding of basic hazardous materials terms.
4. How to perform basic control, containment, and/or confinement operations and rescue injured or contaminated persons within the capabilities of the resources and personal protective equipment available with their unit.
5. How to implement basic equipment, victim, and rescue personnel decontamination procedures.
6. An understanding of the relevant standard operating procedures and termination procedures.

> ■ **NOTE**
> Responders at the Operations level are taught to respond to the emergency and initiate defensive actions to reduce the impact of the spill on people or to limit the spread of the material into the environment. Their actions should not involve product contact.

Although the regulations specify that this level of training must be at least 8 hours, it could be argued that this is not enough time to demonstrate competency in the areas listed as required by the regulations. For this reason, many programs that certify personnel to this level are much longer; it is not unusual to find programs up to 24 hours in length resulting in certification to the Operations level.

The next obvious question that might be asked is just which groups of employees are typically trained to the Operations level. Because this is for those who respond to the release and initiate defensive actions, this group could include fire department personnel who respond from outside the site, or emergency response team members who are part of the site emergency response organization at the facility. Responders at the Operations level are taught to respond to the emergency and initiate defensive actions to reduce the impact of the spill on people or to limit the spread of the material into the environment. Their actions should not involve product contact.

> **■ NOTE**
>
> Technicians are trained to deal with spills by taking offensive action and are protected with the appropriate level of personal protection necessary to handle the types of emergency situations encountered at the release.

Hazardous Materials Technician Level The Hazardous Materials Technician level is designed for those employees whose job it is to respond to a release of a material and do what is necessary to correct the problems encountered. Technicians trained in dealing with spills by taking offensive action are protected with the appropriate level of personal protection necessary to handle the types of emergency situations encountered at the release. The regulations state that they are individuals who "respond to releases or potential releases of hazardous substances for the purpose of stopping the release. They assume a more aggressive role than a First Responder Operational level in that they approach the point of release in order to plug, patch, or otherwise stop the release of a hazardous substance."

Examples of such *offensive* operations that could include entering an area where the material is present, identifying the hazards, taking samples for later analysis, performing field analysis, stopping the flow of material using various plugging, patching, or containment techniques, and even cleaning up or neutralizing the material.

Unlike training at lower levels, there are no restrictions relative to actions at hazardous materials releases for personnel certified to this level. They can wear the full range of chemical protective equipment and perform all types of offensive activities in accordance with the training program that they received.

As with the operational level, a specified minimum number of hours is required for this level of certification. To be certified at this level personnel must receive at least 24 hours of training of which 8 hours shall be equivalent to the First Responder Operations level and they must have competency in the following areas:

1. How to implement the employer's emergency response plan.
2. The classification, identification, and verification of known and unknown materials by using field survey instruments and equipment.
3. Functioning within an assigned role in the ICS.
4. How to select and use proper specialized chemical personal protective equipment provided to the hazardous materials technician.
5. Hazard and risk assessment techniques.
6. Performing advanced control, containment, and/or confinement operations and rescuing injured or contaminated persons within the resources and personal protective equipment available with their unit.
7. Understanding and implementing equipment, victim, and rescue personnel decontamination procedures.
8. Termination procedures.
9. Basic chemical and toxicological terminology and behavior.

Again, while a minimum number of hours is specified for certification to the technician level (24), it is very easy to see that meeting the competencies listed might require a much more extensive training program. Unless the training program is designed for some site-specific hazards of a given industry, and is only training personnel to deal with a specific range of hazardous material (such as those found in a semiconductor firm), it would be understandable to believe that far more than the minimum 24 hours of training would be needed. In fact, very few training programs can be designed that meet the competencies listed above in the minimum number of hours required, especially for those persons who respond to a variety of incidents involving potentially unknown materials. For this reason, programs that certify personnel to this level generally range from 40 hours to as many as 240 hours.

Certification to this level is open to a number of types of employees, including those who are part of a hazardous material team. In some locations, this involves members of the site emergency response team. In other cases, public safety agencies such as fire or police personnel receive this training.

Hazardous Materials Specialist Level The fourth level of response training is the Hazardous Materials Specialist. Although this level is listed as a separate level in the regulations, there is little difference in actual practice at the field level from this and the Hazardous Materials Technician level. In many cases, specialist and technicians function interchangeably in hazardous materials emergencies. Many agencies often do not have anyone certified at the specialist level and only choose to certify their personnel to the technician level because there is little, if anything, that the specialist can do above the level of the technician.

Their duties parallel those of the hazardous materials technician, however, those duties require a more directed and specific knowledge of the various substances that they are called upon to contain. The hazardous materials specialist also acts as the site liaison with federal, state, local, and other governmental authorities in regard to site activities.

Similar to the previous two levels, there is a specified minimum number of hours required for certification to this level plus the participants must have competency in a number of areas. The regulations specify that the training program shall be at least 24 hours equal to the technician level and in addition, have the following competencies:

1. Know how to implement the local emergency response plan.
2. Understand classification, identification, and verification of known and unknown materials by using advanced survey instruments and equipment.
3. Know of the state emergency response plan.
4. Be able to select and use proper specialized chemical personal protective equipment provided to the Hazardous Materials Specialist.
5. Understand in-depth hazard and risk techniques.
6. Be able to perform specialized control, containment, and/or confinement operations within the capabilities of the resources and personal protective equipment available.

7. Be able to determine and implement decontamination procedures.
8. Have the ability to develop a site safety and health control plan.
9. Understand chemical, radiological, and toxicological terminology and behavior.

Once again, we face the problem of minimum hours specified versus the actual time it takes to meet the competencies outlined for the Hazardous Materials Specialist level training. As with the technician, if the persons being certified are working with a limited number of known materials, the programs can come close to the minimum hours listed by the regulations. On the other hand, if the persons receiving the training are to be part of a regional hazardous materials team with multiple responses on unknown types of spills, the training programs can be much longer.

Many of the programs that certify personnel to this level range from 100 to 400 hours, again depending on the complexity of the duties that the specialist will perform and on the range of chemicals that the specialist personnel will be expected to handle.

Because the job responsibilities are similar, the types of personnel who receive this level of certification are similar to those listed for the technician. In this group we find members of hazardous materials teams from both the private and public sector.

Hazardous Materials Scene Manager The last level of training specified for emergency response is the Scene Commander level. This level is for those who "assume control of the incident scene beyond the First Responder Awareness level." In many cases, this is the role of the local law enforcement or fire safety agency who is responsible for public health and safety. In the case of those working at a particular site, it is generally management personnel or key emergency response team personnel.

The training required for certifying personnel to this level is required to be at least 24 hours that it be equal to the First Responder Operational level, and in addition have competency in the following areas:

1. How to implement the employer's incident command system.
2. How to implement the employer's emergency response plan.
3. Understanding the hazards and risks associated with employees working in chemical protective clothing.
4. How to implement the local emergency response plan.
5. The state emergency response plan and the federal regional response team.
6. Knowing and understanding the importance of decontamination procedures.

This last level of certification does not require completion of all of the previous levels. Command level personnel are only required to receive training to the

> **■ NOTE**
> **The Hazardous Materials Scene Commander level is designed for those individuals who will be in charge of the hazardous materials emergency when responders are working at or above the Operations level.**

Operational level and may never have been exposed to any type of experience involving chemical protective clothing. They are managers of the incident, and not the technicians. Because of this, they must often rely on personnel trained to higher levels to assist them in implementing their command and control system. The Hazardous Materials Scene Commander level is designed for those individuals who will be in charge of the hazardous materials emergency when responders are working at or above the Operations level.

Refresher Training—Emergency Responders

Unlike the specific requirements for site worker training programs, the emergency response certification levels do not have specific refresher training hours mandated. Essentially, it is up to the employer to determine the type and relevancy of response-oriented refresher training. The regulations specify for all five of the levels of emergency response training that personnel "shall receive annual refresher training of sufficient content and duration to maintain their competencies, or shall demonstrate competency in those areas at least annually." How this is accomplished is entirely up to the employer.

In some cases, employers may choose to enroll their employees in a recertification class. Because the regulations do not mandate any specific number of hours for each of the levels, recertification classes vary significantly depending on who is providing the programs. In some cases, formal drills and exercises may be a part of the recertification process. In others, the employer might simply administer a written examination or give credit to the employees for the responses that they were involved with in the previous year. Obviously, there is considerable leeway and room to fudge if an employer chooses to.

Determining Emergency Response Levels

In deciding which level to use, it is often wise to utilize a series of questions and answers to help determine the appropriate training levels needed. Following is a short summary of the regulation that helps to establish some parameters as to the training level required. Determining the level of response dictated by the following will lead to establishing the level of training required by the regulations.

1. When a material is released, workers are expected to immediately leave the area and to not take any action to limit the spill. They should initiate an evacuation of the area, and notify the appropriate authorities.

If this is the level of response expected, the personnel should be trained and certified to the First Responder Awareness level.

2. When a material is released, workers are expected to initiate actions to limit the spill of the material without becoming contaminated with it. They should not expose themselves to any of the material at any time. If possible, they should try to minimize the spill by diking the material and keeping it from getting into storm drains, by turning off valves to stop the flow of materials if it is safe to do so, and to initiate evacuation efforts.

If this is the level of response expected, the personnel should be trained and certified to the First Responder Operational level.

3. When a material is released, workers are expected to initiate whatever actions are needed to limit the size of the spill, reduce the consequences of the spill, and clean the spill up when appropriate. These workers will be provided with the required personal protective equipment (respirators and personal protective clothing) to safely work with the materials, and to approach the point of release to control it without becoming contaminated by it.

If this is the level of response expected, the personnel should be trained and certified to either the Hazardous Materials Technician or Specialist levels.

4. When a spill occurs at the facility, I expect someone from my organization to oversee the operation and the handling of that spill including its cleanup. This person would know how to establish a system to manage such an incident and provide a command and control system for the safe handling of the incident.

If this is the level of response expected, the personnel should be trained and certified to the Command level.

Summary

- The regulatory driver for most hazardous materials training programs is the HAZWOPER regulation that was enacted in March, 1990.
- The HAZWOPER regulation came into effect as a result of a number of other environmental laws that dealt with hazardous waste cleanup activities. The most notable was the Superfund Amendment and Reauthorization Act (SARA).
- Federal OSHA has primary jurisdiction for employee safety and health throughout the country. Twenty-five states have their own versions of OSHA and are known as state plan states.
- The Hazard Communication Standard, enacted in the 1970s, outlines basic worker safety training for those workers who are exposed to potentially hazardous materials in the workplace. It does not specify any number of hours of training needed.
- Many of the paragraphs and sections of the HAZWOPER regulation contain information that could be construed as common sense.
- The HAZWOPER regulations cover three separate areas of hazardous waste operations: hazardous waste cleanup sites, treatment storage and disposal facilities, and activities involving emergency response to hazardous materials releases.
- Hazardous waste cleanup site workers receive training in accordance with their expected job functions. This includes general site workers, who receive 40 hours of training plus three days of on-the-job training; occasional site workers, who receive 24 hours of training plus one day of on-the-job training; and supervisors, who receive site worker training plus 8 additional hours.
- TSDF workers are required to receive 24 hours of initial training.
- Emergency response personnel receive training in accordance with their expected job functions. There are five levels including First Responder Awareness level, First Responder Operations level, Hazardous Materials Technician level, Hazardous Materials Specialist level, and Scene Commander level.
- Training programs for emergency response personnel range from a mandated 8 hours up to 24 hours. In practice, many programs can be over 240 hours in length.
- All workers covered by the HAZWOPER regulations are required to have annual refresher training. Refresher training programs for site workers are specified to be at least 8 hours. Other programs do not specify any number of hours for recertification or refresher programs.
- The HAZWOPER regulations comprise seventeen areas from training through sanitation practices.

Review Questions

1. The minimum initial training program required for workers who work with hazardous materials/waste on a *regular* (general site worker) basis:
   A. 4 hours
   B. 8 hours
   C. 24 hours
   D. 40 hours

2. After the initial training, the OSHA regulations specify that workers must receive refresher training every:
   A. 6 months
   B. 2 years
   C. Year
   D. Never

3. Occasional site workers (workers whose job does not routinely involve contact with hazardous wastes but who on occasion, come in contact with hazardous wastes) are required to have a course that is at least how long?
   A. 4 hours
   B. 8 hours
   C. 24 hours
   D. 40 hours

4. Where would I find specific additional information regarding the HAZWOPER regulations?
   A. 29 CFR, Part 1910.120
   B. 39 CFR, Part 1010.120
   C. Title 24 of the Standard Transportation and Commodities Code
   D. SARA Title 3

5. Define HAZWOPER and explain its intent as it relates to worker safety.

6. List the 5 levels of response training according to the HAZWOPER standard.

7. True or False? Operations level training enables a responder to don Level A protection and employ offensive tactics to mitigate a spill.

8. Define CERCLA and explain its intent.

9. Which regulation was designed to encourage the creation of federal and state programs for the land disposal of waste materials?

10. Which regulation is commonly known as "right to know"?

11. True or False? General site worker training is not fully complete until the 24 hours of on-the-job training is provided.

12. True or False? Once hazardous materials technicians and specialists are initially trained, they do not have to have annual refresher training.

13. How many hours of training must a TSDF worker have?

14. List three competencies a hazardous materials scene commander must demonstrate upon completion of initial training.

15. Explain the meaning of "state plan state" and describe the significance of the concept as it relates to environmental regulations.

Principles of Chemistry

Learning Objectives

Upon completion of this chapter, you should be able to:

- Define chemistry.
- Understand the concept of matter and the parts of an atom.
- Define atomic number, atomic weight, and associate twenty-five atomic symbols to their respective elemental name.
- Differentiate between solids, liquids, and gases.
- Understand the difference between pure substances and mixtures.
- Distinguish between homogeneous and heterogeneous mixtures.
- Illustrate and understand the basic gas laws.

INTRODUCTION

Of all the areas of hazardous materials training, chemistry is usually the subject that makes most students cringe. The thought of learning complex formulas or the periodic table has caused many a beginning hazmat student to break out in a cold sweat. Given the choice, most would rather walk into some awful unknown purple cloud than take a chemistry class. Many also believe that to be a good hazmat person, one needs to have a Ph.D. in chemistry or become a chemical engineer. This thinking is unrealistic, and it is time to shed some light on what you really need to know.

First, most hazmat problems can be safely handled by understanding some basic chemical principles and employing clear and concise thinking. It is true that certain instances call for a high level of chemical knowledge, but most day-to-day operations can be safely handled if you just pay attention and think about what you are doing. From a chemical standpoint that means calmly looking at the big picture and understanding the nature of the material. If you understand *what* you are dealing with, it becomes much easier to decide *how* to deal with it.

Second, it is not necessary to commit volumes of technical information to memory. Using a few simple concepts and knowing where to find comprehensive, thorough information is the critical part. Countless hours of research and testing have been performed on many of the chemical compounds you will encounter. Utilize the information that is available in books, electronic databases, and material safety data sheets (MSDSs) instead of becoming a walking encyclopedia yourself. Later chapters will tie in the use of these reference materials but for now, believe that you can work safely around most chemicals if you remember these three important points.

- Always be alert and understand the situation.
- Never take chemicals for granted.
- Take your time and put personal safety first.

Finally, never forget that the chemicals you are dealing with have properties and characteristics that are always constant. In other words, the concentrated sulfuric acid you were working with yesterday is just as dangerous today as it will be tomorrow and next month. It is *your* level of readiness and proficiency that are in question. The good (and healthy!) hazmat person never loses respect for the chemicals or attempts to downplay the situation. Avoid becoming complacent and understand this simple fact: The chemicals will not cut you any slack if you make a mistake. Always take a moment and *think before you act*.

In this chapter we explore several chemical principles that are important for you to understand. Take the time to digest these basics as they will serve as a solid foundation for later sections such as hazard classification, personal protective equipment, and air monitoring.

> **! SAFETY**
> If you understand *what* you are dealing with, it becomes much easier to decide *how* to deal with it.

> **! SAFETY**
> Always take a moment and *think* before you act.

> **■ NOTE**
> Chemistry is the relationship between matter, energy, and reactions.

Chapter 2 Principles of Chemistry

chemistry
branch of study concerning the composition of chemical substances and their effects and interactions with one another

matter
anything that has mass and occupies space

■ **NOTE**
If it exists, it is made up of matter!

proton
basic subatomic particle existing in the nucleus of all atoms; has mass and an atomic weight of 1 amu

neutron
elementary subatomic particle existing in the nucleus having a mass of 1.009 amu's; has no electrical charge and exists in the nucleus of all atoms except hydrogen

electron
subatomic particle having a negative electric charge; electrons orbit the nucleus of an atom

stable atom
an atom that is not in the process of radioactive decay or the formation of an ion

■ **NOTE**
The nucleus is the center of the atom, where all the weight of the particular atom is concentrated.

CHEMISTRY AND MATTER DEFINED

In order to understand what chemistry is all about, it is first necessary to define it. In advance of the definition however, understand that chemistry is a broad field with many disciplines. In this text we explore a few of the many concepts and attempt to narrow the information down to what will benefit you in the field.

There are many interpretations, but the study of **chemistry** is a way of looking at the relationship between matter, how it ultimately bonds together, and the energy involved. Chemistry is the relationship between matter, energy, and reactions.

It is possible to break our study down even further by looking into the concept of **matter**. Matter serves as the foundation of our study of chemical principles and quite frankly is the cornerstone of our life. Simply put, matter is anything that has mass and occupies space. Air is made up of different bits of matter, water molecules are matter, this book and even you are made up of countless tiny pieces of matter. Think of matter as the building blocks that make up everything on the face of the earth. If it exists, it is made up of matter!

Tiny as it is though, matter is not the smallest piece of our chemistry puzzle. Atoms are even smaller divisions of matter and are the key players in the formation of millions of chemical compounds. Atoms are very complex and dynamic but are still not totally understood by modern scientists. Our study however, is centered on the three subatomic particles that make up individual atoms. These particles are called the **proton,** the **neutron,** and the **electron.** Two of the particles, the proton and the neutron, are found in the nucleus of all atoms. The nucleus is the center of the atom and is where all the weight of that particular atom is concentrated. The electrons orbit the nucleus and are very instrumental in how and why certain atoms bond with certain other atoms (more on this later!). At this point however, you must understand that the term *orbit* is simply an expression of the electron activity outside the nucleus. The electrons do not actually go around the nucleus in circles like planets orbiting the sun. Although most textbooks show the action of the electrons in this manner, it is actually much more chaotic and is believed to be an *electron cloud* that surrounds the nucleus. These orbiting electrons in the cloud are essentially weightless, with each one exhibiting its own negative electrical charge.

An important concept to remember is that the number of protons inside the nucleus of a **stable atom** equals the number of electrons orbiting it. Figure 2-1 illustrates a stable carbon atom and a stable chlorine atom with their respective numbers of protons and neutrons. Note that the protons and electrons are equal in number. In a stable atom, protons equal electrons.

The protons have a positive electrical charge, account for approximately one-half of the atom's weight, and are found in the nucleus of every atom. The number of protons are different in the carbon and chlorine atoms shown in Figure 2-1, because the number of protons an atom has is more or less that atom's

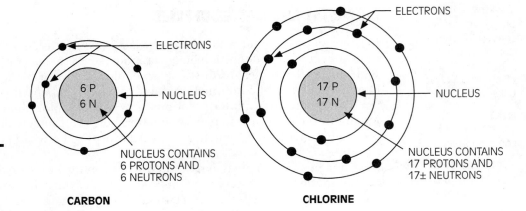

Figure 2-1 *Atomic configurations of carbon and chlorine.*

■ **NOTE**
In a stable atom, protons equal electrons.

■ **NOTE**
The number of protons in the nucleus of an atom is what distinguishes that atom from another.

compound
chemical combination of two or more elements that results in the creation of unique properties and a definite, identifiable composition of the substance

element
one of the 109 recognized substances that comprise all matter at the atomic level; building blocks for the compounds formed by chemical reactions

signature. The number of protons in the nucleus of an atom is what distinguishes that atom from another.

The carbon atom in Figure 2-1 has six protons in the nucleus. All carbon atoms have six protons in the nucleus. Anything other than six protons and the atom ceases to be carbon and becomes something else. If through some major event inside the nucleus you were able to cram in eleven more protons you would no longer have carbon. The sudden addition of the eleven protons would create another element, in this case chlorine. The number of protons would now equal seventeen—the *signature* for chlorine. The proton signature was changed, therefore the atom and all its characteristics were changed. Keep in mind that changing the amount of protons in the nucleus of an atom is no small task. We discuss it like it might be a common occurrence but in reality the nucleus of an atom is one tough cookie and not easily modified by human hands. The point is that the number of protons is very important in establishing the identity of an atom.

Neutrons are essentially weightless, but nevertheless play a key role in keeping the nucleus together. Scientists surmise that the neutrons exhibit what is called the *strong force,* which helps to keep the nucleus intact. When we explore radioactive materials and their characteristics you will find that the number of neutrons a radioactive atom has is usually much larger than the number of protons it possesses. Generally speaking, a large imbalance of neutrons to protons in the nucleus prompts an atom to give off energy. Since electrical stability is the goal of all atoms, the unstable nucleus will emit this energy until stability is achieved. This energy is referred to as radiation and is explored in Chapter 3.

Matter Combined

In nature and the laboratory, microscopic bits of matter bind with other microscopic bits of matter to form the **compounds** we use every day. These bits of matter are referred to as **elements** and are the most pure expression of matter itself.

Figure 2-2 *Sodium and chlorine bonding to form sodium chloride or salt.*

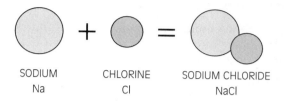

> ■ **NOTE**
> Generally speaking, a large imbalance of neutrons to protons in the nucleus prompts an atom to give off energy.

periodic table
arrangement of the elements in such a form as to emphasize the similarities of their physical and chemical properties

> ■ **NOTE**
> Electrical attraction is the major driving force in atomic bonding.

In other words, elements are the most fundamental substances in nature and are the construction materials for all our chemical compounds. These elements can be found on the **periodic table** and each have unique properties and characteristics. The periodic table is discussed later in the chapter so for now, file away the concept of elements and understand the following example.

A classic expression of a compound is ordinary table salt, which is comprised of one atom of sodium and one atom of chlorine. When these two elements are brought together under the right temperature and pressure, sodium chloride or table salt is formed. This very simple reaction involving two very common elements is illustrated in Figure 2-2.

This combination or bonding is due to an electrical attraction, which enables the two atoms to come together and form a stable compound. Electrical attraction is the major driving force in atomic bonding. Figure 2-3 displays the formation of another common substance we rely on for human existence—water. Water is created by a special type of bonding between hydrogen and oxygen.

When two atoms of hydrogen are bonded with one atom of oxygen under the right circumstances, water or H_2O is created. The formula reflects the constituents of the compound—two atoms of hydrogen (H_2) bonded with one atom of oxygen (O).

These common examples are typical of the reactions that form the millions of chemical compounds we use every day. Although we will not go into the theories behind bonding in this text, it is important to accept the fact that matter bonds together to form many materials. Figure 2-4 should give you an idea of where compounds fit into the big picture of matter and bonding.

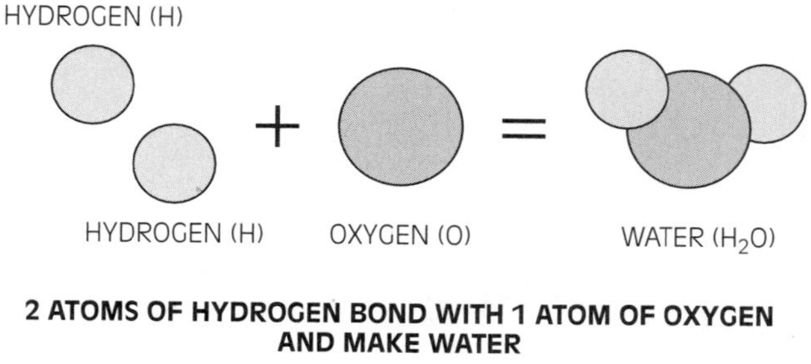

Figure 2-3 *The atomic formation of water.*

2 ATOMS OF HYDROGEN BOND WITH 1 ATOM OF OXYGEN AND MAKE WATER

Solids, Liquids, and Gases

Once bonding occurs, the resulting substance takes on a certain form or state. State of matter refers to the physical nature of a material when it is used, stored, or found in nature. State of matter is a very important piece of the hazmat puzzle to both waste site workers and emergency responders.

Think about how important state of matter is. Would dealing with a heavier-than-air flammable gas be different than facing a pile of combustible metal shavings? Yes! Determining just which of these situations is worse depends on many factors including the actual chemicals involved. What is known however, is that each presents unique hazards and may require different levels of personal protective equipment or **mitigation** tactics. When faced with any chemical situation however, start your investigation by determining the state of the material.

The stuff leaking from the drum, or the tanker truck, or the damaged crate will either be in a solid form, a liquid form, or a gas. Regardless of the particular chemical that may be involved, each state of matter presents a vastly different set

■ **NOTE**
State of matter refers to the physical nature of a material when it is used, stored, or found in nature.

mitigation
operational term used to describe the process of bringing a chemical release under control

SAFETY
When faced with any chemical situation, start your investigation by determining the state of the material.

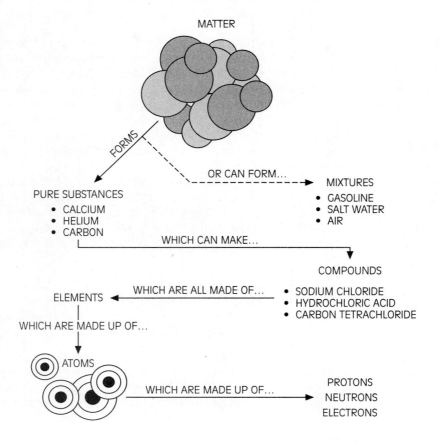

Figure 2-4 *Matter flowchart.*

> **NOTE**
> A chemical may be more or less of a problem based on the form it takes.

of circumstances and potential challenges when it gets out of its container. A chemical may be more or less of a problem based on the form that it takes.

For example, solids may be easy to see and control, but they can be difficult to clean up or clean off tools and chemical suits. Liquids, on the other hand, are much more mobile and have a tendency to permeate into the dirt or run down gutters into drainage systems. A large gasoline spill for example, could send you scrambling to divert the flow from a storm drain or hurrying to get far enough ahead to protect a waterway. It is no small chore to harness flowing liquids and even when they are contained, cleaning them up may present additional headaches. Some liquids may give off lots of flammable vapors, evaporate very quickly, float on water and/or be potentially toxic *or* none of the above *or* some of the above *or* all of the above. You can only be sure of the properties by positively identifying the material involved and looking up the properties in your reference materials.

> **SAFETY**
> A chemical may be hazardous in more than one way. It is usually safest to assume that a chemical has multiple hazards until you can prove otherwise.

On the other hand, gases may be impossible to detect by sight or smell and even more difficult to handle than solids or liquids because they readily leave their container and cannot be recovered or controlled easily. Many gases are odorless and can be fatal if inhaled. Carbon monoxide is a good example. This gas is colorless, odorless, and tasteless and can kill at relatively low doses. Would you guess that carbon monoxide is also flammable? If you said "yes," you are correct. It is actually quite flammable in addition to being lighter than air and present in almost all combustion processes. Once again the issue of a chemical having multiple hazards is present and a reference source may be your best friend. A chemical may be hazardous in more than one way. It is usually safest to assume that a chemical has multiple hazards until you can prove otherwise.

The Makeup of Solids, Liquids, and Gases

The three states of matter are quite different in terms of molecular makeup. Think of solids as being tightly packed at the molecular level with a high degree of molecular order. This tightly packed group results in materials that are quite dense with a definite shape and form. Liquids, on the other hand, have more space between the molecules, which allows them to flow much easier than solids. Gases have even less order and are quite chaotic in their molecular activity. They are very active inside a container and are capable of being compressed or even liquefied in certain cases.

An analogy to illustrate this may be to define the states of matter in terms of a sold-out football stadium. Each person going to the game represents an atom filing into the stadium to take a seat. Just before the kick off, the stadium becomes filled to capacity. All seats are taken and every atom is shoulder to shoulder. The aisles are open of course, and that allows for some movement, but practically speaking the stadium is now a solid.

As the game progresses, we see that things begin to change. It is now the fourth quarter and unfortunately the home team is losing by a jillion points. The fair weather atoms are leaving their seats and heading for the parking lot. A closer look at the stands reveals several open seats between our atoms. There is still a relatively large number of atoms in the stadium, but movement between the seats is now easy and unencumbered. The group is now much less packed than it was at the beginning of the game and we now have the illustration of a liquid.

Now the game is over and there are only a few die-hard atoms around to cheer for their favorite elements. Of a 50,000 seat stadium, maybe 50 atoms are hanging around in the various sections. Movement of those remaining atoms is totally unrestricted and they are free to go anywhere they want. This is much like a gas inside a cylinder. There is total freedom of movement, and the only restrictions are those of the vessel walls. Gases by their nature are very mobile at the molecular level and are characterized by their chaotic nature. Without the walls of a cylinder, gases will go wherever they want and occupy whatever space they are in.

> **SAFETY**
> Without the walls of a cylinder, gases will go wherever they want and occupy whatever space they are in.

Having understood the concepts defining the different states, we can now make the following generalizations:

- A solid is a structure that has a definite shape, is practically noncompressible, is rigid, and has a high degree of molecular order. Solids may be encountered as blocks or rods, fused solids inside drums, or bricks. Figure 2-5 shows an example of a solid material.
- A liquid is a material capable of flowing, has a definite volume, and takes the shape of the container it is in. Liquids are influenced by changes in the terrain and tend to flow easily from cracks and holes in containers. Liquids are virtually noncompressible and either float on water, are heavier than water, or mix with water. Some liquids are thicker than others depending on the **viscosity** of the material. Figure 2-6 shows an example of a liquid.
- A gas is a substance that is compressible, and ultimately takes the shape and occupies the given volume of a container. Gases can be flammable or

viscosity
the ability or the resistance of a liquid material to flow

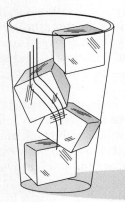

Figure 2-5 *Ice! A common example of a solid.*

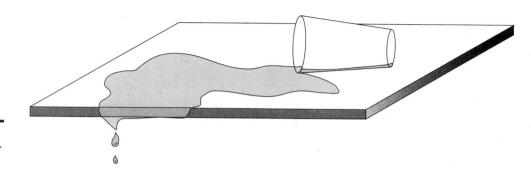

Figure 2-6 *Liquid phase of water after it melts.*

> ■ NOTE
> Look at the oxygen surrounding you right now and you have the best example of a typical gas!

nonflammable, lighter or heavier than air, corrosive, toxic, and even radioactive. Many common gases like propane, butane, and methane are colorless, odorless, and tasteless—properties which make them difficult to detect without the proper monitoring equipment. Look at the oxygen surrounding you right now and you have the best example of a typical gas!

THE PERIODIC TABLE

Practically speaking, no hazmat people walk around emergency scenes or waste sites with a periodic table taped to their forehead. To understand some basic thoughts on matter however, we must explore some information that is contained on the periodic table. Do not fear, it will be fast and hopefully painless!

In order to understand the concept of the periodic table, it may be helpful to discuss some of the background. The periodic table in use today was devised by Dimitri Mendeleev in 1869. Another scientist named Lothar Meyer was doing concurrent research on elements and their properties but Mendeleev gets the credit for arranging the table.

Upon first inspection this jumble of boxes, letters, and numbers may lead you to believe that Mendeleev was completely out of his mind. To the untrained eye it may appear as if he threw a bunch of information on a piece of paper and said "Here it is—I hope it makes sense!" The good news is that there is a whole lot more to the story and it actually *does* make sense. To even his own amazement, Mendeleev realized that there was a systematic way that elements could be arranged. This was big news at the time as no other scientists were of the same opinion. It was determined that by using the **atomic weights** (which we will learn about shortly) of the sixty or so elements recognized at the time, similar properties occurred at *periodic* intervals when arranged properly. In other words, the tabular format emphasized physical and chemical similarities that recur periodically throughout the sequence of elements. Figure 2-7 shows the periodic table that we are using in this text.

atomic weight
the total weight of an atom, the sum of all of its subatomic parts

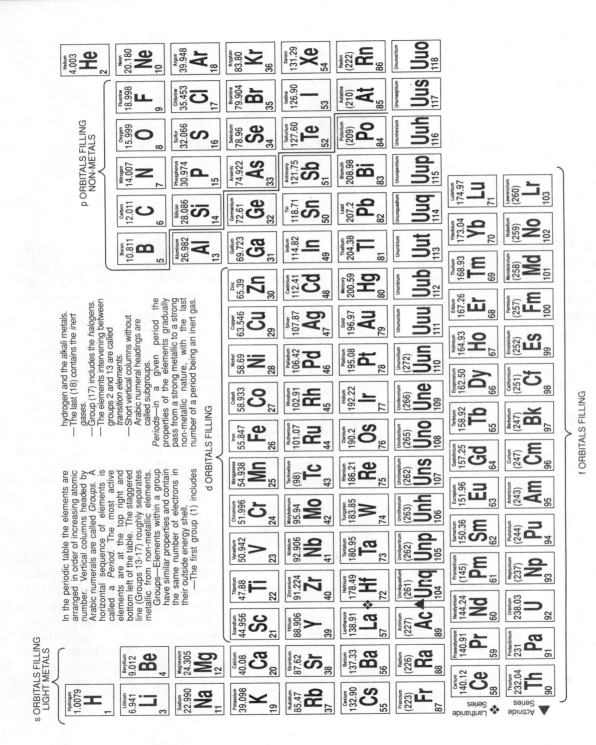

Figure 2-7 *A typical periodic table.*

Atomic Weight

Among the important pieces of information we can learn from the table is atomic weight. Earlier information in the chapter focused on the thought that all matter has mass and therefore has weight. This means that all of the atoms that make up the elements that are bonded together to form compounds have a definite weight and density. The atomic weights found on the periodic table are specific for each atom listed and are expressed in measurements called *atomic mass units* or AMUs. Keep in mind however that these units are incredibly small and may not always be expressed as whole numbers, because the atomic weight listed on the periodic table is an average weight of the atoms that make up the element. For example, carbon has an atomic weight of approximately 12. That value was arrived at by determining the average weight of many carbon atoms. The more important thing to remember here is this—the atoms that make up the elements you see on the table have definite weight. The weight of an atom is primarily centered in the nucleus where the protons and neutrons live and is expressed by the term *atomic weight*. Figure 2-8 represents one method for demonstrating the atomic weight and name of the element carbon. Keep in mind that different periodic tables have different looks but contain essentially the same information. Table 2-1 illustrates the atomic weights of some selected elements.

The principle of atomic weight can be used to your advantage when trying to determine the weight of various gaseous materials. Vapor density is discussed in depth in Chapter 3 but can basically be defined as follows: Vapor density is the relationship of a certain volume of a gas as compared to a like amount of dry air. Essentially this sets the stage for a very important hazmat question—Will the gas that is released be heavier or lighter than air? By using a simple mathematical formula and understanding atomic weight, you can determine the correct answer. Research has found that the weight of a volume of dry air is 29. Do not get hung up on the number 29 and what it represents, just accept the number 29 as the molecular weight of our volume of dry air. If we count up the atomic weights of all

■ **NOTE**
The weight of an atom is primarily centered in the nucleus where the protons and neutrons live and is expressed by the term *atomic weight*.

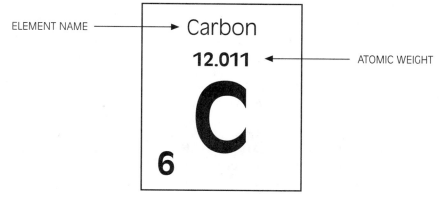

Figure 2-8 *The makeup of a carbon atom as represented on the periodic table.*

Table 2-1 Atomic weights of selected elements.

Element	Atomic Weight
Argon	39.948
Carbon	12.011
Chlorine	35.453
Hydrogen	1.0079
Nitrogen	14.007
Potassium	39.098
Lead	207.2

the elements that make up our sample gas and divide it by 29, we will arrive at a number that is greater than or less than 1. If our number is greater than 1, the gas is heavier than air; if it is less than 1, it is lighter than air. This may appear confusing at first but we will illustrate a computation and it should become very clear.

First, let us find the molecular weight of a heavier-than-air gas like argon. A scan of the periodic table reveals that the atomic weight of argon is approximately 39.948. The next step is to divide the atomic weight of argon by the atomic weight of dry air, as follows:

$$39.948 \div 29 = 1.37$$
$$\text{Argon} \quad \text{Air} \quad \text{Vapor density of argon}$$

As you can see, our resulting number is greater than 1 so we can deduce that when argon vapors are released, they will ultimately sink down and settle in low areas.

To illustrate the vapor density of a lighter than air gas we can look at hydrogen (H_2). This provides a little more complicated computation than argon as hydrogen is a diatomic gas. Diatomic gases require two atoms of the same element to form a stable molecule. Other examples of diatomic gases are oxygen (O_2), nitrogen (N_2), and chlorine (Cl_2). Have you ever wondered why the medical community refers to oxygen as O_2? The answer should be clear now—oxygen is a diatomic gas and the proper reference to it is O_2! Due to the diatomic nature of hydrogen, it is necessary for us to double the atomic weight prior to dividing by 29. Going back to the periodic table we find the atomic weight of a single hydrogen atom to be 1.0079. Doubling this value results in the weight of diatomic hydrogen to be 2.0158. We divide this value by the molecular weight of air (29) to arrive at a vapor density value for hydrogen. The computation for this is as shown.

■ **NOTE**
Diatomic gases require two atoms of the same element to form a stable molecule.

2.0158 ÷ 29 = .0695
Diatomic   Air   Vapor density
hydrogen   of hydrogen gas

As you have no doubt deduced, the vapor density of hydrogen is less than 1, so we now know that hydrogen gas is much more buoyant than air. A final type of vapor density computation deals with more complex formulas involving several different types of atoms. A look at propane gas reveals the formula to be C_3H_8. This means that it takes three atoms of carbon bonded with eight atoms of hydrogen to create propane gas. Figure 2-9 shows the propane molecule drawn out to further define its formula.

Would you suspect that propane gas when released is heavier or lighter than air? The correct answer is that propane is much heavier than air but we can now prove it mathematically. To do this, it is only necessary to add up all the atomic weights of all the atoms that make up the gas in question. This works for all gases no matter how complicated the formula may appear. The following computation shows the breakdown of the atomic weight of propane:

3 Carbon atoms at 12.011 Each	=	36.033
8 Hydrogen atoms at 1.0079 Each	=	8.0632
Total weight of propane	=	44.0962
Divided by the molecular weight of air	÷	29.0000
Vapor density value for propane		1.5205

As you can see, when propane is released, it will sink into low areas as it is approximately one and one-half times heavier than air. Chapter 3 goes into greater detail regarding vapor density and gases and which are heavier or lighter than air.

Atomic Symbols

Another important bit of information derived from the periodic able is that of atomic symbols. These symbols are the one or two letter designator for a given element. Single letter designators for elements such as hydrogen (H), potassium (K), nitrogen (N), and oxygen (O) are always expressed as a single uppercase letter.

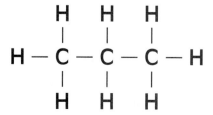

Figure 2-9 *Atomic structure of propane.*

Two letter designators are always expressed as an uppercase letter followed by a lowercase letter. Sodium is a good example of a two-letter atomic symbol and should be written as Na. Other elements with two-letter atomic symbols include magnesium (Mg), chromium (Cr), silicon (Si), and argon (Ar). You will see these symbols over and over as they illustrate chemical formulas. The formula for acetylene is C_2H_2 and can be broken down as follows: The uppercase letter C represents carbon and the uppercase H represents hydrogen. The small subscript number after each is merely stating that in this particular formula, it takes two atoms of carbon and two atoms of hydrogen to form the compound acetylene. This same logic can be applied to virtually any formula you encounter. Simply identify the various elements by atomic symbol then note any subscripted numbers to the symbol it follows. This tells you how many atoms of the particular element are required to form the compound. Hydrochloric acid, for example, has a formula of HCl. You should immediately notice that the compound is made up of hydrogen and chlorine but in this case there are no subscripted numbers. If there are no numbers, you can assume the formula only contains one atom of the listed element. Looking at HCl, you should understand that there is one atom of hydrogen bonded with one atom of chlorine.

The chemical formulas shown above, for example, are a recognized way to express the bonding of matter and are not uncommon at all in the industrial setting. These formulas and millions like them exclusively use atomic symbols to illustrate the constituents of the compound. In other words, the "chemical speak" of formulas is driven by the atomic symbols of the elements they represent. An atomic symbol is the shorthand way to express the name of an element. You will ultimately find that knowing the most common symbols is useful when discussing the chemical compounds. Therefore, it is necessary for you to commit to memory the chemical symbols listed in Table 2-2. You should also know the names and basic properties of the elements that each represent.

Atomic Number

Another important bit of information derived from the periodic table is the **atomic number** of an element. The atomic number refers to the number of protons in the nucleus of a particular atom. For example, magnesium (Mg) has an atomic number of 12. This means that in the nucleus of an atom of Mg, there are twelve protons. The atomic number of an element is an expression of the number of protons in the nucleus of that atom.

Remembering that the number of protons is an atom's signature, an atom with twelve protons in the nucleus can only be magnesium. Figure 2-10 illustrates a stable magnesium atom.

We can now tie all the terms together by looking closely at a stable atom of magnesium as shown in Figure 2-10 and comparing it to the expression found on the periodic table as shown in Figure 2-11. Table 2-3 ties all of the new terms you have learned into one easy to read format.

■ NOTE
An atomic symbol is the "shorthand" way to express the name of an element.

atomic number
the number of protons in the nucleus of a particular atom

■ NOTE
The atomic number of an element is an expression of the number of protons in the nucleus of that atom.

Chapter 2   Principles of Chemistry

Table 2-2 *Selected elements and their physical properties.*

Atomic Symbol	Element Name	Properties
H	Hydrogen	Colorless, flammable Gas
Li	Lithium	Soft, silvery white metal
Na	Sodium	Soft, silvery white metal; very water reactive
K	Potassium	Soft, silvery white metal; less water reactive than sodium.
Mg	Magnesium	Silvery white metal; very light and **ductile**.
Ca	Calcium	Silvery white metal
C	Carbon	Soft, black solid
Si	Silicon	Grayish solid
N	Nitrogen	Colorless, odorless, tasteless gas; nonflammable
P	Phosphorus	Waxy appearance, may be yellowish in tint; also found as a maroon powder
O	Oxygen	Colorless, odorless, tasteless gas; nonflammable
S	Sulfur	Yellowish solid
F	Fluorine	Yellowish gas with a pungent odor; many fluorine compounds are poisonous; **Oxidizer**; nonflammable.
Cl	Chlorine	Acrid smelling gas with possibly a greenish tint; Oxidizer; nonflammable
Br	Bromine	Brownish liquid with a very high **vapor pressure**
He	Helium	Colorless, odorless, tasteless gas; nonflammable and unreactive
Pb	Lead	Soft, heavy metal with a blue or gray tint; lead is the element at the end of the radioactive decay series.
Pu	Plutonium	Silvery white metal; highly radioactive
Hg	Mercury	Very heavy liquid; silver
Au	Gold	Brilliant golden metal, very **malleable**
Ag	Silver	Silvery white metal

ductile
the property of a material that prevents its returning to its original dimension when stress is removed

oxidizer
chemical other than a blasting agent or explosive that initiates or promotes combustion in other materials, causing fire either by itself or through the release of oxygen or other gases

vapor pressure
pressure exerted by a saturated vapor above its own liquid in a closed container

malleable
the property of flexibility; the ability to bend or be hammered into thin sheets

We can gain other information from the periodic table but it is not really pertinent to this text. Hopefully, you now have a good understanding of the makeup of an atom and how to recognize the various elements that may be found in chemical formulas. The atomic symbols and elements listed in Table 2-2 are commonly found in many chemical compounds encountered in waste site work as well as response.

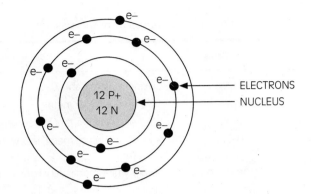

Figure 2-10
Magnesium represented by its atomic structure.

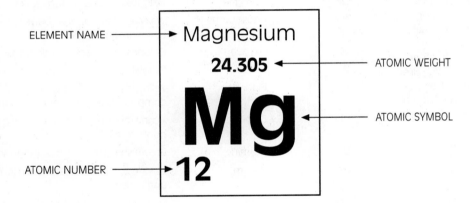

Figure 2-11 *The element magnesium as found on the periodic table.*

Table 2-3 *Technical terms and meanings.*

Technical Term	What It Really Means
The *atomic weight* is 24.305 AMUs	The overall weight of the atom; mostly concentrated in the nucleus
The *atomic symbol* is Mg	Chemical shorthand for magnesium
The *atomic number* is 12	There are twelve protons in the nucleus
The atom is stable	The number of protons equals the number of electrons—the electrical charges balance

COMPOUNDS

> **■ NOTE**
> Compounds are made up of various elements that have come together and formed an entirely new substance.

As mentioned before, new compounds are formed by the process of bonding. Compounds are made up of various elements that have come together and formed an entirely new substance. A compound can be defined as the chemical combination of two or more elements, which results in the creation of unique properties and a definite, identifiable composition of the substance. This means that two substances have come together with the understanding that each is contributing something to the mix. Each has agreed to forfeit some of its original qualities in order to create a new material with unique characteristics.

> **■ NOTE**
> Most of your response work or cleanup operation will deal with compounds.

Remember the earlier example of the bonding of sodium and chlorine? They came together to form sodium chloride, a compound. The distinction of this compound is that sodium, when it is part of this new compound, does not have its original properties. Likewise, chlorine loses its elemental identity, and an entirely new material is formed. The resulting sodium chloride has characteristics independent of the substances that it was originally created from. There are millions of already established compounds with thousands of new ones formed each year. Other examples of compounds include lithium bromide, calcium carbide and sodium bicarbonate. Most of your response work or cleanup operation will deal with compounds.

Mixtures

Mixtures are created when two or more compounds come together but still retain their individual characteristics. Salt water is the classic example of a mixture. To form this mixture, we add normal table salt (sodium chloride) to water. The sodium chloride (NaCl) and the water (H_2O) are mixed together, however, if we choose, the water can be evaporated away, leaving the salt behind. In this case neither substance lost its chemical identity or was altered at a molecular level. Although the water did evaporate, it merely underwent a change in state, not a change in its chemical composition.

> **■ NOTE**
> Temperature makes no difference as the chemical makeup of the substance will remain the same.

Another excellent example of a mixture is even more common than salt water. The air you are breathing right now is actually a chemical mixture of a number of materials. The composition of air is primarily nitrogen, an elemental gas; oxygen, another elemental gas; and trace amounts of argon. While each gas is content to mix with the others in the atmosphere, each retains its own identity and could be separated from the rest by physical or chemical means.

Many hazardous materials situations involve mixtures. Pulling a leaky gasoline tank from a ditch or handling a diesel fuel spill from an overturned tanker truck are examples of situations dealing with mixtures.

homogeneous
having uniform composition throughout

Mixtures can be further broken down as to their type. **Homogeneous** mixtures are those that have a consistent makeup throughout the entire volume. Milk, for example, is a homogeneous mixture. A sample taken at any point in a glass will have identical chemical composition. Temperature makes no difference as the chemical makeup of the substance will remain the same.

Another example may be a 55-gallon drum full of gasoline. This mixture has a consistent blend throughout the entire body which would qualify it as homogeneous. If a sample were drawn from the bottom of the drum, it would exactly match a sample taken from the top or anywhere else from the body of the vessel. Homogenous mixtures make sampling and classification a much easier task as it is not required to capture various stratas within the sample.

Heterogeneous mixtures, on the other hand, are those that have physical and chemical properties that are not uniform throughout the sample. Chocolate chip ice cream is an example of a heterogeneous substance. If you were to take a spoon-sized sample of this, you would find that there is an uneven distribution of chips and ice cream. The next spoon will have a totally different makeup as will the next. The interesting thing about these types of substances is their unpredictable nature. You may never know exactly what the makeup is going to be from one sample to the next. The one thing that is for certain is that the materials comprising the mixture will not mix with one another. If you are able to identify one substance you can rest assured the other will have completely different characteristics.

Think about the following situation to underscore the importance of understanding the two different types of mixtures. Your mythical work site has a drum full of waste oil that has been sitting beside the shop for several years. The cap for the drum has never been in place and the steel 55-gallon drum looks like it has seen better days. You have unfortunately missed the monthly safety meeting and in your absence your best friend nominates you as the site disposal person. Upon your return to work you receive the great news and are charged with the responsibility of disposing of the drum. You obtain a quote from a waste hauler. The quote you receive to remove 55 gallons of waste oil is $500. This seems like a lot of money and you now wonder if there is anything you can do to reduce the cost.

You wonder if there is any way to reduce the amount of material that requires disposal. The answer just might be obvious if you apply some of your newfound information to this situation. For example, the possibility certainly exists that water might have gotten inside the barrel over the years and since it is a fact that oil and water do not mix, we might have a heterogeneous mixture that we can separate. It is also a fact that oil will float on the surface of water creating two distinct layers. With this little bit of information in hand you begin to formulate the following plan.

You will carefully sample the barrel to find out if there is any water or other impurities inside. Knowing that the materials may create distinct layers will allow you to pump off either the oil or the water thereby disposing only what is necessary. Ultimately your plan works and you end up finding that the drum contained 50 gallons of water and only 5 gallons of waste oil. You successfully pump out the water and end up contracting for the disposal of only 5 gallons of waste oil. You were able to reduce disposal costs and the boss thinks you are a genius, because the company did not have to pay lots of money to landfill water. While

heterogeneous
having parts that do not display uniform makeup; having composition that differs from sample point to sample point

PHYSICAL AND CHEMICAL CHANGE

it is not always this simple, using your head and understanding a few concepts can go a long way.

We have discussed the various properties of solids, liquids, and gases. Now it is important to understand that these states may change under certain conditions. The most obvious example of this is through the application of heat or cold. By adding or taking away heat we can change the state of the matter involved.

Think for a moment about what happens to water when it reaches its boiling point of 212°F? It undergoes *a physical* change of state from liquid to vapor. A physical change occurs when a material changes in state but stays the same at the atomic level. The water is still water, chemically speaking, but it has now undergone a change in state or a change in phase. Take away the heat by putting the water in the freezer, and the water undergoes a phase change again; this time from a liquid to a solid. Once more, the molecular makeup of the water is constant, only the phase has changed. Some of the common terminology relating to physical change can be found in Table 2-4.

When doing a sizeup of a chemical release it is important to determine if the release was caused by a material unexpectedly changing state. A liquefied gas such as propane that has been heated and is now expanding from a liquid to a gas within the cylinder can cause some real problems. Eventually the expanding gas within the cylinder may cause catastrophic container failure and propagate a tremendous explosion and fireball. This process typifies a material undergoing a phase change and causing a substantial problem for responders. Always ask yourself if the problem you are facing is based on a simple change in state. Just because you know the reason for the change in state does not necessarily mean you can do anything about the problem but it does help in fully understanding the situation. You must always try to pinpoint the *reason* for the release. Anyone may be able to tell a container

■ **NOTE**
A physical change occurs when a material changes in state but stays the same at the atomic level.

! **SAFETY**
Always ask yourself if the problem you are facing is based on a simple change in state.

Table 2-4 *Physical change terminology.*

What Happens	What It Is Called	Common Example
A solid changes to a gas	Sublimation	Moth balls vaporize, leaving no residue
		Dry ice goes from solid to gas
A liquid changes to a solid	Freezing	Water turns into ice
A solid changes to a liquid	Melting	Ice turns to water
A gas changes to a liquid	Condensation	Moisture forms on the outside of a glass of ice cold soda on a hot day
A liquid changes to a gas	Vaporization	Water boiling

is leaking, but the astute hazmat person strives to figure out why. If it turns out to be as a result of a physical change, you may be able to intervene with tactics that allow you to stop the phase change.

Chemical change or a chemical reaction is different from physical changes in both concept and in the way matter is ultimately affected. When matter bonds, a physical change is not the only thing that may occur as a result of the reaction. With chemical change the materials themselves are subject to change at the atomic level.

In the following neutralization reaction, the by-products are water, a salt compound, and heat.

$$HCl + KOH \rightarrow KCl + H_2O + \Delta\uparrow (\text{heat liberated})$$

When the hydrochloric acid (HCl) is mixed with the potassium hydroxide (KOH), a neutralization reaction occurs. The by-products of this particular reaction are water, a salt compound called potassium chloride (KCl), and heat. This is a chemical change because the two original substances bonded to form completely new products and liberated (generated) heat.

Were you able to see how the bonding occurred? If not, look at the potassium hydroxide (KOH) compound and the hydrochloric acid (HCl) shown to the left of the arrow ($\rightarrow$). The horizontal arrow is chemical shorthand for "reacts to produce." The potassium (K) part of the (KOH) compound bonded with the chlorine (Cl) in the (HCl) and reacted to produce (KCl), which is found to the right of the arrow. The hydrogen (H) leftover from the (HCl), bonded with the hydrogen (H) and oxygen (O) in the (KOH) to form the water compound found to the right of the arrow. Heat is a by-product of this reaction and is always a concern when performing neutralization. Do not worry about how or why this happens, just accept the fact that the bonding occurred and the new substance is formed. These and many other reactions can be very aggressive and generate lots of heat in the process. Some neutralization reactions have actually melted the plastic containers the compounds were in. If a release is being caused by some chemical reaction it may not be possible to intervene until the reaction has ceased. If the reaction is occurring in a closed vessel or by another process, a substantial explosion may result from heat or off gassing. In any case, it is again necessary to really have a good understanding of why the release is happening. This may require the additional help of a chemist but it is well worth the time to consult a higher chemical authority if you are not sure about a possible reaction. Your tactics and ultimate safety may depend on understanding the reason behind the release.

EXOTHERMIC AND ENDOTHERMIC REACTIONS

In order to relate the information on chemical reactions back to our original definition of chemistry, we must now discuss the energies involved when creating new substances. All of the chemicals you will encounter in the field have a

■ **NOTE**
With chemical change the materials themselves are subject to change at the atomic level.

■ **NOTE**
The horizontal arrow is chemical shorthand for "reacts to produce."

! **SAFETY**
Heat is a by-product of this reaction and is always a concern when performing neutralization.

! **SAFETY**
Your tactics and ultimate safety may depend on understanding the reason behind the release.

exothermic
expression of a reaction or process that gives off energy in the form of heat

endothermic
description of a process that ultimately absorbs heat and requires large amounts of energy for initiation and maintenance

■ **NOTE**
Endothermic reactions absorb heat.

■ **NOTE**
kcal is an expression of heat.

certain amount of energy stored within them. This energy is either released as a result of, or required by, a chemical reaction. As it is released, it may ultimately be manifested as electrical energy, heat energy, light energy, or a combination of any of the above. Simply put, if the reaction is giving off energy, it is called **exothermic**.

Fire is a good example of an exothermic reaction. While we all recognize fire as *hot,* it is more technically correct to say that fire is a rapid oxidation of a fuel evolving heat, light, and fire gases. Explosions are also exothermic; mixing acids and bases is generally exothermic; and the action that causes a metal to rust is exothermic (although at very low temperatures).

On the contrary, if the reaction is absorbing energy it is referred to as **endothermic**. Endothermic reactions absorb heat. In other words, there is more heat stored in the products of the reaction than in the matter that creates the reaction. In the example that follows, the products of the reaction are to the right of the arrow and the reactants themselves are to the left. While these examples are a little harder to come by, a common one is the production of something called water gas. A water gas reaction combines water (H_2O) with carbon (C) with the resulting products being carbon monoxide (CO) and hydrogen gas (H_2). The reaction is as follows:

$$H_2O(gas) + C(solid) \rightarrow CO(gas) + H_2(gas) + H = 31.4 \text{ kcal}$$

Kcal is an expression of heat. Although this reaction may appear complicated, the bottom line in the reaction is this; there is more energy stored in the CO and H_2 part of the reaction than on the other side.

Why is any of this important? When chemicals mix together, energy is involved in one way or another. When energy is involved, things happen. Things like heat being released inside a drum which makes pressure go up, causes valves to fail, and/or pipes to burst. These instances often create emergencies or at the very least, cleanup situations.

When handling chemicals that have mixed, it is very important to ask these two questions: Are the chemicals currently reacting and if so, are they going to be confined in any way? If the answer is "yes" to both, it may be time to back off and let nature take its course. It may be easy to see that the tanker truck is leaking, but more difficult to understand why. Oftentimes, we overlook the reason for a chemical release and that can be a dangerous oversight. Once again you may be involving higher chemical authorities on the scene of these incidents. Take time to fully understand the problem before taking action.

! **SAFETY**
Take time to fully understand the problem before taking action.

BASIC GAS LAWS

Many of the situations encountered in hazardous materials response or cleanup involve gaseous materials. Therefore it is important to understand some basic principles of gases and gaseous materials. As we attempt to understand these

■ **NOTE**
Gases fill the space available to them and, more importantly, can be compressed into a smaller volume by adding pressure.

pressure
expression of force measured per unit of surface area

■ **NOTE**
PSI stands for pounds per square inch and is a commonly used measurement when referring to force.

■ **NOTE**
The volume of a gas under constant temperature can change by the addition or subtraction of pressure.

principles, it is sometimes best to look carefully at what is really going on inside a cylinder where the gas is present.

Whether it is a helium tank used to fill balloons at the car lot or an acetylene gas cylinder in the repair shop, there is a lot of activity going on behind the ominous steel walls. Gases fill the space available to them and, more importantly, can be compressed into a smaller volume by adding pressure. The resulting pressure inside may also be affected by heat and cold, thereby changing the molecular activity within the cylinder. In order to understand the concept of pressure, we must first define it.

Pressure is simply an expression of force measured per unit of surface area and may be quantified as PSI. psi stands for pounds per square inch and is a commonly used measurement when referring to force. You may have heard that the atmospheric pressure at sea level is 14.7 PSI. This means that the force of the atmosphere on your body at sea level is 14.7 pounds on every square inch of surface area. Simply stated, it is a matter of how much push is occurring over a given amount of surface area.

To illustrate the force/surface area relationship let us look at snowshoes for just a moment. Imagine that two people of equal mass are placed atop a 10-foot deep snowdrift by a helicopter (might as well be a big example!). They are dressed identically except that one person has on winter boots and the other is wearing big snowshoes. They are far enough apart that one could easily fall through the snow drift while leaving the other behind. The snow has a thin layer of frozen rain on top and when the two people are lowered onto the surface of the drift, it begins to creak and groan.

When the ice breaks through, which of our two volunteers do you think goes to the bottom of the drift and which one gets to go and call for help? Which one would you rather be?

If you answered "The one with the snowshoes" give yourself a gold star. Hopefully you deduced that the greater surface area of the snowshoes would distribute the mass over the weak surface of the snow, thereby keeping a person from falling into the depths of the snow drift. The forces generated by each of them were identical because the masses were the same, but the poor volunteer with the standard boots had a much higher force per unit area. In essence, the mass exerted the force over a more concentrated area hence a higher pressure.

Although the activity of a gas inside a cylinder is much more dynamic, the concept of force per unit area is the same. It is important to keep in mind that the molecules of any gas are in constant motion inside the cylinder. Figure 2-12 illustrates that there are collisions happening constantly between the molecules themselves and the sides of the cylinder.

Another way to look at this is that the expression of the pressure inside the cylinder is simply the frequency of collisions happening over an area of 1 square inch. More collisions mean more molecular activity, which ultimately means higher pressure. This relates to activities that you might be involved in because you should always consider that the properties of a gas are affected by things like

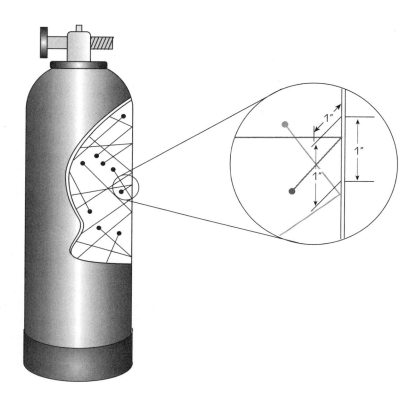

Figure 2-12
Molecular activity occurring within a compressed gas cylinder.

temperature and pressure. The volume of a gas under constant temperature can change by the addition or subtraction of pressure.

If pressure is increased, the volume will decrease. If the gas is cooled while under constant pressure, the volume is reduced and the gas ultimately turns into a liquid. Pressure is also affected by changes in temperature. If the temperature goes up, the activity of the gas within the vessel increases, thereby creating higher pressure inside the cylinder.

Confusing enough? If so, the following explanation of two important *gas laws* should illustrate these relationships and help you to understand what is happening to gases under pressure. If your not confused, congratulations and read on anyway!

■ **NOTE**
Pressure is also affected by changes in temperature.

Boyle's Law

Early in the 1600s, scientists were studying the relationships between pressure, temperature, and volume as they related to gaseous substances. Although the results of those and subsequent experiments were quite technical we can still extract some general information for our use.

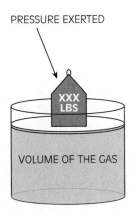 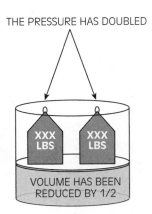

Figure 2-13 *Boyle's law illustrated.*

> ■ **NOTE**
> Boyle's law can be stated as follows: Given that the temperature remains unchanged, the volume of a given quantity of a gas varies inversely with the pressure exerted upon it.

One of the more important revelations about the behavior of gases was discovered by Sir Robert Boyle. In general, Boyle established the compressibility of gases and concluded that if the pressure exerted upon a certain volume of gas was doubled, the resulting volume would be reduced by one-half, assuming that the temperature remained constant. On the other hand, if the temperature remained constant and the pressure is decreased by one-half, the volume of the gas would double. Figure 2-13 illustrates that relationship. Boyle's law can then be stated as follows: Given that the temperature remains unchanged, the volume of a given quantity of gas varies inversely with the pressure exerted upon it.

What you should learn from Boyle's law is that there is a relationship between the pressure exerted upon a gas and the volume inside. To illustrate this, consider the example of air tanks used for self-contained breathing apparatus. These tanks have been improved over the years to reduce weight and increase efficiency but can also demonstrate a clear illustration of Boyle's law.

The older steel tanks contained approximately 47 cubic feet of air under 2,216 psi. The tanks were big and heavy and would cause the wearer to tire very quickly. Recent improvements have been made to the tanks to make them smaller and lighter, yet hold the same amount of air. These improvements have been accomplished by using lighter weight composite materials that are used to store the air at a much higher pressure. The composite tanks are about half the size and weight with pressures in the 4500 psi range. The increased pressure accommodated by the smaller size tank ultimately allowed the same volume of air inside.

> ■ **NOTE**
> Charles's law is as follows: When a gas is kept at constant pressure, the volume of a fixed amount of a gas will be directly proportional to the absolute temperature.

Charles's Law

In 1787 a French physicist named J. A. C. Charles also observed a direct relationship between pressure, temperature, and volume. Charles's law is as follows: When a gas is kept at constant pressure, the volume of a fixed amount of a gas will be directly proportional to the absolute temperature.

His work focused on the change in the volume of a gas when temperature was varied but the pressure remained constant. Simply put, the volume of a gas will go up if the temperature goes up, and down, if the temperature goes down. This law will be true assuming the pressure was not altered by some other external force. We might illustrate this concept further by applying it to a situation that may be encountered in the workplace.

Several compressed gas cylinders of oxygen are stored in the basement of a warehouse next to the boiler room. There is little ventilation and the room is physically uncomfortable due to the heat. The cylinders are warm to the touch and you begin to wonder, "Will the temperature in the room have any effect on the gases inside these cylinders?" The correct answer is "yes" but that is only part of the solution.

In this example the molecular activity will increase to some degree with this referred heat, but not to the extent that it will rupture the cylinders. It would take temperatures as high as 250°F to substantially influence the activity inside the cylinder.

Cylinders involved in fire are a different story. There is certainly enough heat to cause the cylinders to fail and may call for a defensive strategy based on maintaining a safe distance. Letting the incident run its course without committing personnel may be the right call in this instance. At any rate, it is the prudent individual who understands the basic relationships between the pressure, temperature, and volume of compressed gases. More importantly, the prudent and *safe* individual understands the relationships and applies them in the right circumstances.

Summary

- Chemicals can be dangerous. Always be alert and understand the situation, and never take the chemicals for granted.
- Chemistry is the study of the relationship between matter, the energy involved and the reactions that occur when matter is bonded.
- Matter has mass and occupies space.
- Atoms are the smallest division of matter. Atoms are made up of protons, neutrons, and electrons. Protons have a positive charge, neutrons have a neutral charge, and electrons have a negative charge.
- Most of the mass of an atom is centered in the nucleus.
- The number of protons in the nucleus of an atom is its signature. If the number of protons were to change, the atom would be something else. The number of protons an atom has is expressed as its atomic number.
- Elements come together to form compounds.
- State of matter refers to the chemical existing as a solid, liquid, or gas.
- Chemicals can have multiple hazards.
- The atomic symbol is a one- or two-letter designator for the element it represents.
- The atomic weight of an atom is an average weight of like atoms expressed in AMUs (atomic mass units).
- A stable atom is one in which the positive charges (protons) equals the negative charges (electrons).
- Mixtures are a combination of compounds. Each retains its chemical identity and the mixture can be separated by physical means.
- Homogeneous mixtures have a consistent makeup throughout the entire volume of the sample. Material taken from any point in the sample will always be the same.
- Heterogeneous mixtures do not have a consistent makeup. Samples can differ from point to point within the volume.
- Materials can undergo physical change and chemical change. Physical change occurs when the chemical changes state but stays the same at the molecular level. Water changing from a liquid to a vapor via the boiling process is a physical change. Chemical change occurs when materials come together and change at the molecular level. Acid neutralization is an example of chemical change.
- Exothermic reactions give off energy; endothermic reactions absorb energy.

- Boyle's law establishes the *compressibility* of a gas. If the temperature of a gas remains unchanged, the volume of a given quantity of gas varies inversely with the pressure exerted upon it.
- Charles's law deals with temperature and volume. Basically, if the temperature goes up the volume increases. If the temperature goes down, the volume decreases. Charles's law is stated as follows: When a gas is kept a constant pressure, the volume of a fixed amount of a gas is directly proportional to the absolute temperature.

Review Questions

1. Define the following terms:
   Chemistry
   Matter
   Proton
   Neutron
   Electron
   Stable atom
2. True or False? Protons have a positive electrical charge, account for approximately one-half of the atom's weight, and are found in every atom.
3. Why is the number of protons very important in establishing the identity of an atom?
4. What is the major driving force in atomic bonding?
5. What is *state of matter* and why is it so important?
6. Name the three states of matter.
7. Define the periodic table.
8. Show the equations for determining the vapor density for the following:
   Hydrogen
   Nitrogen
   Fluorine
   Chlorine
9. To what does the atomic number refer?
10. Explain the difference between compounds and mixtures.
11. A solid changing to a gas is called _____.
12. A liquid changed to a solid is called _____.
13. A solid changing to a liquid is called _____.
14. A liquid changing to a gas is called _____.
15. A gas changing to a liquid is called _____.
16. True or False? All chemicals have a certain amount of energy stored within them.
17. Explain the difference between endothermic and exothermic reactions.
18. True or False? Pressure is not affected by change in temperature.
19. Briefly describe Boyle's law and state why it is important.

Hazard Classes

Learning Objectives

Upon completion of this chapter, you should be able to:

- List the eight major hazard classes as outlined by the Department of Transportation (DOT).
- Understand the broad hazards of some common explosive materials.
- Define detonation and deflagration.
- Define brisance and describe the potential effects of blast fronts and fragmentation.
- Define ANFO and understand the constituents of the compound.
- Understand the composition of TNT and dynamite.
- Understand the broad hazards of gaseous materials.
- Define the following key terms: compressed gas, liquefied gas, cryogenic liquid, vapor density, diffusion, upper and lower explosive limits, and BLEVE.
- Understand the broad hazards of flammable liquids.
- Define the following key terms: flammable liquid and flash point.
- Define flammable solid.
- Define pyrophoric.
- Understand the broad hazards of oxidizers.

- Understand the oxidation–reduction relationship.
- Understand the significance of the suffixes *-ate* and *-ite* when identifying potential oxidizers.
- Understand the broad hazards of poisonous materials and heavy metals.
- Define the following terms: highly toxic, organophosphate, neuroeffector junction, and synapse.
- Understand the broad hazards of radioactive materials.
- Define radioactive isotope, alpha particle, beta particle, gamma radiation, roentgen, and REM.
- Understand the concept of Time-Distance-Shielding.
- Understand the broad hazards of corrosives.
- Describe the process that creates free hydrogen in acid solutions.
- Describe the difference between acids and bases.
- Explain pH.
- Understand the basic concept of neutralization and give an example.
- Identify the hydroxide ion in caustic solutions.
- Understand the difference between strength and concentration.

INTRODUCTION

■ **NOTE**
A careful understanding of the properties of the material, a sound plan of action, and comprehensive training have been shown to increase the safety margin when handling chemical incidents.

The safe handling of chemicals at the waste site and on the emergency ground is of paramount importance. A careful understanding of the properties of the material, a sound plan of action, and comprehensive training have been shown to increase the safety margin when handling chemical incidents. Whenever handling chemicals, it is important to understand the nature of the materials you are dealing with. Take time to read material safety data sheets (MSDS) or other references regarding the hazards of any chemicals you may come in contact with. Additionally, a good action plan should be in mind when handling chemicals. Determining emergency procedures or basic medical treatments should be considered well in advance of opening or moving any chemical containers. A moment of preplanning can result in a reduction in panic if an unexpected release occurs. During emergency response activities, everyone on the scene should be aware of what to do in case the incident takes a turn for the worse. Even though the event may not have been planned, your response to it certainly can be!

To be as safe as possible when handling chemicals, it is critical that you understand the inherent properties of all of the materials prior to handling them or cleaning them up. This information may be conveyed as part of periodic safety briefings on the scene as well as sound reference and identification efforts combined with your own experience. If you are not sure of what is expected of you or are unclear as to the specific properties of a chemical, ask your supervisor or the person in charge of your operation. Ultimately, it is the scene supervisor's responsibility to ensure that all employees have received the proper equipment and that the action plan is carefully followed.

Although some of the materials you will see in your career may be quite hazardous, they need not pose an unreasonable threat to you. Understanding basic chemical properties and how chemicals can be classified will assist you in many other areas, such as personal protective equipment selection. Proper air monitoring practices and respirator cartridge selection also depend on your understanding of your chemical foe. We covered some basic chemical principles in Chapter 2; now we tie the information together by looking at the different hazard classifications as well as some of the chemicals that exist in each of the different groups.

The information contained in this chapter is designed to assist you with the *recognition* of some common chemical classes encountered in response and waste site work. We also provide some basic safety and handling information and give some broad hazards of the various chemical classes from a big picture standpoint. This is not the definitive text on chemistry or chemical research—it will only serve as a guide for thinking about the various chemical classes. The information on the hazards of the chemicals discussed is not intended to replace the information found in material safety data sheets or other site-specific reference material. Copies of material safety data sheets should be available at fixed facilities or any time a chemical substance can be positively identified. Material safety data sheets provide the most specific and comprehensive information on any given substance and should be used whenever possible.

We focus our discussion on some basic information regarding eight commonly encountered chemical classes or groups. These groups are also recognized by the DOT and are represented by the different identification placards you may have seen on packages and trucks. We discuss those labels and placards in Chapter 5, but for now you should recognize the group of eight to include: explosives, gases, flammable liquids, flammable solids, oxidizers, poisons, radioactive materials, and corrosives.

EXPLOSIVES

Most of us tend to have a great deal of respect for explosive materials. We realize that if things go wrong and we are in the near vicinity of a significant explosion, bad things will happen. This list of bad things can include the disintegration of our bodies from pressure and heat to subsequent impact from projected objects (fragmentation). Explosive events are generally regarded with great care because we realize there may not be a second chance. For this reason we should have a keen understanding of what an explosive is as well as what situations may be conducive to creating an explosive atmosphere.

A layman's definition of an explosive is as follows: A chemical substance that in itself can react to produce a gas at a relatively high temperature and pressure and at such a speed as to damage the surroundings. Heat or friction or pressure can initiate the explosive reaction and the severity of the reaction depends on the particular material and the conditions present. A whole range of blast effects are possible including some of the following:

- *Projection or fragmentation hazard.* Fragments of the explosive container or the physical surroundings are carried outward by the blast front. These fragments can be quite large depending on what is surrounding the explosive material. Protection against fragmentation can best be accomplished by maintaining a safe distance. This decision on distance relies greatly on the materials involved, but should be as conservative as possible.
- *Pressure front propagation.* Explosions are always accompanied by an outward expression of pressure, which is sometimes called a blast front or **brisance.** The pressure created can be enormous (up to several thousand pounds over ambient pressure) and very fast moving. This resulting pressure wave is what causes distant windows to break and the ground to shake following a sizable blast. Incidentally, only a few pounds **overpressure** (5–7 **psi**) are enough to rupture the eardrums of anyone within a short distance of the blast. When explosions occur within buildings, 10 psi overpressure has been significant enough to completely destroy the structure. If you happened to get caught in the pressure front of a serious explosion, a major injury or death would most likely occur.
- *High generation of heat.* Think of an explosion as being an extremely fast fire. The pressure front as well as the fire front move outward very quickly and can cause great damage. Temperatures of fire fronts can be a high as 2,000°F and will cause significant damage to living tissue as well as most common construction materials.
- *Sound wave production.* In some cases the *bang* of the sound wave precedes the *whoosh* of the pressure front and other times it will follow. In either case, sound waves emit from the blast site and are significant, as they will mimic the pressure front in terms of speed.

Explosives are classified by the Department of Transportation (DOT) into various categories depending on the type of explosion that results. The DOT breaks down the class of explosives as represented in Table 3-1. Rather than attempting to digest the entire table it may be more beneficial to break the nature of explosives down into three major categories. These categories are much more broad than the DOT categories, but for field work they are more than sufficient. They are:

1. High explosives
2. Low explosives
3. Blasting agents

brisance
an expression of the shattering effect of a particular explosive material

overpressure
pressure created over and above ambient pressure

psi
pounds per square inch; in this setting, psi is the expression of the pressure a material exerts on the walls of a confining vessel or enclosure

Table 3-1 *Classification of explosives.*

1.1 Explosives with a mass explosion hazard
1.2 Explosives with a projection hazard
1.3 Explosives with predominately a fire hazard
1.4 Explosives with no significant blast hazard
1.5 Very insensitive explosives; blasting agents
1.6 Extremely insensitive detonating substances

High Explosives

detonate
a rapid, self-propagating decomposition of an explosive material evolving pressure and temperature fronts at supersonic speeds

dynamite
an industrial high explosive that is moderately sensitive to shock and heat; main ingredient is nitroglycerin or sensitized ammonium nitrate

SAFETY
Never move or otherwise disturb these sticks of dynamite as the slightest disturbance may cause detonation.

High explosives are those which **detonate** or move at supersonic speed (greater than 1,100 feet per second) throughout the mass of the explosive material. Decomposition occurs instantaneously and self-propagates through the body of the explosive substance. Pressure and temperature waves move outward with the blast front and the pressure developed in the near field (close to the blast) can be extreme. In true detonations, the pressure wave usually out distances the heat wave but it is actually a neck and neck race. **Dynamite** and trinitrotoluene (TNT) are classic examples of high explosives and can have devastating effects upon anything close to the center of the explosion. Although dynamite and TNT are both categorized as high explosive, they are quite different chemically.

Dynamite is a mixture consisting of an absorbent such as sodium nitrate and diatomaceous earth saturated with nitroglycerin. This combination creates a very intense mixture capable of substantial destruction when properly detonated. Freshly synthesized dynamite, however, is actually quite stable and requires a booster charge in order to detonate. Old dynamite is a different story as it may begin to *sweat* if subjected to heat for long periods of time. This sweating is the result of a chemical reaction in which the nitroglycerin breaks down chemically to form nitric acid and glycerol. The resulting mixture becomes shock sensitive and can be recognized by a milky looking substance clinging to the outside surface of a stick of dynamite. *Never move or otherwise disturb these sticks of dynamite* as the slightest disturbance may cause detonation. This type of emergency can become very intense and tension filled for obvious reasons. If sweating dynamite or other explosives are unexpectedly encountered, do not disturb the scene and move to a safe distance with caution. Unfortunately, safe is a relative term here as each incident may vary, but it is not out of line to start your safe zone 1,500–2,000 feet from the material involved. Do yourself a favor—use good judgment and be conservative when setting your safe zones at explosive incidents.

TNT or trinitrotoluene is widely used both in the commercial sector and the military. When newly synthesized, it looks like a yellowish sludge and is somewhat crystalline in structure. It is used as a composite explosive with other materials as well as a bursting charge and standard demolition charge. The chemical backbone of TNT is the common industrial solvent toluene. Toluene is a **flammable liquid** with a chemical structure especially suited for constructing new substances. In the case of TNT, the toluene is nitrated with 3 nitro groups (NO_2), hence the prefix tri-nitro, and then mixed with an acid. The chemical structure for TNT is simple and is shown in Figure 3-1. If you look closely at the structure, the name makes perfect sense—three nitro groups (NO_2) on the toluene backbone—trinitrotoluene. TNT is also a threat in terms of its flammability and may detonate if heated above 450°F. It is also toxic by ingestion, inhalation, and absorption.

Low Explosives

Low order explosives function primarily by **deflagration,** a slower reaction rate than the high order explosives but can be just as devastating. Deflagrations are explosions that move through the explosive material slower than the speed of sound but in which the heat wave precedes the pressure wave. Make no mistake however: Both fronts are propagated within milliseconds and can certainly prove fatal to anyone nearby. Black powder is one of the oldest explosives known and is a good example of a low order explosive. It is comprised of potassium or sodium nitrate, sulfur, and charcoal. Common mixtures include 75% nitrate, 15% charcoal, and 10% sulfur. Black powder looks like crushed up charcoal and is commonly stored in small steel cans. It will deflagrate rapidly when exposed to heat, thereby presenting a dangerous fire and explosion hazard. Black powder is commonly used in timed fuses for blasting as well as in igniters and primers for propellants and detonators.

Blasting Agents

Blasting agents are those materials commonly used to initiate the higher order explosives. Materials like dynamite and C-4 (a military explosive) are not easily detonated under normal circumstances until an explosive train is established. An

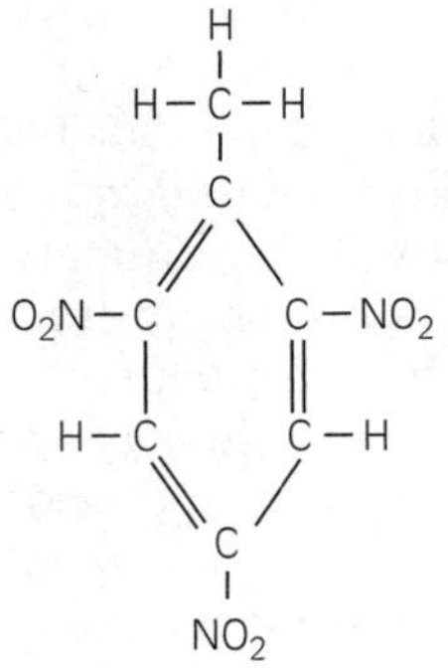

---

flammable liquid
any liquid having a flash point below 100°F (37.8°C), except any mixture having components with flash points of 100°F (37.8°C) or higher, the total of which make up 99% or more of the total volume of mixture

! SAFETY
TNT is also a threat in terms of its flammability and may detonate if heated above 450°F. It is toxic by ingestion, inhalation, and absorption.

deflagration
rapid combustion of a material occurring in the explosive mass at subsonic speeds

■ NOTE
Black powder is commonly used in timed fuses for blasting as well as in igniters and primers for propellants and detonators.

Figure 3-1 *Chemical structure of TNT.*

> **■ NOTE**
> An explosive train is set up to enable low order explosives to detonate higher order explosives.

ANFO
an acronym for ammonium nitrate and fuel oil

explosive train is set up to enable low order explosives to detonate higher order explosives. In other words, a smaller boom must be propagated to initiate the bigger boom created by high explosives. Blasting agents are often used to create these smaller explosions but can be quite hazardous as stand alone materials. A blasting agent that has gained recent attention is a material called **ANFO**. The mixture contains ammonium nitrate (NH_4NO_3) and a fuel oil such as diesel mixed in certain proportions and then subsequently ignited. The ammonium backbone (NH_4) is composed of nitrogen and hydrogen and attached to the reactive ion called nitrate (NO_3). The grade of ammonium nitrate is usually expressed in percent of nitrogen and may be expressed as 20.5% N or 33.5% N. The power of this combination was tragically displayed in the bombing of the federal building in Oklahoma City. The nation was shocked by the event and amazed at the destruction caused by such an easily obtained chemical substance.

This low sensitivity explosive must be initiated by a booster charge to be successfully detonated. It is commonly used in composite explosives where it is combined with a more sensitive explosive partner. Oftentimes aluminum shavings or some other powdered metal is used to make the explosive reaction hotter or more vigorous. Ammonium nitrate is also very **hygroscopic** and must be packed in airtight containers. If exposed to the air for any period of time, its strength as a blasting agent will be greatly decreased. Because of its cratering effect and low cost, ammonium nitrate is primarily used for commercial blasting at construction sites and rock quarrying. This material is not suitable for underwater blasting unless packed in a watertight container or detonated immediately after placement. You may be familiar with ammonium nitrate as a fertilizer widely used in the farming industry. FGAN is fertilizer grade ammonium nitrate that is prilled and often coated to reduce surface to air contact with the ammonium nitrate component. Prills are artificially produced aggregates or nuggets of a material. In the case of ammonium nitrate the prills consist of 94% ammonium nitrate and 6% fuel oil. Figure 3-2 shows an uncontrolled release of FGAN from the underside of a tanker.

hygroscopic
having the ability to absorb and retain moisture from the air

Table 3-2 illustrates the detonation velocity of some of the more common explosive compounds.

> **! SAFETY**
> ● Concentrations of vapor from the surface of a flammable liquid or the body of a flammable gas cloud may build up to form an explosive atmosphere.

Additional Explosive Situations

In addition to the hazard posed by explosive substances such as dynamite, there is a danger of explosive conditions caused by flammable liquids and gases. Concentrations of vapor from the surface of a flammable liquid or the body of a flammable gas cloud may build up to form an explosive atmosphere. These explosive atmospheres could be an especially serious problem if the vapors or gases were somehow confined. Confinement could be accomplished by a buildup of vapors in an underground vault or storage room with poor air movement. These types of materials and situations are not grouped with explosives like TNT, but the end result can be the same.

Figure 3-2 *Release of molten FGAN from a tanker truck.*

> **⚠ SAFETY**
> ● Ignition sources can come from many points including pilot lights, turning on an electrical switch, electrical arcs from machinery turning on or off, or static electricity.

It is not hard to imagine the potential consequences of a prolonged release of propane in an enclosed space. If the leak were contained and the room poorly ventilated, the vapors would begin to collect at the floor. Eventually, the area would be full of propane vapors and reach a point where the vapors and the air were mixed in the right proportions. Once ignition of the gas occurred, the explosion would blow out windows and doors and cause the propagation of a fire and pressure front. Ignition sources can come from many points including pilot lights, turning on a light switch, electrical arcs from machinery turning on or off, or static electricity. Structural damage would occur and anyone trapped inside the room would probably die. For this reason responders and site workers alike should be greatly concerned with explosive atmospheres. Constant air monitoring

Table 3-2 *Detonation velocity of common explosive compounds.*

Name of Material	Use	Velocity of Detonation (expressed in feet per second)
Black powder	Timed blasting fuse	1,300
Nitroglycerin	Commercial dynamites	5,200
Ammonium nitrate	Demolition charge Composition explosives	8,900
TNT	Demolition charges Composition explosives	22,600
Tetryl	Booster charges Composition explosives	23,200
C3	Demolition charge	25,000
PETN	Detonating cord Blasting caps Demolition charges	27,200
RDX	Blasting caps Composition explosives	27,400

!**SAFETY**
Constant air monitoring should be done when working in suspected flammable atmospheres.

flammable range
concentration of gas or vapor in air that will burn if ignited; expressed as a percentage that defines the range between a lower explosive limit (LEL) and an upper explosive limit (UEL)

!**SAFETY**
Flammable atmospheres are not to be taken lightly. Standard chemical protective suits are insufficient protection against explosive forces and flash fire.

should be done when working in suspected flammable atmospheres. Never enter flammable atmospheres unless you have an extremely good reason to do so, and even then it is still risky.

For flammable liquids and gases to behave in this fashion, they must be within their **flammable range.** All flammable liquids and gases have such a point where the mixture of fuel to air is just right for an explosion. For instance, acetone has a lower explosive limit (LEL) of 2.5%. Simply put, this means that the mixture represents 2.5% fuel (acetone) and 97.5% air. Hydrogen has an LEL of 4%. Again, this represents the volumetric measurement of 4% hydrogen and 96% air. You may also think of it as 4 parts hydrogen and 96 equal parts of air mixed together.

Additionally, there is an upper explosive concentration or upper explosive limit (UEL), which is the upper range of flammability for that particular material. The concept of UEL is the same as LEL except it is the upper limit of the explosive mixture. Once again this value is expressed as a percentage and is an important number to be aware of. It is significant because somewhere between the lower and upper explosive limit is where trouble may be possible. If the material is in the right mixture, confined in some fashion, and subsequently ignited, a significant deflagrating explosion will occur. Flammable atmospheres are not to be taken lightly. Standard chemical protective suits are insufficient protection against explosive forces and flash fire. Therefore, it is imperative to consult an

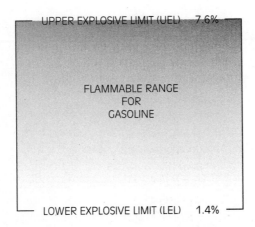

Figure 3-3
Flammable range for gasoline.

MSDS to find out if a material is flammable and if so, what the flammable range is. Figure 3-3 depicts the flammable range for gasoline. The following list shows the flammable ranges for some common liquids and gases:

Acetone (liquid), 2.5–12.8%   Ethyl alcohol (liquid), 3.3–19%
Benzene (liquid), 1.5–8.0%   Hydrogen (gas), 4.0–75%
Carbon disulfide (liquid), 1.0–50% Diethyl ether (liquid), 1.8–48%
Propane (gas), 2.4–9.5%   Methane (gas), 5–15%

When you consider that these ranges are based on the volume of the whole air space available, these numbers may seem fairly high. Keep in mind, however, that an explosive atmosphere need only build up in the immediate vicinity of the source. If ignition occurs, it will almost certainly start a fire in the main body of material, which will spread and also disperse the material, causing it to ignite violently. In other words, explosive vapors may form in pockets rather than evenly throughout the available space. This makes air monitoring very important and very difficult all at the same time. Aside from removing the source, remote ventilation is often the safest and most effective way to deal with flammable atmospheres.

Health Hazards Posed by Explosives

The principal danger posed by explosives is detonation and/or significant fire, however nitroglycerin and dinitrophenol are also toxic. Substances that are toxic, irritants, or sensitizers must be so labeled under the provisions of the Occupational Safety and Health Hazard Communication Standard (effective November 1985). Hazard warning information must be contained on a label and on a MSDS for the material. We are mostly concerned with the big booms that

■ **NOTE**
Substances that are toxic, irritants, or sensitizers must be so labeled under the provisions of the Occupational Safety and Health Hazard Communication Standard (effective November 1985).

may occur, but do not discount a significant fire hazard or toxicity associated with the chemical.

Safe Handling, Use, and Storage of Explosives

When working with explosive substances it is essential to have a sound fire prevention and storage program. Areas where explosives are handled or stored, or any area where explosive vapors may exist, must be free of ignition sources. Nonsparking tools should be used, and electrical circuits should be properly safeguarded against short circuits and sparking. It is vital to understand exactly what you are working with and what the potential hazards may be. Explosive incidents are relatively infrequent but potentially deadly because of the violent nature of the reactions.

Usual personal protective gear is not likely to be effective against an explosive blast. Structural fire fighting gear may be effective however, for fire fighting when significant blasts are not expected. Chemical protective attire may be suitable for protection against toxicity but are ineffective against heat or explosive forces.

Emergency Procedures

Planning and training for emergencies or cleanup operations involving these materials should be carried out before any personnel approach the incident. Anyone dealing with an explosive situation should have a thorough knowledge of the materials involved and the threats that may be present. Do not hesitate to back off and call for more help.

Medical personnel should be available for personal injury and must be in place before anyone goes to work at the scene. Information about particular chemical hazards should be available on an MSDS, which should be made available to assist in prompt and proper medical treatment.

Particular circumstances may require changes in these general procedures. For this reason, it is necessary to be familiar with emergency response procedures for individual chemicals ahead of time.

GASES

This section is designed to encourage an awareness and understanding of the kinds of hazards that can result from working with compressed, liquefied, and cryogenic gases. Gaseous materials may be nonflammable, flammable, toxic, corrosive, oxidizing, or even radioactive. Some gases such as chlorine have multiple properties including a very high health hazard. Chlorine is toxic, corrosive, a strong oxidizer and poisonous. Figure 3-4 shows a typical 1-ton chlorine cylinder that contains the chemical as a liquefied gas.

Generally, gas incidents are difficult to manage successfully for this very simple reason—once a gas escapes its cylinder, it is virtually impossible to get it back in again. You will most likely be dealing with the aftermath of the release rather than the release itself. For the most part, a gas will go where it wants to when released.

Definition of a Gaseous Material

The DOT has several technical definitions of flammable, nonflammable, and poisonous gases. We do not list them here because they do not mean much to the average person who works with or responds to these materials in the field. A much easier way to understand gases is this. A gas is a material in a state of matter that at normal temperature and pressure tends to fill the space available to it. Gases are different from solids and liquids as they are less viscous and less dense at a molecular level. Additionally, if the gas is confined in a cylinder it will expand to fill the whole container. If the gas is released into the air it will diffuse until limited by some confining feature such as the walls of a room or building. **Diffusion** is the action of a gas to move from an area of higher concentration to an area of lower concentration. A common example to illustrate this point may be the pleasant smell of food cooking in the kitchen. In this case the smell is much stronger in the kitchen (area of higher concentration), but you may be able to detect the same odor in an adjacent room (area of lower concentration). This distant detection of the odors is an excellent example of how a typical gas may diffuse when released.

diffusion
the action of a gas to move from an area of higher concentration to an area of lower concentration

Figure 3-4 *A 1-ton chlorine cylinder.*

General Hazards of Gaseous Materials

With a very generic definition of a gas in hand, let us investigate some of the hazards we may encounter as site workers and emergency response personnel. Because many gases are stored in cylinders in excess of 2,000 psi, these materials can pose physical hazards. If a high pressure cylinder is ruptured or the valve stem is knocked off, it can behave like a missile and travel great distances at high speeds. Unfortunately this type of missile has no guidance system and can often change direction radically and cause damage in very erratic patterns. Obviously, anything or anyone in the path of a ruptured cylinder runs the risk of significant injury. Look at Figure 3-5 and see if you recognize the potential hazards.

The cylinders in Figure 3-5 are dangerous from a number of aspects. First, they are unrestrained and could be knocked over very easily. The cylinder at the far right of the photo has no valve cover and is at an even higher risk to have the valve stem knocked off. The piping going through the wall is a gas line, which could also be broken in the event the taller cylinders were to fall over. The small cylinders are stored in a container with some very corrosive cleaning agents. In

Figure 3-5
Miscellaneous compressed gas cylinders in storage.

the event the cleaners leak, the fluid could eat away at the steel cylinders leading to failure. All in all, this is not a good storage scenario but one that could be easily improved. As workers in this field it is incumbent of you to recognize hazards such as these and make corrections where necessary.

Some gases are flammable and can be ignited easily. When these flammable gases are released, they may ignite and flash back to the initial area of the release. Anyone caught in the resulting flash fire could be seriously burned or killed. Chemical protective clothing without special flash fire protection will not protect against fire. Flammable atmospheres must be carefully monitored and only entered under the most extreme circumstances. Most of the time, the area should be ventilated before personnel enter and careful site monitoring should be in place to "watch their backs" while operations are completed. Remember that any mechanical ventilation devices should be remote to the area or intrinsically safe and not subject to creating sparks or other potential ignition sources. Intrinsically safe means that the piece of equipment is not capable of generating an external arc or spark, which is critical when operating in potentially flammable atmospheres.

Examples of some flammable gases that may pose a threat due to fire or explosion are given in the following list. The checkmarks denote gases that are lighter than air.

Acetylene ✓	Hydrogen ✓
Butadiene	Hydrogen sulfide
Butane Ethylene ✓	Methane ✓
Cyanogen	Propane
Diborane ✓	Silane

Many gases are acutely toxic by multiple routes of entry and can kill you even if you suffer the smallest exposure. Arsine for example is a highly toxic gas that can be fatal after only a few breaths. Arsine is widely used in the semiconductor industry as well as silane, germane, and some other very toxic gases. Take care to really understand what these materials are capable of doing and protect yourself accordingly. Some other poisonous gases you may encounter are:

Ammonia, anhydrous	Diborane
Ethylene oxide	Hydrogen chloride, anhydrous
Boron trifluoride	Hydrogen fluoride, anhydrous
Carbon monoxide	Methyl bromide
Chlorine	Germane
Cyanogen	Nitric oxide
Dichlorosilane	Phosgene
Hydrogen sulfide	Phosphine
Fluorine	Sulfur dioxide

> **SAFETY**
> Chemical protective clothing without special flash fire protection will not protect against fire.

> **NOTE**
> Intrinsically safe means that the piece of equipment is not capable of generating an external arc or spark, which is critical when operating in potentially flammable atmospheres.

Gases such as chlorine are quite corrosive and can cause severe lung damage if inhaled. Chlorine irritates the lining of the lungs and creates a fluid buildup that can literally cause you to drown in your own secretions. It is actually labeled as a poisonous gas because its toxicity outweighs its corrosive nature—another example of a chemical having multiple properties.

Ammonia gas or vapor is another flammable respiratory irritant that creates a caustic solution when mixed with water. Most people fail to recognize ammonia as a flammable gas but make no mistake, it is definitely dangerous and can create flammable atmospheres just like propane or acetylene. Take care to avoid skin contact with vapors as sweat or normal body moisture can react with the vapor to form the corrosive compound called ammonium hydroxide. Ammonium hydroxide or aqua ammonia has a relatively high pH (it is caustic) and may cause localized skin irritation.

Compressed, Liquefied, and Cryogenic Gases

The gaseous materials used and stored in the workplace are found in three basic forms: compressed, liquefied, and cryogenic. Each of these forms can present a variety of hazards and problems for those who use, work with, or respond to emergencies involving gases. The next sections discuss each of these different types of gases and give some specific chemical examples of each classification.

Compressed Gases Compressed gases are stored in cylinders under relatively high pressure. The concept is that a gaseous material such as oxygen or helium has been forced into the cylinder by applying lots of pressure. Remember Boyle's law from Chapter 2? It basically states that as the pressure exerted upon a gas is increased the volume decreases. This act of compression is a good thing as it enables us to put a lot of material into the cylinder. One of the potential downsides however, is the heat generated by compressing a gas. This form of heat is called the **adiabatic heat** of compression and is generated anytime a gas is forced into a cylinder under pressure.

This type of heat generation occurs when filling self-contained breathing apparatus (SCBA) cylinders. If you have ever filled one of these cylinders, you noticed that as the air is forced inside, the outside of the bottle begins to feel warm. This is due to the fact that the air is being forced into the tank and is normal during this type of operation. Old steel cylinders were actually filled in a water bath to absorb the resulting heat, to counteract the heat of compression, and to reduce the loss of pressure after the filling process was completed. You may have seen that after the bottle was filled and the heat dissipated, the pressure went down slightly. This is because heat also increases molecular activity, which then raises the pressure to a certain degree. As the molecular activity stabilizes due to cooling, the gas actually condenses slightly and the pressure drops. For this

---

! **SAFETY**

Gases such as chlorine are quite corrosive and can cause severe lung damage if inhaled.

adiabatic heat of compression
descriptive of a system in which no net heat loss or gain is allowed

reason you should fill any compressed gas cylinder slowly and then top it off again after it has cooled to maximize the available volume.

As mentioned previously, high pressure cylinders are dangerous in the event they are compromised. Therefore, when working with compressed gas cylinders it is wise to remember:

- Never move the cylinders without putting the valve cover on.
- Use cylinder dollies whenever possible.
- Do not expose compressed gas cylinders to temperatures above 250° Fahrenheit.
- Do not modify valves or threads.
- In the event cylinders are involved in fire, treat them like explosives—maintain safe distances.
- Understand the chemical hazards associated with the gas including signs and symptoms of exposure and emergency actions. Read the MSDS to find out what the health and fire hazards are. Be aware of any special handling instructions.
- Keep cylinders secured when in use or in storage.
- Never assume a cylinder is empty if it is unmarked. Develop a system to designate full and empty cylinders.
- Do not try to figure out what the gas is by smell. You might be wrong!

Liquefied Gases Gases can also be used and stored in the liquefied state. Liquification is accomplished by subjecting the gas to a certain amount of pressure while lowering the temperature inside the storage vessel. The ease with which this is done depends on the molecular weight and the chemical structure of the substance. Some gases, such as propane and butane, can be easily condensed into a liquefied state by using pressure and temperature reduction. Other gases, such as oxygen and helium, are **permanent gases** and can only be liquefied at extremely low temperatures. We will not go into great detail about the process of liquefying a gas but you should understand that the gas exists inside the cylinder differently than a compressed gas does. Liquefied gases are stored in the cylinder just as the description implies—in the liquid phase. The primary advantage to liquefying gases is that the container can hold significantly more potential volume when stored in a liquid state. Imagine that you have taken a huge volume of gas and condensed it into the liquid state. Liquids are more tightly packed together at the molecular level than gases so in essence you get more volume of a given material when it is in a liquid phase.

Propane is a classic example of a material that is gaseous at normal temperature and pressure but can be easily liquefied. Once propane is stored inside a cylinder in the liquid phase it becomes a highly expansive gas when released. This means that once the liquid propane is released from the cylinder it instantly

---

permanent gases
gases that cannot be liquified by pressure alone

■ NOTE
Liquefied gases are stored in the cylinder just as the description implies—in the liquid phase.

■ **NOTE**

This characteristic of changing from a condensed liquid to a gas is called expansion ratio and is important to understand.

BLEVE
(boiling liquid expanding vapor explosion) a major failure of a closed liquid container into two or more pieces

boiling point
temperature at which a liquid changes to a vaporous state at a given pressure; usually expressed in degrees Fahrenheit at sea level pressure (760 mmHg, or one atmosphere)

changes phase from a cold liquid to a colorless, odorless (unless artificially odorized), and tasteless gas that is heavier than air. This characteristic of changing from a condensed liquid to a gas is called expansion ratio and is important to understand because a liquefied gas cylinder actually contains a lot more potential gas than you may think. It is much like trying to contain the genie inside the lamp of Aladdin. The genie is many times larger than the volume of the tiny lamp and once the cork is pulled—watch out!

Liquefied propane is in many ways similar to the genie in the lamp. It has an expansion ratio of over 270 to 1. This means that for every one volume of liquid stored in the cylinder, there are approximately 270 corresponding volumes of propane *vapor* available. If a given vessel, say a 1,000-gallon propane tanker, were on fire, there is a release potential of 270,000 vapor volumes. Figure 3-6 shows the tanker and the potential volumes contained inside.

No liquefied gas cylinder can be expected to contain that kind of volume and pressure increase without failing. In the event the container does fail under these circumstances, a **BLEVE** is said to have occurred. BLEVE is an acronym that stands for:

Boiling
Liquid
Expanding
Vapor
Explosion

Another key property of propane that makes it easy to liquefy but also susceptible to a BLEVE is a very low **boiling point** of −44°F. At any temperature above −44°F, propane will instantly convert to the gas phase. The heat impingement then is a critical point in our example of the 1,000-gallon tanker mentioned previously. Once the temperature caused the pressure to rise to dangerous levels, the BLEVE was imminent because a chain of events was set into motion that could

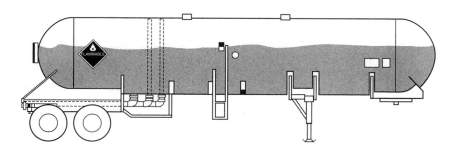

Figure 3-6 *An illustration of the expansion ratio of propane.*

LIQUID VOLUMES = *1000 GALLONS*
@ A 270:1 EXPANSION RATIO
YIELDS *270,000* VAPOR VOLUMES.

not necessarily be reversed. First the propane was heated above its boiling point, which caused it to become a boiling liquid inside the vessel. Due to the 270 to 1 expansion ratio, the now heated propane started turning into an expanding vapor. In the event the fire overwhelms the ability of the container to vent the resulting pressure, the entire contents will vaporize upon release. In our example, this means that 270,000 volumes of flammable vapor become airborne and immediately burst into flame. This gas explosion occurs with incredible force and will vaporize anything proximal to the blast. Since the tanker could not hold the ensuing pressure, a catastrophic container failure occurred and caused our resulting BLEVE. In essence a BLEVE occurs when materials with high expansion ratios are heated above their boiling points inside closed vessels. Figure 3-7 illustrates the tanker exploding due to its inability to contain the growing pressure.

Confused? Let's use a more common example that we can all relate to—popcorn. Popcorn pops for the same reason that the 1,000-gallon propane tanker explodes. Inside each kernel of corn is a little water. Water has a boiling point of 212°F, which is much higher than that of propane. Water also has the characteristic of expansion ratio just like propane. Would you think the expansion ratio of water is higher or lower than that of propane? The correct answer is higher. The expansion ratio of water is approximately 1,700 to 1. For every one volume of water there are 1,700 corresponding vapor volumes (steam in this case). So the stage is set for a BLEVE just like in our example of propane. In this case we put the popcorn in the microwave and turn it on. The microwaves heat the water until it reaches its boiling point of 212°F. The water now becomes a boiling liquid inside the kernel and instantly expands according to its 1,700 to 1 ratio. Because the kernel of corn does not have a relief valve, it suffers a catastrophic container

> ■ NOTE
> In essence a BLEVE occurs when materials with high expansion ratios are heated above their boiling points inside closed vessels.

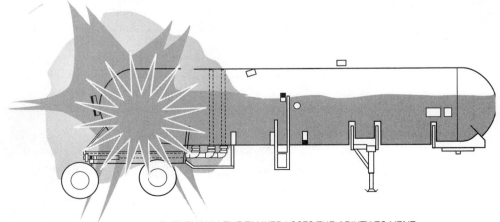

EVENTUALLY, THE TANKER LOSES THE ABILITY TO VENT PRESSURE. ONCE THAT OCCURS, CATASTROPHIC CONTAINER FAILURE IS IMMINENT.

Figure 3-7 *Propane tanker explosion.*

Chapter 3   Hazard Classes

> **⚠ SAFETY**
> BLEVEs are not limited to just liquified gases like propane and butane. Any liquid capable of expanding once it has been taken above its boiling point is susceptible to a BLEVE.

failure due to the BLEVE. We achieved the same results with two totally different chemicals! BLEVEs are not limited to just gases like propane and butane. Any liquid capable of expanding once it has been taken above its boiling point is susceptible to a BLEVE.

Establishing safe zones will be critical, as huge chunks of the container may be hurled hundreds of yards from the blast site. Figure 3-8 shows the aftermath of a BLEVE in a cylinder used to fuel a backyard BBQ grill. Notice the complete destruction of the cylinder, which was hurled about 50 yards from the point of the blast.

Cryogenics The third form that gases are typically used and stored in is called **cryogenic.** The word *cryogen* was derived from the Greek word *kryos* which means very cold. This is somewhat of an understatement though as these gases are liquefied by cooling the material below its respective boiling point. Once below the boiling point, the gas is in a liquid state and can only be maintained by ensuring a low temperature. Many cryogens originate from the atmosphere and are cooled to the necessary temperature by an interesting process. Initially, the gas

cryogens
a classification of gases that are cooled to a very low temperature, usually below −150°F (−101°C), to change to a liquid

Figure 3-8
Remnants of a failed propane cylinder.

is drawn into a compression chamber via a dryer to remove as much moisture as possible. The gas is also filtered as much as possible to remove all the crud and particulates before it moves through the system. The compressed gas (air) is cooled inside a chamber by a relatively cold circulating liquid. Liquid ammonia is often used as it can drop the temperature to approximately −30°F. Remember that as a gas is cooled, the pressure drops. In order to keep a significant pressure (usually 2,000 psi) constant, more air gets pumped in and cooled until we have a full chamber of gas at 2,000 psi and cooled to −30°F. The cold compressed air is then allowed to escape into another vessel, which again lowers its temperature. This process is repeated and different size containers are used as catch vessels to reduce the temperatures to the desired levels. Once the air condenses into a liquid the desired gases (nitrogen, hydrogen, helium, etc.) are boiled off or distilled at the prescribed temperatures and captured elsewhere. This is not an easy process and is routinely used when the gas cannot be liquefied by other methods. To qualify as a cryogenic material, the gas must have a boiling point no greater than −150°F. The benefit of using cryogens is that at these temperatures, it is possible to get an even larger amount of gas into a given cylinder. Expansion ratios of up to 1,445 to 1 are possible, which makes these cryogens economical to use and store. Table 3-3 lists numerous cryogens. Notice that many have boiling points much lower than −150°F.

Figure 3-9 shows the vapor cloud released by a cryogenic nitrogen tanker. The vapors are extremely cold and are certainly hugging the ground.

One of the key characteristics of cryogenic materials is their unique storage vessels. While the pressures inside are low (under a few psi), keeping a constant temperature is of utmost importance. Cryogens are therefore stored in cylinders,

Table 3-3 *Examples of cryogens.*

Substance	Boiling Point (°F)	Expansion Ratio
Liquid argon	−302	840:1
Liquid fluorine	−306	980:1
Liquid helium	−452	700:1
Liquid hydrogen	−423	848:1
Liquid krypton	−243	695:1
Liquid natural gas	−289	635:1
Liquid neon	−411	1,445:1
Liquid nitrogen	−320	694:1
Liquid oxygen	−297	857:1
Liquid xenon	−163	560:1

Figure 3-9 *Cryogenic nitrogen gas cloud.*

> **SAFETY**
> Because of their extremely cold temperatures, cryogens are capable of causing significant damage to anything that comes in contact with them.

which act like thermos bottles. These containers have silvered linings and are designed to use the inherent temperature of the material inside to help maintain the extremely cold atmospheres. In the event temperatures are not maintained, cylinder failure will be rapid as most cryogenic containers are not at all designed to handle significant pressures. Figure 3-10 shows a typical dewar cryogenic container.

Because of their extremely cold temperatures, cryogens are capable of causing significant damage to anything that comes into contact with them. Never touch the liquid and do not walk into visible clouds as it may freeze whatever personal protective equipment you may be wearing, as well as tools, equipment, and humans. If you come in contact with cryogenic materials, severe frostbite and tissue damage can result. Care should always be taken when working around these materials as no conventional chemical suit or gloves is sufficient to protect you against these extreme temperatures.

Health Hazards Posed by Gases

Some health hazards that may be posed by gaseous materials are:

- Carcinogens
- Corrosive materials
- Toxic materials

Figure 3-10 *Large cryogenic dewars in storage.*

- Irritants
- Sensitizers
- Substances affecting target organs

> **SAFETY**
> Whenever working with gases, it is important to consider that some might have more than one property.

If a gaseous substance presents one of these health hazards, OSHA standards require that information about the hazard be listed on a hazard label and also be available on an MSDS. Whenever working with gases, it is important to consider that some might have more than one property. An example of such a gas is anhydrous ammonia, which is a flammable, corrosive, and toxic gas.

> **NOTE**
> Remember that you breathe about 35 pounds of air a day, an enormous amount of exposure for your lungs.

It is also important to remember that a substance in the gaseous state automatically increases the danger of exposure because of the possibility of inhalation and eye and skin exposure. Remember that you breathe about 35 pounds of air a day, an enormous amount of exposure for your lungs. Reduce your exposure and protect that route of entry! Cryogens carry the extra hazard of causing severe thermal burns to any exposed skin. Take care not to walk in vapor clouds or breathe the vapors. The pipes in Figure 3-11 are covered with ice frozen onto them from the moisture in the air. If the pipes are cold enough to do that to the air, think of how your skin would be affected.

Many gases are heavier than air and thus tend to form layers at the lowest level they can reach. Concentrations at floor level or in equipment wells may reach unexpectedly high levels.

Figure 3-11 *Ice formation on cryogenic nitrogen piping.*

vapor density
the weight of a vapor or gas compared to the weight of an equal volume of air

■ **NOTE**
Vapor density is the weight of a given gas compared to an equal amount of dry air.

Even inert or nontoxic gases can displace air and collect in concentrations high enough to be suffocating. Additionally, flammable gases can also cause asphyxiation through the displacement of oxygen.

When reading MSDSs you may encounter a term called **vapor density**. It is important to understand this term as it is a description of how much the gas weighs when compared to air, which is important to know if you are monitoring for a particular gas or working in a confined space or trench and a certain gas may be present. Vapor density is the weight of a given gas compared to an equal amount of air. Figure 3-12 is a graphic illustration of a heavier-than-air gas escaping from a cylinder.

Vapor density is therefore defined as the weight of a vapor or gas compared to the weight of an equal volume of dry air. This is simply an expression of the density of the vapor or gas measured against the dry air, which is assigned a numerical value of 1.0. Materials lighter than air have vapor densities less than 1.0. Ammonia vapors for example have a vapor density of .59—about half the weight of an equal volume of air. Propane, hydrogen sulfide, butane, chlorine, and sulfur dioxide all have vapor density values greater than 1.0, therefore they are heavier than air. Propane vapors are approximately twice as heavy as an equal volume of air.

All vapors and gases will eventually diffuse or mix with air, but the lighter materials will initially dissipate (unless confined) while heavier vapors and gases

Figure 3-12 *Vapor density illustrated.*

are likely to concentrate in low places, along or under floors; in sumps, sewers, and manholes; or in trenches and ditches where they may create fire or health hazards.

Safe Handling, Use, and Storage of Gases

Working with pressurized substances presents its own set of problems. The proper use of connectors, valves, and reducing valves should be demonstrated. Specifications for compressed gas cylinders and their filling limits are set by voluntary standards and regulations. Cylinders should not be dragged, and care should be exercised to avoid shocks and blows.

Safe handling and use of hazardous materials is always based on no contact, and modern engineering practice including ventilation systems should be used if at all possible. If there is a possibility of skin contact or inhalation, the activity should be reviewed and analyzed for correct personal protective equipment. The danger of toxic vapor inhalation, reaction with or absorption through moist mucous tissues and the eyes, or absorption through the skin should always be minimized.

Leakage of flammable gases may result in the formation of explosive atmospheres, and precautions should be taken to eliminate static and electrical sparks in the workplace.

The use of protective gear, including location, storage, need for use, maintenance, cleaning, and decontamination should be discussed.

The principle of segregation of hazards applies to gases as well as other materials. Ideally, flammable gases should be stored separately and away from oxidizers and explosives. Poisons and corrosives should also be stored away from each other and from any material that might cause a breach in their containers. In the event poisonous gases are encountered, care should be taken to eliminate any potential for nearby fire. Poison A cylinders are not equipped with relief valves so catastrophic failure is the only option for extreme overpressure.

Emergency Procedures

Move any exposed victims to a source of uncontaminated air. Rinse skin or eyes exposed to corrosives or poisons with large quantities of water (the location and use of safety showers and eye baths should be discussed). Quick action is essential, because the longer the contact, the greater the damage. It may be necessary to support the victim's breathing in the event of respiratory and or cardiac arrest. Make sure any victims are thoroughly decontaminated before such measures are taken. You do not want to become another victim by allowing a careless exposure.

Particular circumstances may require changes in these general procedures. For instance, some substances may react with water to create extremely hazardous conditions, so that drenching with water is not recommended. For this reason, it is necessary to be familiar with the encountered chemicals before action is taken.

Proper emergency response procedures should be given on an MSDS for the material. It is important to know about and to follow this information. Medical assistance should be obtained quickly if needed and it is also important to provide the physician with the MSDS so that treatment can be quickly provided.

Other emergency response measures include isolation and containment of the material. Special cleanup may be required, particularly because it may be unsafe to vent toxic gases to the atmosphere. Proper procedures must be followed to make sure that no harmful residues are left in the workplace or are released to the environment.

FLAMMABLE LIQUIDS

This section discusses the hazards associated with flammable liquids. The hazards of these types of materials might seem obvious, but a number of issues must be understood when dealing with this class of substances. In addition to posing the threat of fire, most of these materials also present a health hazard for the worker or emergency responder. When working with flammable liquids, it is usually safe to assume that they possess more than one property. A typical flammable liquid label is shown in Figure 3-13.

Figure 3-13
Flammable liquid container label.

flash point
the minimum temperature at which a liquid fuel gives off a vapor in sufficient concentration to ignite—if an ignition source is present

combustible liquids
any liquid having a flash point at or above 100°F (37.8°C)

> **SAFETY**
> The lower the flash point, the greater the danger of combustion.

Flash Point and Flammability

The flammability of a material is measured by the particular material's **flash point.** Flash point is important to understand because it is the vapors of a material that burn and not the material itself. It would then make sense to know at what point those vapors are emitted from the surface of the liquid. In the case of flammable or **combustible liquids,** we use the term *flash point* to denote the minimum temperature at which the material gives off flammable vapors. The lower the flash point, the greater the danger of combustion. For example, gasoline has a flash point of −43°F. This temperature is quite low and means that at any temperature above −43°F gasoline will give off flammable vapors.

This concept is important because almost any reference material will list the flash point for a flammable liquid and you can almost always get an accurate temperature reading of the environment. If the flash point is lower than the prevailing temperature, you should be concerned with the vapor production from the body of the liquid spill and potential ignition sources if the vapors are flammable. Materials with low flash points like gasoline almost always pose a threat of vapor ignition.

Most regulatory agencies classify flammable and combustible liquids on the basis of their flash points. In 1994, the Department of Transportation identified flammable and combustible liquids as follows: A flammable liquid is one with a flash point at or below 140°F, and a combustible liquid as one with a flash point greater than 140°F.

Many agencies use a variety of definitions for flammable and combustible liquids, but this text uses the one quoted by OSHA, the National Fire Protection Association, and most fire codes. It lists flammable liquids as having a flash point under 100°F, and a combustible liquid as one with a flash point at or above 100°F.

> **■ NOTE**
> Flammable and combustible liquids are further subdivided based on their relative hazard.

Flammable and combustible liquids are further subdivided based on their relative hazard. The low flash point group consists of substances having a flash point below 0°F (–18°C). These materials present a high degree of danger because the material almost always gives off flammable vapors. Some of these substances include

Acetaldehyde	Diethylamine	Heptene
Acetone	Diethyl ether	Hexane
Acrolein	Diisopropyl ether	Iso-pentanes
Allyl chloride	Dimethyl ether	Methyl pentane
Amyl nitrate	Ethyl amine	Octane
Carbon disulfide	Ethyl ether	Petroleum spirit
Cyclohexane	Ethyl mercaptans	Propylene oxide
Cyclopentane	Furans	Tetrahydrofuran
Diethoxymethane		

The intermediate flash point group consists of materials having a flash point from 0°F (–18°C) to 73°F (23°C). These materials are also considered hazardous but are not as high a hazard as with the first group. Some substances found in this intermediate flash point group include

Acrylonitrile	Diethyl ketone
Alcohols (most)	Dioxane
Allyl ethyl ether	Ethanol
Amyl acetates	Ethyl methyl ketone
Bromopropanes	Heptane
Butyl acetates	Methyl ethyl ketone (MEK)
Butyl methyl ether	Naptha, petroleum
Chloromethyl ethyl ether	Propanol
Cycloheptane	Toluene
Cyclohexene	Xylenes

The high flash point group consist of substances having a flash point from 73°F (23°C) to 141°F (61°C), which are regarded as presenting minor danger. Some high flash point liquids include

Anisole	Diethylbenzene	Hydrazine
Bromobenzene	Ethyl butyrate	Nitroethane
Butanol	Formalin	Turpentine
Chlorobenzene	Furfural	

Substances having a flash point above 141°F (61°C) like the various oils, are not considered to present a fire hazard, according to this classification scheme.

> **! SAFETY**
> Remember—flash point is not the sole measure of the degree of hazard for a given chemical.

vapor pressure
the pressure exerted by a saturated vapor above its own liquid in a closed container

The thinking is that these materials would require significant preheating in order to ignite. Remember—Flash point is not the sole measure of the degree of hazard for a given chemical.

Substances with high **vapor pressures** evaporate rapidly and form combustible atmospheres. Vapor pressure is the pressure exerted by a saturated vapor above its own liquid *in a closed container*. Essentially, the liquid particles are vaporizing until the space above the liquid is saturated. The vapor then condenses and returns some of the vapor particles back to the liquid phase. Figure 3-14 illustrates the concept of vapor pressure.

This evaporation-condensation cycle repeats itself until an equilibrium is obtained. Once that happens, the optimum vapor pressure for the liquid has been reached. This concept may be confusing but it applies practically because if the liquid has a high vapor pressure inside a closed container, it will surely evaporate quickly if released.

If the concept of vapor pressure is still confusing, think of it this way. Vapor pressure is the *wantability* of a liquid to become a vapor. Certain liquids like methyl alcohol, ethyl alcohol, and isopropyl alcohol have high vapor pressures. They give off lots of vapors inside a closed container and evaporate quickly when released. This property is evident if you have ever opened a bottle of rubbing alcohol. You immediately smell it and if the lid is off for any period of time, the whole room will smell like a medicine cabinet. Imagine if 55 gallons of this stuff were to be spilled or the lid were off of a full 1-gallon container? How much flammable vapor production do you think that here would be? The nontechnical answer is "lots" and could be proved quantitatively with careful air monitoring.

Two additional facts you should know about vapor pressure.

1. The vapor pressure of a substance at 212°F (100°C) is always higher than the vapor pressure of the substance at 68°F (20°C). This means that heated

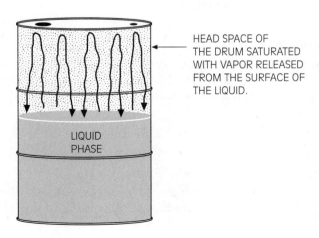

Figure 3-14 *Vapor pressure illustration.*

chemicals are more active at the molecular level than cold chemicals and the eventual vapor production once released will be higher.

2. Vapor pressures reported on MSDSs are expressed in mmHg. For purposes of a benchmark, 760 mmHg is equivalent to 14.7 pounds per square inch, which is equal to 1 atmosphere. Materials with vapor pressures above 760 mmHg are capable of exceeding ambient pressure and can be considered vaporous. Remember this term applies to liquids inside their container but you can be sure that if the material has a high vapor pressure it will readily evaporate when the lid is removed.

Isopropyl alcohol (IPA) is an excellent example of a material that evaporates rapidly. Alcohol spills in general create great volumes of flammable vapor. While that characteristic presents us with a certain challenge, it also tells us that the problem is essentially taking care of itself: The liquid product is quickly turning to vapor that will ultimately diffuse into the atmosphere. High volatility liquids like IPA evaporate quickly on heated pavement and may simply go away before we roll out all our hazmat toys. Liquids such as diesel fuel and oil do not evaporate so quickly and the only way to clean them up is to take offensive action. The point here is that some liquids are volatile and give off lots of vapor and are gone rather quickly. This may be a good thing under the right circumstances if we fully understand the prevailing conditions. Read MSDSs or other reference sources to get the technical information on a given substance. Once completed, you may then evaluate your situation and use the inherent properties of the chemical to your favor if it is safe to do so.

Solvents

No single chemical class can be defined as a solvent. When looking at site operations in the chemical industry it is natural to consider hydrocarbon based solvents as a unique group because of the many common characteristics found in chemicals such as toluene, acetone, methyl ethyl ketone, xylene compounds, paint thinners, and countless other organic compounds, used to break down other substances. Figure 3-15 shows a storage situation of ether, a common solvent compound.

For the purpose of our discussions, solvents are defined as materials used to dissolve other materials, primarily other hydrocarbon-based substances. Some are used as cleaning fluids or for degreasing and others are used for extracting soluble substances from raw materials.

However, water is a very common solvent and dissolves a wide range of substances, including alcohol compounds and solids such as sodium chloride, calcium fluoride, and potassium hydroxide pellets. Water is often regarded as the universal solvent because it dissolves so many things. A true solvent is deemed to be the dissolving media or the substance that will ultimately break down what is called the *solute*. Making a top quality glass of chocolate milk will clearly

Figure 3-15
Examples of flammable liquid containers.

> **SAFETY**
> Many solvents are flammable liquids, but may also present a health hazard. There can be health effects of all kinds—solvents may be toxic, corrosive, or irritating to the skin, eyes, and mucous membranes.

> **SAFETY**
> It is essential to wear the proper personal protective equipment when handling solvents.

illustrate the solvent–solute relationship. The nice cold glass of milk is the solvent or the dissolving media, and the chocolate powder is the solute or what is dissolved. There you have solvent–solute in a nutshell.

Many solvents are flammable liquids but may also present a health hazard. There can be health effects of all kinds—solvents may be toxic, corrosive, or irritating to the skin, eyes, and mucous membranes. Remember that these solvents will break down a fatty substance very near and dear to you—Your skin! You are covered by a lipid (fat) layer that is nonsoluble in water (a good thing), but highly soluble in hydrocarbons. This is bad in the event you expose unprotected skin to something like gasoline. The gasoline will migrate right through your skin and end up circulating through your entire body. It will end up stored somewhere in your fatty tissue because it is not soluble in water and subsequently eliminated with other water soluble compounds. If you have ever cleaned grease or paint off your hands with paint thinner or gasoline you most likely noticed that your skin felt dry for one or two days, because the solvent (thinner or gasoline) was successful in breaking down your lipid layer and essentially defatting your skin. Additionally, some of the chemical was absorbed into your body and is probably hanging out in some chemical lounge chair deep in the fat cells of your body!

It is essential to wear the proper personal protective equipment when handling these types of compounds. It is also important to protect your respiratory system because inhalation of the vapors of many of these substances can cause decreased mental capacity, loss of motor skills, nausea and vomiting, and various long-term health effects. Large exposures to the lungs may even cause a condition called *chemical pneumonitis*. This chemically induced pneumonia causes fluid build up in the lungs and can be fatal or in the very least life threatening.

For this reason you should understand all information on labels and material safety data sheets.

Health Hazards Posed by Flammable Liquids

Various flammable liquids can present health hazards or combinations of health hazards such as carcinogens, corrosives, highly toxic substances, irritants, sensitizers, toxic substances, and substances having target-organ effects. A substance posing one of these health hazards should have a health hazard warning label, and information should also be available on an MSDS. Some examples of toxic and corrosive compounds are as follows:

Toxic Flammable/Combustible Substances

Aniline	Dinitrobenzenes	Nitrobenzene
Benzonitrile	Dinitrotoluenes	Nitrotoluenes
Carbon tetrachloride	Ethylene dibromide	Tetrachloroethylene
Chloropicrin	Hexachlorobenzene	Toluidines
Dichlorobenzenes	Nitroanilines	

Corrosive Flammable/Combustible Substances

Acetic anhydride	Alkyl amines	Diethylenetriamine
Acetic acid (50% to 80%)	Benzoyl chloride	Ethanolamine
Acetic acid, glacial	Benzyl bromide	2-Ethylhexylamine
Acrylic acid	Butyl amine	1-Pentol

Again, warnings of these hazards should be available on safety labels and material safety data sheets.

Solvents should be used with adequate ventilation, because almost all organic solvents have an anesthetic effect, leading to dizziness and sleepiness, or unconsciousness. Prolonged exposure could eventually lead to death.

Skin exposure to solvents should be avoided as a matter of common sense. Solvents can remove protective oils in the skin, and some can even penetrate the skin. Many solvents can cause systemic or target-organ damage, particularly to the liver and blood-forming organs. Even if the substance is not corrosive or toxic, chemical exposure is to be avoided.

Safe Handling, Use, and Storage

The consistent physical danger from flammable liquids and solvents is the fire hazard. Vapors may form combustible or explosive atmospheres and ignite from uncontrolled ignition sources. Flammable liquid fires occurring in the presence of water increases the surface area of the fire and makes extinguishment difficult. Foaming agents should be utilized for these types of incidents and great care must

Figure 3-16 *Fire in a bulk flammable liquid container.*

be taken to properly protect responders from heat and flame. Figure 3-16 illustrates a floating flammable liquid fire in a bulk storage facility.

Precautions and monitoring for flammable atmospheres should always be in place before work begins. Figure 3-17 shows workers monitoring for a potential flammable atmosphere.

Protect against electrical sparks and other accidental sources of ignition. Prohibit the use of any nonintrinsically safe instruments within any potentially flammable atmosphere. The use of protective gear, including location, storage, need for use, maintenance, cleaning, and decontamination should be discussed and understood by all on the scene or at the site. Remember that standard chemical protective suits are worthless in a fire situation!

Principles of storage and segregation should be based on the recognition of the fire hazard. The storage area should be constructed to be free from ignition sources and able to contain accidental releases by the use of secondary containment. Flammable liquids should be stored separately from substances that could react with them, such as oxidizers.

Emergency Procedures

Planning for emergencies at an emergency scene may sound redundant but is absolutely critical. Even the most thought out plans go awry occasionally and

> **SAFETY**
> Protect against electrical sparks and other accidental sources of ignition. Prohibit the use of any nonintrinsically safe instruments within any potentially flammable atmosphere.

> **SAFETY**
> Flammable liquids should be stored separately from substances that could react with them, such as oxidizers.

Figure 3-17 *Workers monitoring a potentially flammable atmosphere.*

■ **NOTE**

Many flammable liquids and organic solvents do not mix with water, and hosing down a spill with water will just spread the hazard.

create horrible situations. One sure way of turning an emergency into a catastrophe is by having something go wrong. Always have a backup plan involving fire fighting procedures, secondary escape routes, and medical treatment. The main requirements with flammable liquids and solvents are ensuring adequate ventilation and taking fire precautions. Many flammable liquids and organic solvents do not mix with water, and hosing down a spill with water will just spread the hazard.

FLAMMABLE SOLIDS

The good news about this class of chemicals is that they are not frequently encountered on the emergency scene or the waste site. The bad news is that because they are uncommon in those settings, the comfort level and understanding of flammable solids may be low. When something like white phosphorous comes into contact with air in train load quantity, it is going to be high profile event. So what is a flammable solid? Rather than tell you what they are, let us look at what they are not. This class actually excludes the traditional Class A flammable materials such as wood, paper, and plastic. These materials are solids and they are indeed flammable, but we do not include them in this otherwise specialized class of substances. Flammable solids do include substances such as calcium carbide, naphthalene, white and red phosphorous, and ammonium picrate. You may not know the properties of these chemicals, but you can rest assured that they do not behave like a log burning in your fireplace. And because the fire does not behave like a normal organic combustion process, the extinguishment procedures are different.

> ■ **NOTE**
>
> Most flammable solids can be handled easily unless they are on fire or otherwise reacting in some fashion.

Most flammable solids can be handled easily unless they are on fire or otherwise reacting in some fashion. A pile of magnesium shavings for example can be swept up using a broom and any old dustpan. It is a minor event that can be accomplished by anyone with a working knowledge of a broom. Ignite the shavings, however, and the game changes big time. Now we have an extremely hot fire that cannot be extinguished with water, because the magnesium fire burns so hot that it actually rips the water molecule apart to get to the oxygen (no small task, by the way). In some cases, the application of water causes an explosion, which sends bits of flying molten magnesium into the air. Other fires may occur as well as significant skin burns to anyone nearby. All in all, it is very important to acquaint yourself with the properties and extinguishment methods of any flammable solid before attempting to handle it.

Flammable Solids Defined

flammable solids
a solid, other than a blasting agent or explosive as defined in 24 CFR 1910.109(A), that is likely to cause fire through friction, absorption of moisture, spontaneous chemical change, or retained heat from manufacturing or processing, or which can be ignited readily and when ignited burns so vigorously and persistently as to create a serious hazard

pyrophoric
describes a chemical that ignites spontaneously in air at a temperature of 130°F (54.4°C) or below

It may be easier to give the broad classifications of **flammable solids** based on how they behave. You may then learn the nature of various materials in the workplace and classify them into one of the following groups:

- Flammable metals such as zirconium, titanium, sodium, and lithium.
- Spontaneously combustible materials such as white and yellow phosphorus. These substances are classified as **pyrophoric.**
- Metals that burn intensely or are difficult to extinguish such as magnesium and aluminum. Sulfur also falls into this category although chemically speaking it is a nonmetal.

- Water-reactive solids such as metallic sodium, lithium, and rubidium as well as zinc powder and calcium carbide. When water contacts calcium carbide, acetylene gas is created.
- Flash point solids such as camphor and naphthalene. These solids do not burn but give off flammable vapors when heated. Camphor has a flash point of 150°F while naphthalene has a flash point of approximately 175°F.

Flammable solids can be found as powders, shavings, or filings from a milling process, solid bars or rods, and chips or chunks. Fused sodium for example, may come in 55-gallon drums and subsequently heated and pumped into a chemical process. The bottom line is that these materials may come in a wide array of forms and you should be prepared for any of them.

Examples

There are numerous examples of flammable solids, which should be obvious to even a casual observer in the field. Materials such as sodium, magnesium, lithium, sulfur, and carbon are all classified as flammable solids. Some other examples may not be so obvious and include such materials as phosphorous, picric acid, and calcium carbide. Phosphorous has several allotropes that may need to be investigated in order to fully understand a particular hazard. An allotrope is one of several possible forms of a substance. Phosphorus can be found in red, yellow, and white. Each is an allotrope and has unique properties.

Other unique flammable solids include

Barium azide

Methyl parathion

Nitrotoluene

Paraformaldehyde

Sodium hydride

Trichlorosilane

Trinitrobenzene

These few examples of flammable solids should give you an idea of the types of substances that fall into this category. Remember that each may have very unique properties so stay aware and research all materials before taking any cleanup or emergency actions.

Health Hazards Posed by Flammable Solids

It is relatively easy to reduce exposure to flammable solids as a group unless they are reacting. The primary hazard is inhalation of dusts and finely divided powders as they are easily inhaled. Some dusts such as sodium metal react with the moisture in the lungs to form a caustic solution that causes burning of sensitive tissues. All attempts should be made to avoid exposure with skin, eyes, and mucous membranes. Again, the mixture of some metallic dusts with moisture on the body may result in chemical burns. It is also imperative to avoid all clouds of smoke as they may include toxic by-products of the burning metal.

Safe Handling, Use, and Storage

Because some flammable solids are water reactive, it is often necessary to take special measures in designing containers for them. The prospect of water contact should always be questioned no matter what flammable solid is involved. Figure 3-18 shows some suspect storage practices of a water-reactive flammable solid.

A good rule of thumb for these materials is to not add water until you are certain it will not cause an adverse reaction. Picric acid, on the other hand, is stabilized by adding water. In transit it is commonly found to contain about 10% water in the solution. This addition of water reduces the possibility of explosion when shocked or heated. Contrast that with dry picric acid which is classified as a high explosive closely resembling trinitrotoluene (TNT). Make sure you know what the addition of water will do to the substance.

Both acids and bases may react adversely with flammable solids and cause the release of toxic gases. Do not allow flammable solids to mix with other materials unless you are certain of the resulting reaction.

!**SAFETY**
Do not allow flammable solids to mix with other materials unless you are certain of the resulting reaction.

Figure 3-18 *Poor storage of water-reactive flammable solids.*

The use of protective gear in fire situations may be questionable depending on the solid. For example, magnesium burns at approximately 1,200°F and even firefighter's turnouts are insufficient. The flying chunks of magnesium could land on your body and basically lodge in the skin and burn for quite some time. Some antipersonnel devices used in the military are called "Willie Petes," a pseudonym for white phosphorus. These devices explode and send showers of burning white phosphorus on any unlucky person in the near vicinity. The chunks burn through the clothing and deep into the skin just like the magnesium example above. In non-fire scenarios, little or no chemical protective equipment may be needed. If dusts can be controlled, the risk of cleaning up spilled solids is minimal. Again, this is product-specific so research it before you touch it. Some flammable solids are very active chemically, so great care must be taken to ensure compatibility with other stored materials and their containers.

> ■ **NOTE**
> Some flammable solids are very active chemically, so great care must be taken to ensure compatibility with other stored materials and their containers.

Emergency Procedures

There are recommended general actions for handling flammable solids but they are very brief. Some solid spills can be covered with a tarp or heavy plastic to minimize dust blowing around or water contact from light rain. Regardless of the material involved, hold off adding any water until you are sure the situation will not worsen. Most emergency actions are based on letting the incident stabilize itself through fire or just sweeping up the stuff on the ground. Some of the more exotic metals require advanced techniques but for the most part, these incidents are easily handled.

One exception to the use of water on flammable solids is in the case of contact with people. In such cases it is usually best to brush off as much powder as possible followed by the application of large amounts of water to reduce the tissue damage and prohibit the creation of a toxic paste being formed on the victim's body. Safety showers and eye wash stations should be located in close proximity to the areas where flammable solids are used, stored, and handled.

In most cases you will be well served by following the initial basic procedures of chemical spills: Isolate the area and deny entry. This certainly buys you time and reduces the exposure of personnel. Many of these products are highly reactive and can mix with other materials to release toxic, flammable, or explosive vapors.

OXIDIZERS AND ORGANIC PEROXIDES

oxidation
chemical reaction that involves a transfer of electrons from a fuel to an oxidizer

A commonly held perception of **oxidation** is that oxygen has combined with an element and created an oxide. Unfortunately, this is not the whole picture, chemically speaking, so we have to expand on it slightly. Oxidation really occurs when electrons have been transferred from the outer electron shells of a reducing agent or fuel to the outer shells of an oxidizing agent. Chlorine, fluorine, and oxygen are all oxidizers by the fact that they accept electrons in chemical reactions. Sodium, potassium, and calcium are examples of reducers or fuels as they all donate electrons to the oxidation-reduction reaction. Oxygen is sometimes gained from the process of oxidation but it is not the sole way we define oxidation. Understanding that oxidizers are materials capable of reducing fuels should now expand your thinking. Oxidation is a complicated chemical reaction and one in which heat is generally liberated. Oxidizers as a group are very aggressive from a chemical reaction standpoint and should be treated with great respect.

We have already introduced oxygen, so let us expand on its role as an oxidizer. The main source of this important gas is the atmosphere, which contains about 21% oxygen. Common oxidation reactions using atmospheric oxygen are respirations, combustion, or a slow oxidation process like rusting. The rate at which reactions proceed depends on the concentration and availability of oxygen present, among other factors. The oxygen in air is diluted by inert nitrogen in a ratio of approximately 4:1, so the rate of oxidation is much less than it would be with pure oxygen. As responders and/or waste site workers we should be quite concerned with our surrounding oxygen content. If enough oxygen is present with fuel, and an ignition source ties all three together at the chemical level, fire can occur. In order to have combustion we must bring together heat, fuel, and oxygen and have them participate with a chemical chain reaction. In this case the oxygen will do its job as an oxidizer and reduce the fuel.

Oxidizers are also substances that can release oxygen when they react and thus provide a self-contained source of energy to fuel a reaction. In a broader technical sense, oxidizers may contain no oxygen; chlorine is an example. This property is very important in promoting useful chemical reactions, but it can also be a hazard.

Oxidizers are very reactive compounds, and the kinds of reactions that can occur depend on the prevailing conditions and the other reactants involved. When combining with combustible materials, oxidizers will burn fiercely. Look at Figure 3-19 and determine if the storage practice is a good one. Would you store oxidizing agents on wood shelves?

Keep in mind that oxidizers may cause spontaneous combustion when mixed with fuels. A number of factors can influence this reaction, such as strength of the oxidizing agent, concentration of solutions, and nature of the fuel, but in general the mixture of oxidizers and fuels should be avoided. A common oxidation-reduction reaction that illustrates this point is the mixing of pool chlorine and brake fluid. Calcium hypochlorite, or pool shock treatment contains a large amount of available chlorine. We have already mentioned the oxidizing tendencies of chlorine and you should be aware that it is a powerful oxidizing agent. Some pool chlorine in the granular form has as much as 65% available chlorine in the compound. This means that it is a strong oxidizer and should be treated with care. Brake fluid is a common reducing agent and is quite susceptible to

■ **NOTE**
Common oxidation reactions using atmospheric oxygen are combustion and a slow oxidation process like rusting.

■ **NOTE**
In a broader sense, oxidizers may contain no oxygen; chlorine is an example.

Figure 3-19 *Storage of technical grade (30%) hydrogen peroxide.*

being mixed with the pool chlorine. Many a homeowner's garage contains both of these substances. Either and/or both can be purchased at most hardware or grocery stores. If a little brake fluid is poured or mixed into a small pile of the mentioned pool chlorine, a vigorous reaction begins. Within a minute or so, the slow oxidation causes heat to build up and ultimately results in combustion of the fuel (brake fluid). A large, toxic cloud of chlorine gas is liberated as well as substantial heat production. All this from two common materials.

This same reaction may occur if hydrogen peroxide in high concentrations is mixed with wood shavings. Nitric acid above 72% concentration is also classified as a strong oxidizer. If this compound comes into contact with anything organic it will cause it to combust very intensely and rapidly. There are countless other examples to support this concept and you should definitely be aware of the potential hazards when dealing with oxidizers. Some common industrial oxidizers are the following:

Aluminum nitrate	Fluorine	Permanganates
Ammonium nitrate	Hydrogen peroxide	Potassium peroxide
Calcium hypochlorite	Oxygen	Sodium hypochlorite
Calcium perchlorate	Ozone	Sodium peroxide
Chlorates	Perchlorates	

Some oxidizers are stronger than others and concentration has a lot to do with how reactive an oxidizer may be. It takes a solid chemical background to know this simply by looking at the name or the structure but there is some good information that can be derived by workers in the field.

Do you notice anything consistent in the chemical names in the list? Look closely at the endings of the names and you should see that many end in *-ite* and *-ate*. This point is important to remember when attempting to determine if a particular material is an oxidizing agent. If a chemical name ends in *-ate* or *-ite* it could indicate that the material is an oxidizer! The ate/ite endings do not always indicate oxidizers but it is a good place to start if you have no idea as to the nature of the material. Additionally, notice that some of the compounds in the list have "peroxide" in their name. This is also an indicator that a material may be an oxidizer because of the meaning of the word *peroxide*. In chemical speak, *per* means many and the oxide suffix tells you that the compound contains oxygen. Combining the two meanings gives you *many oxygens*. Because we know that oxygen is a strong oxidizer, we can assume that the compound is oxidizing in nature. Many compounds fit this bill such as potassium peroxide, lithium peroxide, and sodium peroxide. Remember that this rule is not absolute and should only be employed as a preliminary way to classify a given substance. Always refer to the appropriate MSDS or other reference source to verify the exact nature of the substance.

> **SAFETY**
> If a chemical name ends in *-ate* or *-ite*, it could indicate that the material is an oxidizer!

Organic Peroxides

A special class of oxidizers very useful for their chemical properties is organic peroxides. They, too, can react explosively when involved in a fire. The transportation hazard label normally says "Organic Peroxide" and should make you immediately take notice. The reason these substances pose such a high danger is found in their chemical structure. With organic peroxides, an organic compound is chemically bonded with some sort of oxidizing agent. In other words, the oxidizer and the reducer are already bonded together and waiting to react. The concept is much the same as the nitrated toluene discussed in the explosives section. A common organic solvent was beefed up with an oxidizer and a powerful explosive was the result. As you know, these substances can react in a big way if involved in a fire or other heat-producing reaction.

Methyl ethyl ketone peroxide is a good example of an organic peroxide. In this case a common organic solvent has many oxygens (remember the definition of peroxide?) attached to it. Methyl ethyl ketone peroxide is a very aggressive solvent itself used extensively in plastic fabrication. It is irritating to the skin and is susceptible to combustion. All in all, methyl ethyl ketone peroxide is a mild example of how aggressive an organic peroxide can be. At the other end of the spectrum is a material called benzoyl peroxide. You may be acquainted with this substance from the acne medicine commercials. A classic example of the oxidizer–fuel mixture on the same compound, benzoyl peroxide is a white granular powder that can be toxic if inhaled and prone to explosion if dry. Treat organic peroxides with care and understand the nature of the specific material that you are handling.

Health Hazards Posed by Oxidizers

> **■ NOTE**
> Oxidizers present more of a physical hazard due to the dangers of fire and explosion either from the compound itself or when reacting with other materials.

Oxidizers as a class are not defined as a health hazard. They present more of a physical hazard due to the dangers of fire and explosion either from the compound itself or when reacting with other materials. If one of these substances does pose a health hazard, and some oxidizers do, information about the hazard should be contained on a health hazard label and also be available on an MSDS.

Apart from these possible dangers, unnecessary exposure to any chemical should be avoided. Oxidizers are highly reactive and sometimes react with moisture to release toxic fumes. Most oxidizers are skin irritants and dry agents should not be allowed to contact skin, eyes, and mucous membranes.

Safe Handling and Storage of Peroxides

> **! SAFETY** If there is a possibility of human contact, the substance should be reviewed and analyzed for possible routes of exposure and compatibility of protective equipment—the danger of toxic vapor inhalation or reaction with moist mucous tissues may exist.

Safe handling and use of oxidizers clearly should be based on no contact as strong oxidizers can degrade all but the toughest personal protective equipment (PPE). There is no "oxidizer PPE" available and the selection of any protective gear should be product specific. In addition, absorbent materials and storage vessels should be carefully selected. Many absorbents are organic based and can react vigorously and burn if exposed to strong oxidizers. For example, hydrogen peroxide is a strong oxidizing agent at high concentrations. At concentrations around 70% liquid hydrogen peroxide can ignite organic substances such as wood and/or paper. Nitric acid above 72% should also be given special consideration. Picking up these spills with organic matter may result in a fire or at the very least a vigorous chemical reaction. If there is a possibility of human contact, the substance should be reviewed and analyzed for possible routes of exposure and compatibility of protective equipment—the danger of toxic vapor inhalation or reaction with moist mucous tissues may exist.

Because of the high reactivity of oxidizers it is important to make sure that containers do not leak. They should also be stored away from materials with which they could react violently. Figure 3-20 shows a poor storage scenario of a whole range of chemicals. Notice the oxidizers on the top shelves in close proximity to some incompatible materials.

Always note the concentration of the oxidizers in solution. It is usually safe to say that the higher the concentration, the higher the hazard. Solutions above 30% are hazardous and solutions above 70% should be regarded as extremely hazardous.

Emergency Procedures

> **■ NOTE**
> Planning and training for emergencies involving these materials should be carried out before the emergency occurs.

Planning and training for emergencies involving these materials should be carried out before the emergency occurs. Oxidizers involved in fires may react with explosive violence. In almost all cases oxygen and/or heat is released, so the resulting fire will be self-sustaining. Carbon dioxide or other inert gas fire

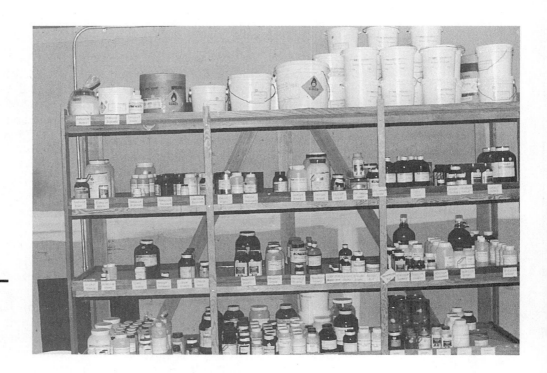

Figure 3-20 *Poor storage practices of a variety of materials. (Notice the oxidizers.)*

extinguishers are usually ineffective as they absorb little heat. For small fires, dry chemical extinguishing agents may be effective. One should be careful, however, because a fire involving these materials is not likely to be small for very long. Probably the most effective means of dealing with large fires is to flood with large amounts of water, providing the water is not reactive with the chemical released. Extinguishing action would result from dispersion and cooling rather than reducing the supply of oxygen.

Oxidizers spilled on skin must be rinsed off immediately. The location and use of safety showers and eye baths should be discussed. Quick action is essential, because the longer the contact the greater the hazard. Usually a 15-minute rinse is the minimum amount of time for washing off these chemicals.

> **! SAFETY**
> Oxidizers spilled on skin must be rinsed off immediately.

Particular circumstances may require changes in these general procedures. For instance, the substance may react dangerously with water. In such cases, if a dry material contacts the person, a quick brushing off of the material might be indicated prior to the application of water. For this reason, it is necessary to be familiar with emergency response procedures for individual chemicals ahead of time.

Further stages of emergency response include isolation of the spilled material to prevent spread of the damage. These reactive substances can react with other chemicals and may release toxic vapors and cause fires or explosions. After the hazard is contained, there is still the cleanup to consider. Common inorganic chemicals may be neutralized, but more complex substances must be thoroughly tested to make sure that no harmful residues are left in the workplace or released into the environment.

POISONS, PESTICIDES, AND CARCINOGENS

This section discusses the hazards associated with substances classified as poisons and carcinogens. This category is quite broad and can be discussed at great length and detail, but we will attempt to keep it brief. The intent is to discuss different subcategories of poisonous materials such as pesticides and some specific dangerous chemicals that you should be aware of. You are definitely encouraged to have a heightened awareness and understanding of the kinds of hazards that can be posed by the materials we present.

poisons
substances that are likely to cause death or serious injury if swallowed, inhaled, injected, or come in contact with the skin, eyes or mucous membranes

Definition of Poisons Numerous terms are used to describe really bad chemicals, among them *toxic, poisonous*, and *lethal*. Do they all mean the same thing or is each unique in meaning and application? The answer is both yes and no. How can this be? The reason is that all things can be toxic to a certain extent depending of the dosage. You may have heard the axiom "The dose makes the poison" issued by Paracelsus, a Swiss alchemist from the 1500s who came up with this brilliant observation about chemical exposures. In reality almost any chemical can be poisonous; it just depends on the dose, length of exposure, and other things like your own physiology.

We will approach this broad topic by saying that **poisons** are substances that are likely to cause death or serious injury if they are swallowed, inhaled, injected, or come in contact with the skin, eyes, or mucous membranes. Once again, this definition can include many different chemicals but we can narrow it down further. Poisons are substances whose main function or primary hazard is to kill or cause serious health effects. In some cases, the poison interferes with the proper functioning of the body by affecting oxygen distribution in the bloodstream or by blocking nerve impulses. These health effects are usually acute and may ultimately result in death. Pesticides should come to mind as a group of chemicals that can be considered poisonous. The suffix *-cide* means to kill and can be applied to various groups of chemicals. The word *pesticide* literally means "to kill pests." Rodenticides kill rodents, and fungicides kill fungus. So how do you know if a material is poisonous? Reading the MSDS or other reference information is the best way to get health information.

■ **NOTE**
These health effects are usually acute and may ultimately result in death.

From an occupational safety and health point of view the definition of a toxic substance results from its effects on test animals. From a responder's

> **SAFETY**
> The goal of working with chemicals should always be to suffer no unplanned exposure.

standpoint, a toxic or poisonous substance will hurt or kill you if you suffer a lethal exposure. Choosing the wrong personal protective equipment in this case could prove to be a fatal error. It is therefore critical to totally understand the chemical before you take action. Remember that no amount of water washing will be effective on chemicals that are affecting you internally. This is not to say that you should not perform external decontamination, because you most certainly should, but once the chemical is in your body and doing damage you are well beyond the benefits of washing your skin. The goal of working with chemicals should always be to suffer no exposure. Figure 3-21 shows students practicing an emergency decon of a fellow worker.

Again, the damage a chemical causes may be inside your body where a water wash will do no good. How effective will washing your skin be if the chemicals are affecting nerve transmissions? The answer is zero. In this setting you could only be helped by invasive therapy and aggressive treatment. This means needles and drawing lots of blood and other highly unpleasant procedures. So avoid all this headache (and other significant aches) and watch yourself.

Most systems, including OSHA, define two categories of toxic materials: Toxic and Highly Toxic materials.

Figure 3-21 *Workers simulating emergency decon procedures. Courtesy Safety Compliance Management.*

Highly toxic materials are those substances falling within any of the following categories:

1. A chemical that has a median lethal dose of 50 milligrams or less per kilogram of body weight when administered orally to albino rats weighing between 200 and 300 grams each.
2. A chemical that has a median lethal dose of 200 milligrams or less per kilogram of body weight when administered by continuous contact for 24 hours (or less if death occurs within 24 hours) with the bare skin of albino rabbits weighing between two and three kilograms each.
3. A chemical that has a median lethal concentration in air of 200 parts per million by volume or less of gas or vapor, or 2 milligrams per liter or less of mist, fume, or dust when administered by continuous inhalation for one hour (or less if death occurs within one hour) to albino rats weighing between 200 and 300 grams each.

Toxic materials are those substances falling within any of the following categories:

1. A chemical that has a median lethal dose of more than 50 milligrams per kilogram but not more than 500 milligrams per kilogram of body weight when administered orally to albino rats weighing between 200 and 300 grams each.
2. A chemical that has a median lethal dose of more than 200 milligrams per kilogram but not more than 1,000 milligrams per kilogram of body weight when administered by continuous contact for 24 hours (or less if death occurs within 24 hours) with the bare skin of albino rabbits weighing between two and three kilograms each.
3. A chemical that has a median lethal concentration in air of more than 200 parts per million but not more than 2,000 parts per million by volume of gas or vapor, or more than 2 milligrams per liter but not more than 20 milligrams per liter of mist, fume, or dust when administered by continuous inhalation for one hour (or less if death occurs within one hour) to albino rats weighing between 200 and 300 grams each.

Other classes of substances that have a physiological action and are defined as health hazards by the OSHA standard are Irritants and Sensitizers.

Irritants are chemicals that are not corrosive but cause a reversible inflammatory effect on living tissue by chemical action at the site of contact. The definition of an irritant is different in the shipping regulations of the Department of Transportation; an irritant is defined as a substance that in contact with fire or air gives off dangerous or intensely irritating fumes. Because this is a more severe hazard, care should be taken to distinguish between shipping labels and health effects labels.

Sensitizers are chemicals that cause a substantial proportion of exposed people or animals to develop an allergic reaction in normal tissue after repeated exposure to the chemical.

> **NOTE**
> The degree of danger posed by a poison depends on its physiological action, its route of attack, and its reactivity.

Health Hazards Posed by Poisons The degree of danger posed by a poison depends on its physiological action, its route of attack, and its reactivity. The amount of damage that can result from exposure depends on factors such as the concentration of the chemical and the duration of exposure. In some cases, transient contact with a chemical such as benzene may have little effect. On the other hand, an inhalation or skin exposure of a cyanide compound may kill you.

If there is contact with a poison, a whole range of physiological effects could occur. Some poisons, such as carbon monoxide, can interfere with the oxygen distribution system of the body. The red blood cells that normally carry oxygen to the cells are unable to do their job due to their interaction with the CO (carbon monoxide). The hemoglobin in the red blood cells prefer carbon monoxide 200 times more than they prefer oxygen. The result is that the normal oxyhemoglobin created inside the RBC (red blood cell) turns into carboxyhemoglobin and prohibits normal cellular respiration. In case you did not know it, normal healthy people breathe in order get rid of excess CO_2 and to oxygenate the cells of the body. This process keeps the tissues alive, which in turn keeps the organs alive, which in turn keeps you alive. If this system fails or becomes compromised, you are in deep trouble in a real hurry.

> **NOTE**
> Neuroeffector junctions are where nerves meet organs and synapses are junctions between nerve bundles.

Other substances, such as the organophosphate group of pesticides, may paralyze muscles and affect nervous system activity. Parathion, an organophosphate compound, creates havoc at the cellular level inside your body. It interferes with nervous system transmissions at what are called *neuroeffector junctions* and *synapses*. Neuroeffector junctions are where nerves meet organs and synapses are junctions between nerve bundles. This whole system is like the wiring in your house. The main breaker panel is like a central brain with multiple runs of wiring going outward in all directions. Some of these runs of wiring are interrupted by switches that can be turned on or off depending on the need for power. Your body is a lot different in makeup but very similar in concept. The wiring of your body originates in your brain and spinal cord and spreads throughout your body via an intricate framework of nerves. These nerves are also interrupted by "switches," which allow certain body functions and muscular activity to be turned on and off. These switching stations are the synapses and neuroeffector junctions as mentioned previously. This on-off switching is done by chemical reactions that occur in fractions of a second. At both junctions, there is a tiny space between the nerve itself and the opposing section. This gap is bridged by chemical impulses that finish the job of nervous transmission. When the brain says "jump" (meaning any kind of organ activity, moving your muscles, or tearing of your eyes), a lightning fast impulse is sent down the nerve much like electric current through your household wiring. When the impulse gets to the junction, a chemical is secreted by the nerve, which then travels across the gap to find its specific receptor site. This receptor site is the lock, which only accepts a very specific chemical profile or key. This lock and key relationship is vitally important to the overall success of the nervous system as it is the final step in impulse transmission. Figure 3-22 illustrates the physiology of a typical synapse.

All nerve terminals within major organs, muscles, arteries, and so forth have a host of locks that accept certain keys. Some locks called alpha 1 receptors are stimulated by *go* keys, which cause constriction of arteries and the bronchioles inside the lungs. Other locks are stimulated by *go slow* keys such as beta 2 receptors, which oppose the alpha 1's and cause dilation of the same arteries and bronchioles. It is not important to memorize the various receptors but you should be aware of the concept. Chemical reactions are a key part of nervous system impulses and organ function. In the event these transmissions are compromised, total body systems may be affected. This is physiologically critical as the body always needs to be in balance.

> ■ NOTE
>
> Chemical reactions are a key part of nervous system impulses and organ function.

Organophosphates fit into the picture by inhibiting the breakdown of certain chemical compounds that stimulate the action of the intrinsic eye movement, some heart activity, the action of the salivary glands, and most of the organs of the abdominal cavity. Without going into great detail, these body systems suffer unyielding stimulation and create a condition called SLUDGE. This condition is typical of organophosphate overexposure and results in the victim displaying the following signs:

<u>S</u>alivation

<u>L</u>achrymation

<u>U</u>rination

<u>D</u>efecation

<u>G</u>astric disturbance

<u>E</u>mesis

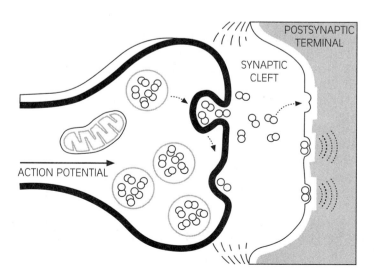

Figure 3-22
Mechanism of action for a neurotransmitter impulse.

The victim's heart also becomes bradycardic and results in serious hypotension and possible seizures. This condition is very serious and requires the administration of a drug called *atropine* to offset the action of the organophosphate. The atropine is not a sure antidote and many patients ultimately die from the exposure. All this disturbance is caused by an interruption of the nervous system. The lock and key relationship is interrupted by the organophosphate resulting in serious and potentially fatal consequences.

Other poisonous materials may take a period of hours or days to affect you. Arsenic is an example of a poison that takes longer to act. If you ate a small amount of arsenic once, it would probably not noticeably affect you. However, repeated small doses over a long period of time can impair health and even be life threatening. Lead poisoning or repeated exposures to **PCBs** cause similar reactions. Often limited, single exposures are usually not significant. Most everyone has washed their hands in gasoline or paint thinner at some point in their lives. This sporadic exposure sequence carries only a remote possibility of causing any health problems at all. In most cases it is repeated dosing over the years that ultimately proves to be problematic.

Poisoning symptoms also vary widely. A victim may suffer from convulsions or cessation of breathing to no noticeable symptoms. It is possible for a poison to enter the system with no noticeable effect and then produce profound symptoms some time later. Low concentrations of hydrofluoric acid fit this profile. Hydrofluoric acid (HF) is known as "the bone seeker" and easily penetrates the skin. At concentrations less than 20%, a victim may not even notice that there has been an exposure. There may be no redness or burning or itching that is characteristic of an acid exposure. Only after the HF has migrated into the muscle tissue will it break apart to form free hydrogen (H+) and fluorine (F−). The creation of free hydrogen is what causes the pain in an acid exposure. Because the compound does not readily **ionize,** the pain caused by the creation of free H+ is delayed. Once the compound has broken apart, the fluorine tends to bond with calcium wherever it is found. Because calcium levels are high in the bones, the fluorine chooses to bond there. This fluorine interaction with the calcium in the bones is also quite painful and may cause substantial damage. Large dermal exposures of HF could result in death. Exposure of over 25 square inches of body surface area with a significant concentration of HF (>10%) could be fatal. HF may also cause systemic effects like **hypocalcemia** and **hyperkalemia.** These conditions may ultimately affect the heart and/or kidneys, which could lead to significant systemic disturbances and death. Figure 3-23 shows HF stored at a fixed facility.

A number of chemicals are regarded as particularly dangerous because of their high toxicity or threat of exposure. Some are easily recognizable while others may not give any hint as to their toxicity. The substances in the following list are considered to be toxic to some degree:

PCBs
polychlorinated byphenyl, a highly toxic chemical compound previously used as a cooling agent in electrical transformers

■ **NOTE**
The creation of free hydrogen is what causes the pain in an acid exposure.

hypocalcemia
medical condition characterized by abnormally low levels of calcium in the blood

hyperkalemia
medical condition characterized by an abnormally high level of potassium in the body

Acetone cyanohydrin
Acrylamide
Aldol
Ammonium fluoride
Aniline
Barium cyanide
Benzonitrile
Bromoform
Calcium cyanide
Chloroform
Cyanogen bromide
Dichlorobenzenes
Dichlorodimethyl ether
Dichloromethane
Dinitrotoluenes

Epibromohydrin
Ethyl bromide
Hexachlorobenzene
Hydrocyanic acid
Lead arsenates
Nickel carbonyl
Nitroanilines
Nitrotoluenes
Phenol
Phenyl mercaptan
Potassium fluoroacetate
Sodium cyanide
Tetraethyl lead
Toluidines

Figure 3-23
Concentrated HF containers.

Some toxic gases that pose an inhalation hazard are:

Ammonia, anhydrous

Arsine	Hydrogen chloride, anhydrous
Boron trifluoride	Hydrogen fluoride, anhydrous
Bromine chloride	Hydrogen sulfide
Carbon monoxide	Methyl bromide
Carbonyl sulfide	Methyl chloride
Chlorine	Nitric oxide
Diborane	Phosgene
Dichlorosilane	Phosphine
Ethylene oxide	Selenium hexafluoride
Fluorine	Sulfur dioxide
Germane	

The OSHA standard is based only on toxicity data without regard to subsidiary factors such as volatility. Whether the compounds listed above would be classed as Highly Toxic or Toxic would depend on the results of toxicity testing. The information should be given on a warning label and on an MSDS for the substance.

Carcinogens

■ **NOTE**
Carcinogens or carcinogenic substances are those which are capable of causing cancer.

Carcinogens or carcinogenic substances are those which are capable of causing cancer. Harmful health effects of chemicals can range from gross destruction of tissue (as with corrosives), to interference with normal body functioning (as with toxic materials), to disruption of processes within body cells that affect their ability to divide. In cancer, modified body cells start reproducing and continue to do so in an out-of-control fashion. This leads to tumor formation and is a severe assault on a normally functioning body. For this reason carcinogens can be thought of as poisons.

Cancer is not a single disease. Many types of body cells can show cancerous activity and result in different symptoms. The causes of cancer are not definitely known, although many are thought to be dietetic or environmental. Medical research is making continued progress in isolating the origin, but to date has not been able to provide any definitive cause of cancer. Methods of treatment are also improving and showing a higher rate of success. Despite these small advances, cancer still remains one of the most common and feared diseases of our time.

It has been known that exposure to some substances is correlated with the development of cancer. With regard to dose levels and the duration of exposure, the range of effects seems to be broad. Some authorities hold that there is no safe

> **■ NOTE**
> From an occupational safety and health point of view, the distinction of a carcinogen results from its effects in animal testing.

exposure level, and this conservative view is certainly the safest one to take. From an occupational safety and health point of view, the distinction of a carcinogen results from its effects in animal testing. Specific exposure tests have been developed and performed on test animals. The tests take about two years to run and analyze, so they are expensive and the amount of information obtained does not accumulate rapidly. There is also an argument about the applicability of test data obtained with regard to humans. The question that is always asked of animal research is "If the chemical causes cancer in lab rats, does it really mean that it will cause cancer in humans?" This is a big question that is not easily answered. A simple review of the big saccharin scare years ago brings up a valid point. In this test, **heroic doses** of saccharin were found to cause cancer in the rat population. There was no evidence that cancer was caused in humans but saccharin was essentially "blacklisted" as a sweetener. Was there a link between the animal testing and the human element? It will never be known but one thing is for sure—it certainly raised some interesting questions.

heroic doses
an expression of exposure levels used in animal testing; referred to as the maximum tolerated doses of the test chemical

According to the OSHA standard, a chemical is considered to be a carcinogen if it has been

- Evaluated by the International Agency for Research on Cancer and found to be a carcinogen or potential carcinogen.
- Listed in the Annual Report on Carcinogens published by the National Toxicology Program as a carcinogen or potential carcinogen.

Furthermore, OSHA specifically regulates the following eighteen substances as carcinogens:

2-Acetylaminofluorene	4-Dimethylaminoazobenzene
Acrylonitrile	Ethyleneimine
4-Aminodiphenyl	Methyl chloromethyl ether
Arsenic, inorganic	alpha-Naphthylamine
Benzidine	beta-Naphthylamine
bis-Chloromethyl ether	4-Nitrobiphenyl
Coke oven emissions	N-Nitrosodimethylamine
1,2-Dibromo-3-chloropropane	beta-Propiolactone
3,3-Dichlorobenzidine	Vinyl chloride

> **■ NOTE**
> According to the OSHA standard, information on whether a compound is a carcinogen must be included on a safety label and a material safety data sheet for the substance.

It is important to keep the problem of carcinogens in the proper perspective. In nature and the laboratory there are millions of chemicals known and/or created. Of these, perhaps 40,000 find some use in the chemical industry and commerce. Any given facility is not likely to use more than 2,000 at the most, yet OSHA specifically regulates 18 compounds. According to the OSHA standard, information on whether a compound is a carcinogen or not must be included on a safety label and a material safety data sheet for the substance.

Health Hazards Posed by Carcinogens Again, the problem of cancer initiation is extremely complex. Not only are the causes uncertain, there is also a possibility of a combination of factors, including individual susceptibility that influence one's chance of getting cancer in the first place. Most often the effects are not immediate, with some cancers not becoming active for 20 years. For these reasons alone it is usually impossible to track down any one exposure and label it as the trigger for the type of cancer.

Further complications arise because of the existence of promoters, substances which are not themselves carcinogenic but which can ultimately allow other substances to have a carcinogenic effect. These promoters are extremely difficult to isolate because the list is so lengthy. Additionally, a promoter may not have the same effect from one individual to another. This is why cancer is such a baffling disease. Determining exactly what caused the cancer from one case to another is like finding a needle in a haystack. The best course of action for workers in the field is to identify any and all carcinogens and take the appropriate precautions.

> ■ **NOTE**
> The best course of action for workers in the field is to identify any and all carcinogens and take the appropriate precautions.

Safe Handling, Use, and Storage of Poisons and Carcinogens

A very stringent and mandatory system of controls has been established for regulated substances. Some of the provisions are:

- Establishment of a controlled area where carcinogens are processed, used, handled, or stored.
- Access by authorized employees only.
- Use of closed vessel operations.
- Requirement that employees wash hands, forearms, face, and neck on each exit from the area.
- No drinking fountains in the controlled area.
- No food, beverages, cosmetics, smoking materials, or chewing gum in the controlled area.
- Maintenance of reduced air pressure in controlled areas. Ensure that the airflow is always inward.
- Prohibition of dry sweeping or mopping in the area.
- Use of proper labels and warning placards.
- Establishment of proper training programs.

Although these measures are required by rule only for regulated substances, they form a comprehensive safety program that could be applied for any suspected cancer-causing material. The program is based on elimination of contact with the material.

Because poisons are so specific in their harmful effects, it is necessary to make sure that they are properly contained and provided with warning labels.

Safe handling and use of toxic materials should be based on no contact, and work practices should employ closed reaction systems where possible. If there is a possibility of contact, practices should be reviewed and analyzed for possible routes of exposure. The danger of toxic vapor inhalation, reaction with or absorption through moist mucous tissues, or absorption through the skin may exist. On the emergency scene, medical personnel should have a clear understanding of the potential effects before any workers enter the site. Contingency planning for unforeseen exposures should be given a high priority. Unfortunately, you may be required to evacuate large numbers of people during these incidents. A fire involving vinyl chloride for example creates huge clouds of thick black smoke. Any persons downwind could be exposed to inhalation of a known carcinogen. Such a fire may require a large-scale evacuation, which is no easy task for any agency. Evacuations can present huge obstacles as they are very time consuming as well as labor intensive.

Emergency Response Procedures

There are recommended general actions, including removal of the source of contamination and diluting or neutralizing the product. These actions are easier said than done in many cases because of the high degree of concern with the released substance. Other emergency response measures include isolation of the spilled material. After the hazard is contained, there is still the cleanup to consider. Proper procedures must be followed to ensure that no harmful residues are left in the workplace or are released to the environment.

The best plans include great measures to ensure that contact with the substance is minimized. If it is possible to let the incident stabilize on its own with little impact, choose this option! Remember that success in the hazmat business is where the problem did not get a whole lot worse after you arrived and no exposures to personnel occurred.

Rinse liquid spills on the skin with large quantities of water (the location and use of safety showers and eye baths should be discussed). If the poison has been swallowed, medical attention must be found. It is not always the best course of action to induce vomiting and administer a universal antidote. For example, corrosives may cause as much damage coming up as they did going down, so inducing vomiting is not recommended. Petroleum distillates and most hydrocarbons should receive special attention in the event of ingestion. Do not induce vomiting with these materials as it may cause further health problems. Aspiration of vomit contaminated with hydrocarbons may cause *chemical pneumonitis*. This chemical pneumonia has a high mortality rate and can often cause more fatalities than the original exposure. Understanding the exposure potential is critical to providing the proper care. In the case of powder exposures to the skin, brush off as much as possible prior to a water wash. Give emergency medical personnel and hospital staff as much information about the chemical as possible, which can easily be accomplished by sending the MSDS to the hospi-

■ **NOTE**
Understanding the exposure potential is critical to providing the proper care.

tal along with the victim. The more the doctors know, the quicker they can treat the exposure.

RADIOACTIVES

What image comes to mind when you consider the possibility of handling a radiation incident? The truly honest person will most likely answer "fear." Plain and simple, the prospect of handling a radioactive substance makes most of us highly uncomfortable. Thoughts of glowing skin and flippers growing out of our heads are usually what we think of in the event of a radiation exposure. Your fears are not unfounded and radiation incidents are definitely to be respected. But the good news is that these incidents are few and far between. Furthermore, if we understand what radiation is and how to protect ourselves, we should be able to stay safe no matter what the circumstance. In order to do this however it is necessary to understand the following information:

- What radiation is and where it comes from.
- Atomic structure and the origin of radioactive energy.
- Some terms and definitions relating to exposure.
- Time-distance-shielding.

It is also necessary to realize that this section covers only basic information and will not go into great technical detail. Volumes of information on this topic are available and if you are interested please do some research on your own.

Radiation

radiation
energy emitted from matter in the form of electromagnetic waves

Radiation has been around since the "big bang" or divine creation (or whatever you believe) brought something out of nothing. Somewhat of a deep concept, but the reality is that as soon as there was matter, there was nuclear radiation. This natural phenomenon did not originate with the nuclear age or Three Mile Island or Chernobyl, but we did not really understand its existence until the late 1800s. A German physicist, W. K. Roentgen, was doing some research in 1895 and ultimately discovered X-rays. They were called X rays because "X" is the mathematical symbol for an unknown and Roentgen did not fully understand where these rays were coming from. Later he began to understand their nature and ultimately won the Nobel prize in 1901 for his efforts. He believed that some atoms had unique properties that caused them to emit energy different than their stable counterparts. Through testing and research he isolated this strange energy and was able to determine that it was incredibly intense. Since then, scientists from all generations have completed exhaustive research on this topic and have given

> **■ NOTE**
> Unstable atoms are radioactive and they in turn emit radiation.

radioactive isotope
a radioactive isotope of any element

> **■ NOTE**
> All elements on the periodic table have at least one radioactive isotope.

electromagnetic radiation
propagation of waves of energy of varying electric and magnetic fields through space

us volumes of incredibly technical information. Nuclear radiation in fact, is the most studied environmental hazard in the entire world. It affects life as we know it and will continue to do so until the end of time.

So what is radiation and where does it come from? First of all we need to understand the difference between radiation and radioactive. Unstable atoms are radioactive and they in turn emit radiation. Essentially, radioactive is a description of the atom and the spontaneous emission of electromagnetic particles from its nucleus. On the other hand, radiation is a description of the energy emitted. This is critical to the understanding of a **radioactive isotope** and the ensuing radiation from its instability. An isotope is a form of a stable atom. The atom has all the properties of the stable form except that it differs in mass and has a different number of neutrons in the nucleus. An example of this can be found by comparing the stable carbon atom and its radioactive isotope called carbon-14. The carbon-14 has more neutrons in the nucleus and an atomic weight greater than its stable counterpart. A radioactive isotope then, is a variation of its stable counterpart that emits radiation. All elements on the periodic table have at least one radioactive isotope.

All atoms have the potential to become unstable and decay. This decay is the atom's way of attempting to achieve stability. In essence it is dealing with an internal instability usually resulting from an imbalance of protons and neutrons. The nucleus is described as undergoing decay since its decomposition is in the process of emitting energy. When this imbalance is substantial as in the example of uranium (atomic weight of 238, atomic number of 92), an atom releases its excess energy from the nucleus thereby creating electrically charged particles and/or **electromagnetic radiation**. Electromagnetic radiation is a massless, nonelectrical force that travels from the nucleus in "waves." The resulting energy is characterized by wavelength—the shorter the wavelength, the more intense the energy.

The charged particles are called alpha and beta particles while the pure energy (electromagnetic radiation) is called gamma rays. Other types of electromagnetic radiation include microwaves and visible light, but gamma radiation is different as it is of higher energy and originates in the nucleus of unstable atoms. For a review of atomic stability read the section that discusses the components of an atom in Chapter 2.

Alpha, Beta, and Gamma Radiation

We have touched on the different types of radioactive decay, but it is important to understand the differences between them and the potential hazards they present. Alpha radiation comes from a charged particle that is emitted from the nucleus of an unstable atom. This particle is quite heavy atomically speaking and is the most highly charged of the three forms. The good news is that the alpha particle uses up most of its energy before it travels very far from the nucleus. Safe distances from alpha radiation are accomplished by staying just a few feet away from

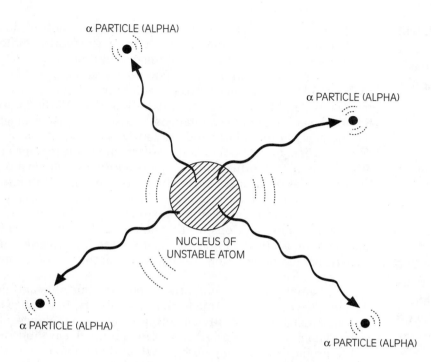

Figure 3-24 *Example of alpha particle emission.*

alpha particles
small charged particles emitted from the nucleus of an unstable atom

half-life
the time required from the decay process to reduce the energy production to one-half of its original value

■ **NOTE**
Elements with atomic numbers greater than 82 are regarded as heavy and can be suspected radioactive elements.

the source. Figure 3-24 is a graphic illustration of **alpha particles** being emitted from a nucleus.

Remember, this particle is only emitted from the nucleus of an unstable atom. It struggles for stability and the only way this stability can occur is for the atom to release some energy. It then becomes much like a cat trying to cough up a fur ball. Both the cat and the atom have something in common—a real need to release something in order to feel more stable. This change in energy state, or atomic fur ball release, can occur over a few minutes or could last hundreds of years depending on the isotope. Many reference sources will refer to this energy release duration as an isotope's **half-life.** The half-life is the time it takes for a radioisotope substance to undergo enough changes in energy state to reduce its original mass by half. Some half-lives are as short as millionths of a second or as long as billions of years. Carbon-14 has a half-life of approximately 5,700 years whereas cobalt 60 has a half-life of approximately 25 years. Half-lives are unaffected by external factors such as ambient temperature or atmospheric pressure.

Alpha particles are generally emitted from elements with heavy nuclei such as radium and plutonium. Elements with atomic numbers greater than 82 are regarded as heavy and can be suspected radioactive elements. These particles can become airborne, however, and inhaled by the unsuspecting worker. Unfortunately, you will be intimately involved with this particle as it will be sitting in some deep corner of your lungs bombarding everything within a few inches of it.

Chapter 3 Hazard Classes

beta particles
by-product of radioactive decay from an unstable nucleus; the process changes a neutron into a proton and subsequently emits an electron from its orbit

gamma radiation
electromagnetic radiation

> ⚠ **SAFETY**
> In almost all cases an internal hazard is worse than skin exposure, because the exposure occurs inside the body where the source may have access to a wider range of organs or tissues.

> ■ **NOTE**
> This pure energy is sometimes called a *photon* and is a short wavelength, high frequency type of wave that can travel approximately a mile from its source.

> ■ **NOTE**
> Fission is the splitting of an atom's nucleus, which results in an incredible amount of energy release in the form of neutrons, pure energy, and propagation of some other atoms.

ionizing radiation
energetic, short wavelength rays emitted by elements and isotopes, strong enough to remove electrons from atoms

High efficiency particulate air (HEPA) filters are usually adequate to protect your lungs from alpha particles and should be used whenever contaminated dusts are suspected.

Beta particles are much smaller than alpha particles and therefore travel greater distances from the nucleus. They are also faster-moving but have less of a charge than alpha particles. Depending on the strength of the source, a beta particle can travel several millimeters into your skin. Beta particles can be inhaled or even ingested just like alpha particles, but they do not go far enough into the body to reach vital organs where the damage could be life threatening. Drinking water contaminated with beta particles would result in a serious internal hazard and could possibly cause radiation sickness or cancer or some other type of fatal organ dysfunction. In almost all cases an internal hazard is worse than skin exposure, because the exposure occurs inside the body where the source may have access to a wider range of organs or tissues.

Atomically speaking, beta particles are much like electrons except that they are not in a "normal" orbit around a nucleus. They are sometimes captured from their orbits into the nucleus and catapulted outward in the atom's quest for stability. This is a complicated process that we do not have to worry about in this text. Suffice it to say that beta particles can be positively or negatively charged and more intense than alpha particles. Some common beta emitters include carbon-14, phosphorus-32, and sulfur-35.

The third kind of radioactive decay is **gamma radiation.** This radiation gets all the press and creates instant panic because it is so dangerous. You would be wise to understand what kind of hazards this decay presents. First let us look at what gamma radiation is. It differs from alpha and beta particles in that it is not a particle—it is pure energy traveling through space in the form of electromagnetic waves. This pure energy is sometimes called a *photon* and is a short wavelength, high frequency type of wave that can travel approximately a mile from its source. It moves at the speed of light (186,000 miles per second) and can travel deep into the body, affecting organs and body tissues with extreme exposures causing severe burns and rapid death. You will not be protected by chemical suits, firefighters' turnout gear, the standard walls of a building, or any other common materials of construction. Gamma radiation can be caused by several activities but chiefly it is from a fission reaction. Fission is the splitting of an atom's nucleus, which results in an incredible amount of energy release in the form of neutrons, pure energy, and the propagation of some other atoms.

Most of us do not hang around places that accomplish nuclear fission so we will probably never see its destructive nature. We may have an X ray taken some time in our lives and that is about as close as you can get to the potential power of gamma radiation. The best protection is distance but it may be way too late for that once you detect its presence. Figure 3-25 shows the relative skin penetration power of the three forms of decay.

All three forms of radiation we are discussing are categorized as **ionizing radiation.** Ionizing radiation is far worse than nonionizing radiation as it interacts

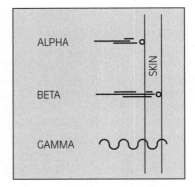

Figure 3-25
Examples of skin penetration by radioactive isotopes.

with the atoms of the somatic cells within the body. Somatic or body cells are neutral in charge unless they are hit by an alpha or beta particle. This collision may knock an electron out of its natural orbit thereby creating an electrical imbalance. If this happens, an **ion** is created inside your body cells. An ion is a neutral atom with an imbalance in electrical charge due to the gain or loss of an electron. If the atom gains an electron it is called a **cation;** if it loses an electron, it is called an **anion.**

Your body generally deals with repairing this kind of damage on a regular basis and life goes on unchanged. In the event the body responds negatively to the newly created ion, or does not respond at all, biological changes occur. These changes may manifest themselves as radiation sickness or various types of cancer. Radiation sickness is a group of symptoms resulting from an excessive radiation exposure to large parts of the body. Some of the symptoms include nausea and vomiting, hair loss, exhaustion, and hemorrhage. Length of exposure, intensity of the source, biological factors, and a whole host of other conditions influence just how serious the exposure is. You should always limit or avoid an exposure because you just may not know the consequences for a long, long time.

Sources of Radiation

We are bombarded daily by any number of radioactive sources. Cosmic radiation, primarily from the sun, penetrates our atmosphere and rains down upon each and every one of us. Although the dose is relatively low (30 mrem per year), it is still present and is part of the background radiation you cannot escape. mrem is a unit of absorbed dose into body tissues; numerically it is one-thousandth of a rem. The rocks and soil that make up the earth are also sources of radiation. The uranium, thorium polonium, and such present in our minerals are always producing radiation energy. This is also a low amount, approximately 40–50 mrems per year but

---

ion
an atom or molecule that has acquired a positive or negative charge by gaining or losing an electron

cation
a positively charged ion created when an atom loses an electron from its orbit

anion
a negatively charged ion created when an atom gains an electron in its orbit

■ **NOTE**
mrem is a unit of absorbed dose into body tissues; numerically it is one thousandth of a rem.

once again, it is another source. Since 5 rems is the maximum yearly dose for occupational exposure, background radiation is insignificant for most people. Your own body emits radiation due to elemental isotopes within our soft tissues and organs. We certainly cannot escape this source unless we can figure out how to survive without carbon and potassium and other vital elements.

In addition to these naturally occurring sources we can receive radiation through many sources produced by humans as well. Therapeutic radiation for cancer and other disease may bring us up close and personal with the effects of constant radiation. X rays and natural fallout from weapons testing are also artificial sources from which we cannot escape. The really important source is the potential radiation exposure from occupational incidents. This can be from a daily job that you do to responding to some unknown chemical release. In these settings, the dose absorbed could be potentially fatal (300–650 rems) without the possibility of a second chance. These are the big events that we all dread and hope never to encounter. You should not discount, however, the lower dose events that may occur. One of the big hazards of radiation exposure is from a constant dosing over a long period of time. It is therefore critical to keep track of all exposures and never trivialize any events involving radioactive substances.

Safe handling procedures and storage practices are an extremely complicated process with radioactive substances. There are so many laws, regulations, and handling procedures that it would be impossible to include them all. The best advice here is to study the procedures and practices used at your work site. There will be many, and it is important to be acquainted with them well in advance of beginning work.

Emergency Procedures

Responding to radiation events can be quite tense and worrisome. The typical responder may never see a transportation incident in his or her entire career, because the packaging and transportation requirements for radioactive substances is very strict. Containers for these materials are essentially bombproof and have never been involved in a serious transportation release.

Routinely working with radioactive substances is another matter. There is always a possibility of spilling liquids or container breakage with the spread of radioactive contamination. When a source is identified, or thought to be leaking, all personnel should be evacuated and the **Radiation Safety Officer** notified immediately. Some very basic safety actions can then be taken:

- Prevent radioactive material from getting on clothes or skin.
- Guard against ingesting any radioactive material.
- Contain loose radioactive material from spreading, especially liquids.
- Clean up contamination and monitor personnel and the area for radiation levels.

> **SAFETY**
> One of the big hazards of radiation exposure is from a constant dosing over a long time period.

Radiation Safety Officer person in charge of the program that concerns radiation-related issues at fixed facilities

Time-Distance-Shielding

Another commonly accepted method for reducing radiation exposure is to employ the principles of time, distance, and shielding.

When working with radiation sources, the more time you are exposed to a source, the more radiation exposure you will receive. Radiation exposure should be limited whenever possible. Since all radiation you receive is significant to you, always keep your exposure to a minimum by working with radiation sources only when necessary.

Distance is also very effective and, in many cases, the most easily applied method of radiation protection. Technical research shows that radiation intensity falls off very rapidly as the distance increases. The inverse square law for reduction of radiation applies to most sources of radiation. The law goes like this: If you double the distance from the source of radiation you decrease the intensity of the radiation by a factor of four. Therefore, it is important to stay as far from the sources of radiation as possible. This is not always possible when working directly with radioactive materials but can definitely be employed in emergency situations.

Shielding should be designed so that no large amount of radiation is released except through the desired portal. This shielding is useful at fixed facilities but once again may not be practical in emergency response. For these situations, it may be necessary to protect personnel with physical barriers. The more material used, and the greater its density, the more radiation it will stop. It is impossible to eliminate all emitted radiation, but it can be reduced to acceptable levels.

Acute whole body exposure may be encountered from atomic weapons, reactor accidents, or commercial facilities. The basic term used when describing radiation exposure is the roentgen. Sometimes abbreviated with the letter R, it is the international unit of measurement for X rays and gamma rays but has been expanded to include alpha and beta particles. This measurement is only an indication or attempt to quantify an amount of radiation and is not intended to project possible health effects. Exposures are usually expressed in units of measurement per hour. As an example, a reference material may caution you to avoid any exposure of more than 5Rs per hour. This basis is important to understand because it is used in Table 3-4. Metric system logic applies to radiation exposure in this way—1 roentgen (R) equals 1,000 milliroentgens (mR). Table 3-4 lists the health effects of a possible radiation exposure.

> ■ **NOTE**
> Metric system logic applies to radiation exposure in this way—1 roentgen (R) equals 1,000 milliroentgens (mR).

corrosives
chemicals that cause visible destruction of or irreversible alterations in living tissue by chemical action at the site of contact

CORROSIVES

This class of chemicals will most likely represent a large portion of your hazmat activity. Along with flammable liquids, **corrosives** are the most widely used substances in industry. In fact, sulfuric acid is at the top of the list of the most

Table 3-4 *Health effects of radiation exposure.*

0–25R	No detectable clinical effects. Delayed effects may occur.
25–100R	Disabling sickness is not common and exposed individuals should be able to proceed with usual duties. Delayed effects may be possible, but improbable.
100–200R	Nausea and fatigue is most likely. Higher doses may cause vomiting. Delayed effects may shorten life expectancy but probably less than 1%.
200–300R	Nausea and vomiting within the first day. There may be delayed health effects of up to 2 weeks. Once symptoms appear they may include loss of appetite and general malaise, sore throat, pallor, diarrhea, moderate emaciation.
300–600R	Nausea, vomiting, and diarrhea are probable within the first hours of exposure. There may be no other significant signs for up to one week. Once the symptoms occur they may include loss of appetite, general malaise, and fever during second week, followed by hemorrhage, and inflammation of the mouth and throat. There may be diarrhea, and emaciation in the third week. Death may occur in 2 to 6 weeks. Death occurs in approximately 50% of the individuals exposed to more than 450 roentgens.

common industrial chemicals produced and used in the United States. Because sulfuric acid is used in just about every segment of industry, some economists base how technologically advanced a nation is by how much sulfuric acid it uses. Petrochemical plants, semiconductor fabricators, biotechnology, and plastic manufacturers all use large amounts of this chemical. Chances are that if a facility makes a product, sulfuric acid is probably involved. Figure 3-26 shows several containers of sulfuric acid.

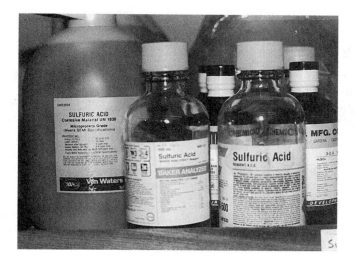

Figure 3-26
Concentrated sulfuric acid storage.

We will look at this class of chemicals from a broad perspective and focus on some common corrosives you should be familiar with. The section also encourages an awareness and understanding of the kinds of hazards that can result from working with corrosives. Like many other hazardous materials, corrosives can present a variety of challenges to the site worker or emergency responder. Some corrosives like hydrofluoric acid are poisonous but nonflammable whereas others like acetic acid are less corrosive but combustible. Corrosives also include caustic materials such as sodium hydroxide and potassium hydroxide, which may aggressively defat the skin upon exposure. Corrosives may be the broadest category of chemicals you will ever come into contact with. Countless chemicals within this class pose multiple hazards so watch out and look at the big picture of the substance you may be facing.

> **SAFETY**
> Countless chemicals within this class pose multiple hazards so watch out and look at the big picture of the substance you may be facing.

Definition of Corrosives

Corrosives are defined by the DOT as materials that can cause visible destruction to tissue and/or steel during a predetermined period of time. For most cases this testing occurs over a 4-hour period.

From an occupational safety and health point of view the definition of a corrosive results from its effects on living tissue rather than on metal plates. Specific tests have been prescribed to be performed on test animals, and corrosives are compounds that cause, in the words of OSHA a "visible destruction of, or irreversible alterations in, living tissue by chemical action at the site of contact" during the 4-hour test period. Seems pretty gruesome, but the reality is that this class of materials has some very dangerous chemicals in it.

Corrosives are thus defined in terms of their chemical and physiological action, which does not have any particular relation to the chemical's structural nature. In other words, the health effects that corrosives cause are the chief reason these substances are grouped together.

Acids

polar
description of a molecule where the positive and negative charges are permanently separated; differs from nonpolar substances in which the electrical charges may coincide; polar molecules will ionize in water and conduct an electrical current

Attempting to define an acid may be difficult for the nonchemist but the definition should be understood. Generally speaking, an acid may be defined as a material that has loosely held hydrogen. Now that probably makes no sense at this point, but keep this in mind; acids are substances that are combined of two parts—the acid group and hydrogen. Hydrochloric acid is an excellent example of this two-part concept. The formula for hydrochloric acid is HCl and if we dissect it, and look for the loosely held hydrogen, we find the formula to be quite telltale. Figure 3-27 represents the splitting of H^+ and Cl^- when hydrochloric acid (HCl) is placed in water.

This loosely held hydrogen is a key factor in the strength of the acid. It is also an indicator of how much potential damage could be caused by a skin exposure. Figure 3-27 shows that when HCl is placed in water its **polar** nature is to interact

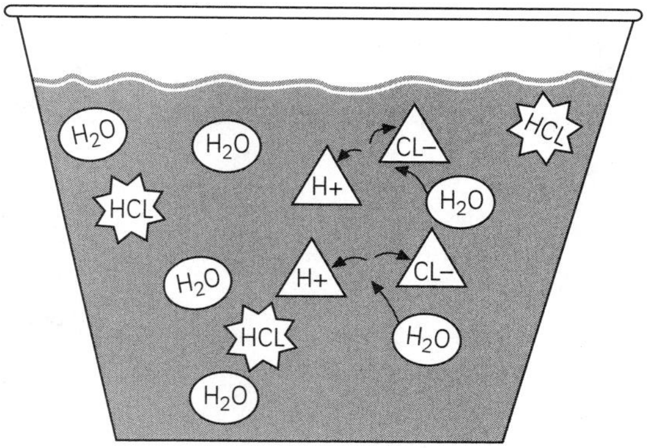

Figure 3-27 *Water splitting the hydrochloric acid into free H+ and Cl–.*

> **SAFETY**
> It is prudent to regard all acids as strong and capable of damage until proved otherwise.

with the hydrogen–chlorine bond. If water is successful in splitting the hydrogen from the chlorine, a free hydrogen ion (H+) is created. This free hydrogen ion then bonds with water. There is more on this process in the following section.

You should also know that all things called acids are not strong or necessarily bad for you. Boric acid, for instance, is used as an eyewash. Carbonic acid makes your soda fizz and dilute acetic acid is also known as vinegar. Information on whether a compound is a true corrosive capable of skin destruction must ultimately be based on experimentation. It will also be recorded on a safety label and the MSDS for the substance. It is prudent to regard all acids as strong and capable of damage until proved otherwise.

Acidic Strength

pH
the symbol relating to the hydrogen ion concentration of a solution; a convenient way to express the relative acidity or alkalinity of a solution

ionization
the formation of ions that occurs when a neutral molecule of an inorganic solid, liquid, or gas undergoes a chemical change

The relative strength of a corrosive material is expressed by a scale known as the **pH** scale. In simple terms, pH stands for the "power of hydrogen." Shown in Figure 3-28, this scale has a range of 0 to 14. Acids go from 0 to 6 on the scale whereas bases have a range from 8 to 14. A pH of 7 is considered neutral and is the point we strive for when neutralizing. Strong acids are considered to have a pH of 2 or less. Strong bases have a pH of 12.5 or more.

The relative strength of an acid is determined by how easily the material releases its hydrogen ion. The term for this process is called **ionization** and inorganic acids (those not containing carbon) tend to be stronger than organic acids, because inorganic acids release their hydrogen ions easily. Some examples of common inorganic acids include hydrochloric acid, phosphoric acid, and sulfuric acid. This action of ionization is then our basis for defining pH as the power of hydrogen.

The technical definition of pH is quite a bit more involved but not important for us to know at this level. Just remember that when an acid is mixed with water, hydrogen ions are released from the acid group and subsequently bond

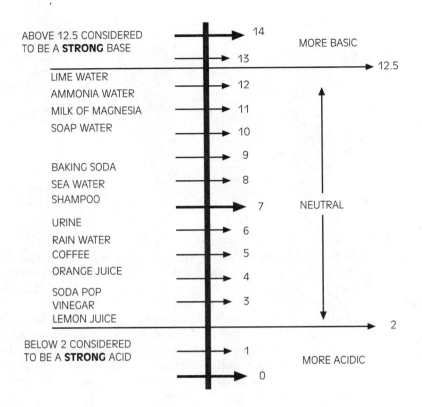

Figure 3-28 pH values for some common substances.

■ NOTE

The strength of an acid is based on how readily the compound gives up its H+ to water.

concentration
the relative amount of a substance when combined or mixed with other substances

with some of the water molecules. When this happens, the H_2O gets an extra hydrogen ion and becomes H_3O. Impossible you say? Actually it is very possible and the process even has a very descriptive definition. It is said that the extra hydrogen ion (H+) has solvated to the water (H_2O) creating a hydronium ion (H_3O). Figure 3-29 shows the process of solvation.

Look at the chemical formulas in Table 3-5 and notice the presence of hydrogen in all of them. This H+ is ultimately released to the water and is what causes the pH paper to turn color (usually red for acids and blue for bases). The free H+ also causes the damage to skin and metals. The H+ is the bad guy here and the more H+ that goes into a solution, the stronger the acid. The strength of an acid is based on how readily the compound gives up its H+ to water.

Inorganic acids such as those in Table 3-5 ionize readily and are considered strong. Don't forget—the less complete the ionization, the weaker the acid.

Acidic Concentration

Another concept that must be understood is **concentration.** Concentration is often confused with how strong an acid is, but it really refers to how much of the

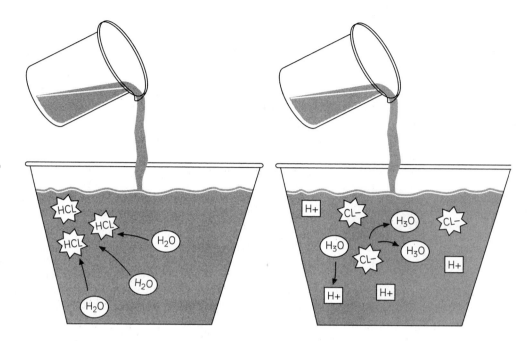

Figure 3-29 *The process of solvation: Left: When HCl is placed into the water it is broken into H+ and Cl–. Right: Some of the free H+ remains and solvates with the water molecules to form H_3O.*

acid is mixed with water. The strength of an acid is identified by how readily it ionizes. Sulfuric acid is traditionally used in 97% concentration. This simply means that the other 3% of the volume is water.

Water is the other part of an acid solution and its presence is always implied when discussing concentration. Concentration is the relative amount of a substance when combined or mixed with other substances. Examples include 2 ppm of chlorine gas in air or a 50% hydrofluoric acid solution.

Table 3-5 *Examples of inorganic acids.*

H_2SO_4	Sulfuric acid
HF	Hydrofluoric acid
HCl	Hydrochloric acid
HNO_3	Nitric acid
H_3PO_4	Phosphoric acid
$HClO_4$	Perchloric acid
HBr	Bromic acid
HI	Hydriodic acid

■ NOTE
Strength is not related to concentration as it does not affect the ability of an acid to ionize.

At the 97% concentration, we could say that we have a concentrated, strong acid. If the concentration of the sulfuric was dropped to 10% (the other 90% being water) we would now have a strong, dilute acid. Strength is not related to concentration as it does not affect the ability of an acid to ionize. Do you think it is possible to have a concentrated weak acid? The answer is yes and an example is the chemical compound called acetic acid. This acid does not easily give away its H+ to water and is considered to be a weak acid. Even if the concentration is 70% it is still considered weak because of its poor ionization potential. So in this case we would have a concentrated, weak acid. Again, if we drop the concentration to 10%, we would have a dilute, weak acid. Remember, ionization potential is not affected by a change in the concentration of an acid.

bases
any chemical substance that forms soluble soaps with fatty acids; also referred to as alkalis

alkali
any chemical substance that forms soluble soaps with fatty acids

caustics
alkalis

Bases

Another group of materials within the corrosive classification are **bases.** We have touched on them throughout this section but it is important to know that they are essentially a stand-alone classification. These substances are also known as **alkalis** or **caustics,** but in reality they are all descriptions of the same thing. Bases are common chemicals and you most likely have several around your own home. Most drain cleaners contain some degree of sodium hydroxide. Baking soda (sodium bicarbonate) is a base as well as your bottle of ammonia and most oven cleaners. To chemically define a base, one has to look at things from the molecular level. We defined an acid as a substance that gives up its H+ to water; we define a base as a substance that dissociates its OH− from the compound and into the surrounding solution. The process is conceptually the same as was described with acids. The main difference here is that we are talking about the OH−, which is called the hydroxide ion. The OH− is negatively charged and naturally attracted to anything positively charged.

! SAFETY
Bases characteristically cause worse skin damage than acids because they actually work beneath the surface of the skin.

Bases characteristically cause worse skin damage than the acids because they actually work beneath the surface of the skin. The slippery feeling you get between your fingers when they are exposed to something like drain cleaner is due to the fact that you are *melting*. This slick feeling is the subcutaneous fat being brought to the surface by the action of the base. The process is called **saponification** and is a common occurrence with caustic exposures. Do not think that this is a surefire weight loss program, but it does melt away the fat!

saponification
a reaction between an alkali and a fatty substance that forms water soluble byproducts and a soap

Bases are mostly used and stored in the liquid or solid phase. They are common substances, and if you work in the hazmat field for any length of time you will certainly encounter a few. Table 3-6 shows some common bases.

Sodium and potassium hydroxide are very prevalent in industrial settings. They are generally used and stored as concentrated liquids or solid pellets. The solid pellets pose problems if they are allowed to become airborne dusts. These dusts are **anhydrous,** and only need to find water to get back into solution. This water can be found in your eyes, nose, armpits, or any other wet or sweaty part of the body.

anhydrous
containing no free water

Table 3-6 *Examples of common bases.*

NaOH	Sodium hydroxide
KOH	Potassium hydroxide
NH$_4$OH	Ammonium hydroxide
Mg(OH)$_2$	Magnesium hydroxide
CaCO$_3$	Calcium carbonate
NaHCO$_3$	Sodium bicarbonate

You will also notice the consistent listing of OH in most of the formulas in Table 3-6. This is the same type of signature for most of the bases as the H+ was to the acids. Seeing this in the formula, or reading "hydroxide" in the chemical name, will almost always tip you off to the presence of a base. The formulas and names are not always the most reliable however as evidenced in the last two examples in Table 3-6. Here you see the names *carbonate* and *bicarbonate*. You should readily notice that neither has the telltale OH in the formula or hydroxide in the name. Furthermore, both chemical names end in -ate which was a tip off to a possible oxidizer. Clearly these two substances are not oxidizers and may tend to confuse us in our quest to categorize chemicals based on loose chemical rules. You may also rest assured that these are not the only examples of substances that may be difficult to categorize based on name alone. The most reliable way to identify a substance is to get its name then look up the properties in some reliable reference source.

Bases are classified as strong or weak, and concentrated or dilute, just like the acids. The same rules apply, except we substitute the OH– for the H+ in our definitions. Concentration varies depending of the need, but remember that it does not relate to strength.

Neutralization

■ NOTE
The most reliable way to identify a substance is to get its name, then look up its properties in some reliable reference source.

neutralization
a reaction between an acid and a base

■ NOTE
The goal in neutralization is to reach a neutral pH so the material can be handled and disposed of safely.

Neutralization becomes important when we must handle a corrosive release. Neutralization is a process of deactivating strong acids using an appropriate amount of a base (caustic) at an appropriate concentration. Neutralization also describes a process employed to deal with caustic spills. Bases can be neutralized with the appropriate amounts of an acid at an appropriate concentration. Essentially neutralization is based on the antagonistic relationship of the two classes—acids and bases. The goal in neutralization is to reach a neutral pH so the material can be handled and disposed of safely. Compounds containing fluorine are a little different because of their inherently toxic properties, but other than that, neutralization is very effective and widely used to take care of corrosive spills.

A great benefit to neutralizing a corrosive is that it does not appreciably add to the overall spill volume, thereby reducing disposal costs. Dilution of strong acids and bases is usually not feasible as it takes thousands of gallons of water to change the pH of an acid or base by one pH value, because dilution does not work at the atomic level by interacting with those pesky hydrogen ions—it just reduces the concentration by diluting the overall volume. Strong, concentrated, corrosives are more reactive and evolve more heat when neutralized, so that plastic tools or containers may melt or become deformed and fail. Remember—Dilution may not always be the solution!

The anatomy of a typical neutralization decision and effort may go as follows:

- A corrosive is spilled on a flat surface where a puddle may form. The material is identified and referenced for chemical properties.
- The appropriate neutralization agent is decided upon. If the spill is acidic, choose a moderate pH base to perform the neutralization. If the spill is caustic, choose a mild acid. Commercial neutralizing agents are also available. They are more user friendly than making your own but still operate under the same premise.
- Slowly add the neutralizing agent, taking care to do frequent pH checks. It takes quite some time to move the pH a little bit, but once it gets close to 7 slow down. It is very easy to slam past 7 one way or the other and have to totally change your neutralizing agent.
- The resulting liquid will consist of water and a salt compound and will be easily disposed of.

A neutralization reaction is simple in concept and is shown in Figure 3-30. This particular reaction illustrates the interaction between hydrochloric acid and sodium hydroxide. Do you see where the water and the salt compound come from? The sodium bonded with the chlorine and formed the salt compound called sodium chloride. The hydrogen left behind by the loss of chlorine bonded with the OH on the other side of the equation and formed water. Once complete neutralization is achieved the hazard of the material is removed and the spill can be easily absorbed and disposed of.

Health Hazards Posed by Corrosives

The degree of danger posed by a compound primarily depends on the strength, concentration of the chemical, and the duration of exposure. Some substances

■ **NOTE**
The degree of danger posed by a compound primarily depends on the strength, concentration of the chemical, and the duration of exposure.

Figure 3-30 *A typical neutralization reaction involving hydrochloric acid and sodium hydroxide.*

$$HCL + N_aOH \rightarrow N_aCl + H_2O + \Delta \uparrow (HEAT\ LIBERATED)$$

have a slower corrosive action (resulting in skin destruction from 3 to 60 minutes), but are regarded as particularly dangerous because the vapor is toxic. Other substances are extremely dangerous as they cause skin destruction in less than 3 minutes exposure. Some particularly hazardous corrosives are found in the following list:

- Allyl chlorocarbonate
- Bromine
- Chromium oxychloride
- Chromosulfuric acid
- Fluorosulfonic acid
- Hydrofluoric acid
- Nitric acid
- Nitrohydrochloric acid
- Selenic acid
- Selenium oxychloride
- Sulfur chlorides
- Sulfur trioxide
- Sulfuric acid
- Sulfuryl chloride
- Thionyl chloride
- Trifluoroacetic acid
- Vanadium tetrachloride

Substances are regarded as presenting a medium danger if they can cause skin destruction from 3 minutes to 60 minutes under test conditions but have no particular additional hazard. Examples of some of these chemicals include the following:

- Acetic acid (50% to 80%)
- Acetic acid, glacial
- Acetic anhydride
- Acetyl bromide
- Acetyl iodide
- Acrylic acid
- Alkyl, aryl, or toluene sulfonic acid
- N,N-Diethyethylene diamine
- Diethyldichlorosilane
- Diphenyldichlorosilane
- Dodecyl trichlorosilane
- Ethyl chlorthioformate
- Ethylphenyldichlorosilane
- Ethylsulfuric acid
- Fluoroboric acid
- Fluorosilicic acid
- Formic acid
- Hexafluorophosphoric acid
- Hexyl trichlorosilane
- Thiophosphoryl chloride
- Trimethylacetyl chloride
- Vanadium oxytrichloride

Substances regarded as presenting a minor degree of danger (but still meeting the definition of skin destruction in a 4-hour test) include the following:

- N-Aminoethylpiperazine
- Amyl acid phosphate
- Butyric acid
- Cyanuric chloride
- 3-(Diethylamino)-propylamine
- Ethanolamine
- 2-Ethylhexylamine
- Hydroxylamine sulfate
- Methacrylic acid
- Phosphorus trioxide
- Tetraethylenepentamine
- Triethylenediamine
- Vanadium trichloride
- Zirconium tetrachloride

Some materials can pose multiple threats. The possibility of inhalation toxicity has been mentioned, and some substances are not only corrosive but are readily absorbed through the skin, which may result in further systemic damage or poisoning. Hydrocyanic acid is a good example and you should be especially careful with this compound. The skin exposure and localized burning is usually the least of your worries as the real danger lies below the skin. Hydrocyanic acid is one of those substances that is usually fatal in large doses and cannot be effectively treated in the field.

Many corrosives are especially reactive and can mix with other chemicals to result in a complicated situation. Heat is often the by-product of these reactions and fire could be the result of the mixing. Some substances, such as sulfuric acid, can react with water or even moisture in the air to release toxic gases or react with organic material to generate heat. Others are not only corrosives but also flammable liquids or oxidizing materials. Some acids may also react with certain metals to evolve toxic or flammable gases. Information on these added dangers should be obtainable from an MSDS.

Safe Handling, Use, and Storage

Because many corrosives are so reactive, it is often necessary to take special measures in designing containers for them. Acids and bases corrode metals, therefore they are usually shipped and stored in plastic and glass containers. Figure 3-31 shows a typical polyethylene drum used to store corrosives.

The use of protective gear, including location, storage, need for use, maintenance, cleaning, and decontamination, should be discussed. As corrosives are

Figure 3-31 *Typical polyethylene 55-gallon drum used to store corrosives. Sulfuric acid is commonly shipped like this.*

very active chemically, great care must be taken to ensure chemical compatibility. In other words, make sure your equipment will not fail if it is exposed to a heavy corrosive.

Emergency Procedures

There are recommended general actions, including dilution or neutralization of the corrosive, but they are very product specific and should be clearly understood prior to action. Because they can be easily neutralized and therefore quickly disposed of, spills involving corrosives are often easy to resolve. Only those trained in proper procedures should attempt to neutralize the materials. In many cases, the process of neutralization can create a significant reaction that liberates heat. In such cases, severe consequences could result. Figure 3-32 shows a student adding a base to an acid spill.

In cases of spills, many people assume that they can dilute the material with water to make the corrosive less concentrated. As mentioned previously, this may not be the greatest idea because of the increased spill volume. Flooding with large quantities of water can reduce the degree of hazard by making the product less corrosive, but it also creates larger amounts of a less concentrated material. In addition, significant local heating and spattering can occur when water is added to concentrated materials.

Figure 3-32 *A student practicing neutralization in a mock spill scenario.*

One exception to the use of water on corrosive materials is in the case of contact with people. In such cases it is always best to immediately apply large amounts of water in an attempt to dilute the product and stop the tissue damage. Safety showers and eye wash stations should be located in close proximity to the areas where corrosive materials are used, stored, and handled. Quick action is essential, because the longer the contact, the greater the damage.

Further stages of emergency response include isolation of the spilled material to prevent spread of the damage. In the case of corrosive materials it is essential to follow the basic rules of chemical spills—isolate the area, and deny entry. Remember that many of these products are highly reactive and can mix with other materials to release toxic, flammable, or explosive vapors.

After the hazard is contained, there is still the cleanup to consider. Common inorganic chemicals can be neutralized, but the fate of more complex substances must be tracked to make sure that no harmful residues are left in the workplace or are released to the environment. In all cases, only properly trained and protected personnel should be involved in such activities.

Summary

- Whenever handling chemicals, it is important to understand the nature of the materials you are dealing with.
- Have a good action plan in mind when handling chemicals.
- The importance of preplanning can never be underestimated.
- There are eight commonly encountered classes or groups of chemicals: explosives, gases, flammable liquids, flammable solids, oxidizers, poisons, radioactive materials, and corrosives.
- It is imperative to consult Material Safety Data Sheets once a chemical has been identified. If a victim is being transported to a medical facility, the MSDS should be sent along with the prehospital care providers.
- The DOT breaks down the classification of explosives into six classes which can then be broken down into three major categories: High explosives, Low explosives, and Blasting agents.
- When working with explosive substances it is essential to have a sound fire prevention and storage program.
- Gaseous materials may be nonflammable, flammable, toxic, corrosive, oxidizing, or even radioactive.
- **B**oiling, **L**iquid, **E**xpanding, **V**apor, **E**xplosion occurs when materials with high expansion ratios are heated above their boiling points inside closed vessels.
- If you contact cryogenic materials, severe frostbite and tissue damage will result.
- It is important to remember that a substance in the gaseous state automatically increases the danger of exposure because of the possibility of inhalation and eye and skin exposure.
- Most regulatory agencies classify flammable and combustible liquids on the basis of their flash points.
- The consistent physical danger from flammable liquids and solvents is the fire hazard. Vapors may form combustible or explosive atmospheres and ignite from uncontrolled ignition sources.
- Flammable solids may come in many forms and you should be prepared for any of them.
- Oxidation occurs when electrons have been transferred from the shells of a reducing agent or fuel to the outer shells of an oxidizing agent. Oxidation is a complicated chemical reaction in which heat is generally liberated.

- If a chemical name ends in -*ate* or -*ite*, it could indicate that the material is an oxidizer.
- The selection of any personal protective equipment (PPE) should be specific to the incident and/chemical.
- Depending upon the dose, length of exposure and individual physiology, poisons are substances that are likely to cause death or serious injury if they are swallowed, inhaled, injected, or come in contact with the skin, eyes, or mucous membranes.
- The suffix -*cide* means "to kill" and can be applied to various groups of chemicals.
- OSHA (and most systems) defines two categories of toxic materials: Toxic and Highly Toxic.
- Poisonous materials may take hours or days to affect you and may present in a wide range of signs and symptoms.
- Contingency planning for unforeseen exposures should be given a high priority.
- Radioactivity is a description of the process of atomic decay. Radiation is a description of the energy emitted.
- Gamma radiation is different from alpha and beta radiation in that it is not a particle but a form of pure energy traveling through space in the form of electromagnetic waves and is the most dangerous, with extreme exposures resulting in severe burns and rapid death.
- Radiation sickness is a group of symptoms resulting from an excessive exposure to radiation.
- You should always limit or avoid a radiation exposure and take into consideration that the effects and consequences may not be evident for a long time.
- Technical research shows that radiation intensity falls off very rapidly as the distance from the source increases.
- Corrosives are a broad category of chemicals and pose multiple hazards.
- A key factor in the strength of an acid is the loosely held hydrogen.
- The strength of an acid is identified by how readily it ionizes, whereas concentration simply refers to how much of the acid is mixed with water.
- Bases are classified as strong or weak, and concentrated or dilute, just like the acids.
- A great benefit to neutralizing a corrosive is that it does not add appreciably to the overall spill volume, thereby reducing disposal costs.

Review Questions

1. Explain the difference between detonation and deflagration.
2. Which of the following most closely defines brisance?
   A. A sound wave that permeates nearby buildings.
   B. A pressure wave that emits outward in all directions from the site of the blast.
   C. Tiny fragments carried from the blast by air pressure.
   D. The origin of the explosion.
3. Which of the following most closely defines the acronym ANFO.
   A. A mixture of ammonium nitrate and ferric oxide.
   B. A mixture of ammonia and fuel oxides.
   C. A mixture of ammonium nitrate and a fuel oil such as diesel.
   D. A mixture of amines and nitrated fluorine.
4. True or False? Dynamite and TNT have the same chemical composition.
5. What is vapor density?
   A. How much of a substance will mix with water.
   B. The weight of a vapor compared to the weight of a liquid at the same temperature and pressure.
   C. The weight of a liquid compared to the weight of water.
   D. The weight of a gas or vapor compared to that of an equal volume of dry air.
6. True or False? The health effects of chemicals are always immediate, letting you know that they are hurting you.
7. In order to find the vapor density of any gas, it is necessary to take the _____ of the sample gas and _____ it by 29.
8. True or False? Caustic chemicals are capable of causing saponification.
9. A pH of 1.6 indicates that the material is which of the following:
   A. A caustic material
   B. A mild acid
   C. A strong base
   D. A strong acid
10. What does BLEVE stand for?
    A. Burning Liquids Expanding Violently
    B. Boiling Liquid Exploding Vapor Event
    C. Boiling Liquid Expanding Vapor Explosion
    D. Burning Liquids Expanding Vapor Explosion
11. What is a carcinogen?
    A. The level at which 1/2 of the population of rats will die.
    B. A material that causes death at low doses.
    C. A disease-producing material such as anthrax.
    D. A substance known or suspected to cause cancer.
12. A material with a vapor density of 2.5 will most likely do which one of the following?
    A. Rise
    B. Mix with air
    C. Sink into low areas
    D. Mix with water
13. What is the basic principle behind flash point?

A. It defines the temperature at which a substance will auto ignite without a flame source.

B. It is an expression of a liquid fuel with a flash point at or above 100°F.

C. It describes the nature of a liquid fuel with a flash point less than 100°F.

D. It is an expression of the minimum temperature at which a liquid fuel gives off flammable vapors.

14. What is specific gravity?

A. It is a definition of the effects of gravity in a specific application.

B. It tells us the weight of a specific human on the surface of the planet.

C. It is a way to express the weight of a liquid compared to the weight of water.

D. It represents the weight of a vapor relative to air.

15. What is the flammable range?

A. The temperature at which vapors will auto ignite without a flame source.

B. The range above which a material will burn.

C. The right mixture of vapor to air needed to burn.

D. The minimum temperature at which a liquid fuel gives off flammable vapors.

16. The hazard(s) associated with cryogenic nitrogen is that the material is: (Choose all that apply.)

A. Toxic

B. Capable of displacing O_2 in closed areas

C. Flammable

D. Extremely cold and may cause frostbite

17. Gases are typically stored and transported in which of the following states? (Choose all that apply)

A. Pressurized

B. Synthesized

C. Liquefied

D. Cryogenic

18. True or False? Hazardous materials rarely, if ever, possess more than one hazardous property.

19. Which of the following gases/vapors are lighter than air? Use the mathematical method for determining vapor density if necessary. (Check all that apply.)

_____ Nitrogen

_____ Propane

_____ Hydrogen

_____ Argon

_____ Neon

20. True or False? Regarding acids, concentration and strength mean exactly the same thing.

21. IPA (isopropyl alcohol) has a flash point of 53°F. The outside temperature is 68°F. While unloading this product from the truck, several cases fall to the floor and the product spills. Several large pieces of machinery and a welding operation are going on nearby.

Is there a potential danger from fire in this scenario?

Yes   No

Why?

A. The material is not considered flammable.

B. The material is giving off flammable vapors and the ambient temperature is above the material's flash point.

C. The material would need to be heated to its autoignition temperature to be a hazard.

22. Which of the following types of radioactive materials are particles?

   A. Alpha only
   B. Alpha and beta
   C. Beta only
   D. Gamma and alpha

23. Explain the difference between radiation and radioactive.

Chapter 4

Principles of Toxicology

Learning Objectives

Upon completion of this chapter, you should be able to:

- Define toxicology.
- Describe the branches of toxicology.
- Understand the difference between toxic and poison.
- Identify the two categories of chemical effects: acute versus chronic.
- Define the modes of action: local versus systemic.
- Define routes of entry into the human body.
- Define which route produces the highest potential hazard to most workers.
- Describe the three time frames of toxic exposure.
- Define LD_{50}, LC_{50}, LDlo, LClo, LChi, LDhi, odor threshold, PELs, TLVs- TWA, STEL, ceiling limits, and IDLH.
- Define asphyxiant.
- Define carcinogen and teratogen.
- Describe additive, potentiation, antagonistic, and synergistic health effects.

INTRODUCTION

For many of us, a chemical exposure may invoke panic, which may stem from the fact that chemicals are often associated with their possible terrible health effects. When we hear about hazardous chemicals, we are bombarded with terms such as *toxic, highly toxic*, and *dangerous*. Whenever we read something reported about chemicals, it generally is related to some of the harmful effects that they have or some new study that proves the material is toxic or poisonous. For those of us who work in a field where we contact chemicals, these messages can be very disconcerting. The good news, however, is that working with chemicals is not all doom and gloom. Although there are some truly deadly substances on the market, a large component of chemical safety lies in good information. The true goal of our toxicology study then becomes one of sorting out the facts from the fiction.

This chapter discusses hazardous materials toxicology in a practical way. The chapter is not intended to replace a more extensive discussion of toxicology as would be found in a course specifically on the subject. It is, however, intended to provide a brief overview of the subject for those involved in emergency response and waste site work.

As a way to illustrate some of the misconceptions that relate to hazardous materials, let us review the following statements about **carcinogens**. Examine the list and try to determine which are true and which are only myths.

- Cancer is the leading cause of death in the United States.
- Over one-third of all cancers are caused by exposure to chemicals on the job.
- Almost all chemicals that are produced will cause cancer in people. Stated another way, there are thousands of chemicals that are known to cause cancer in people.
- There is an epidemic of cancers in the United States, caused mostly by exposure to synthetic (human-made) materials.
- Air pollution appears to cause only about 1% of all cancers.
- About half of all chemicals that have been tested have been shown to cause cancer in rodents when they are exposed to high doses.
- Chronic infections can cause cancer and actually contribute to about one-third of all of the cancers reported worldwide.
- Most of our exposure to toxic and/or potentially carcinogenic chemicals occurs inside buildings.

By the end of the chapter, you will know which statements are true or false as well as understanding other relationships to the toxicological effects of chemical exposure.

■ **NOTE**
Although there are some truly deadly substances on the market, a large component of chemical safety lies in good information.

carcinogen
a material that either causes cancer in humans, or, because it causes cancer in animals, is considered capable of causing cancer in humans

BACKGROUND

The study of harmful effects of chemicals has been going on for a long time. In prior years, the study involved scientists who were involved in working to find medicines and other materials to cure diseases. As they began to work with the various materials, it was found that many were as capable of causing harm as they were of healing. As this type of study expanded, an entire subdivision of scientists, called *toxicologists* was organized and began to investigate the harmful health effects that chemicals were capable of causing.

In broad terms, toxicology means the study of the adverse or harmful effects of chemicals on living organisms. Toxicology is far reaching in its scope when we consider that adverse effects can mean a number of things from causing rashes to causing death. It is important to note that living organisms involve a whole host of subjects from single cell amoebas to human beings. Looking at the field of toxicology as a whole, we find that a number of disciplines involve many different types of toxicologists. For example, the field of toxicology has a number of major branches including the following:

Clinical toxicology. This branch of toxicology is concerned with studying the diseases caused by, or uniquely associated with, the various toxic substances. This area of toxicology is often involved in the study of drug overdose treatment.

Descriptive toxicology. This branch of toxicology is concerned directly with toxicity testing, that is, actually determining the effects that a substance has.

Mechanistic or biochemical toxicology. This branch of toxicology is concerned with understanding how chemicals exert their toxic effects on living organisms and tries to answer the question of why something causes harm.

Forensic toxicology. This branch of toxicology deals with the medical and legal aspects of the harmful effects of chemicals on man and animals. It applies techniques of analytical chemistry to answer medical questions about the harmful effects of chemicals.

Environmental toxicology. This branch of toxicology is dedicated to developing an understanding of chemicals in the environment and of their effect on man and other organisms. It is broad in its scope because the environment covers a variety of media, which may include air, groundwater, surface water, and soil. Additionally, the study within this branch also involves a whole host of subjects in the environment, ranging from the very lowest form of animal life through humans. Ultimately this branch is involved with determining the environmental fate of the materials.

Regulatory toxicology. This branch of toxicology deals with the responsibility of deciding what safe levels of chemicals may be allowed in the various environmental media, including the workplace. The safe levels are then refined and adopted as standards, guidelines, or policies, and are developed on the basis of the data from studies provided by the other branches of toxicology.

Industrial toxicology. This relatively new branch of toxicology deals with the disorders produced in individuals who have been exposed to harmful materials during the course of their employment.

It is from the work of these toxicologists that we now have a good understanding of the harmful effects of some of the materials that we might contact and the manner in which they exert these effects on us.

EXPOSURE MECHANISMS

Now that we have discussed the basics of toxicology, we should feel confident that there is considerable study underway that deals with the materials we work with. Let us now direct our attention to what some of the study of toxicology has determined relative to our working with hazardous materials.

As we review what is involved in chemical exposure, we first learn that the actual process whereby we are exposed is quite complex. Common sense tells us that simply working with or around a toxic material does not mean that we have been exposed to its harmful effects. In order for a material to hurt us, a number of factors must come into play. These factors can be summarized by the following:

$$\text{TOXICITY} + \text{EXPOSURE} + \text{YOU} = \text{HAZARD or RISK}$$

Toxicity Factors

■ **NOTE**
Toxicity refers to the inherent properties of the specific materials.

The easiest of these concepts to quantify, or know specifically, is the toxicity of the substance. Toxicity as we define it in this case refers to the inherent properties of the specific materials. This property for most materials has been studied and provided to us in the form of the material safety data sheet. This information should be available to all site workers and/or made available through hazard communication programs. Emergency responders may have a more difficult time obtaining this information until a positive identification of the substance is made. Once identification is made, the relative toxicity of the material should be easily referenced. These factors include such characteristics as lethal dose, possible health effects, and signs and symptoms of exposure.

Exposure Factors

The second variable in our equation is far less quantifiable, somewhat variable, but in some cases more easily controlled. This is both good news and bad.

Chapter 4 Principles of Toxicology

exposure factors
conditions of a chemical exposure that determine how damaging the event may be

Although we cannot fully quantify all of the **exposure factors** easily, and although they do vary, the ability to limit exposure is well within our control. Let us now look at some exposure factors and discuss how they affect our overall risk as we work with or respond to chemical incidents.

Concentration of the Chemical An important question to answer regarding chemical exposure is, "What was the concentration of the material that caused the exposure?" In some cases, the exposure potential may be low because the concentration of the material is low. The concentration of certain chemicals directly impacts the damage that occurs as a result of the exposure. This line of thinking is useful when considering the effects of exposure to acids or other corrosive materials. In most cases, the more concentrated the material, the more damage that will occur. Higher concentrations can result in worse damage, which ultimately results in more significant effects. Yet low concentrations of various acids are found in the soft drinks that we regularly consume.

> **SAFETY**
> The concentration of certain chemicals directly impacts the damage that occurs as a result of the exposure.

Duration of the Exposure How long did the exposure actually last? The duration of the exposure will give us information relative to the effects that we can expect. Short-term exposures are often those that our bodies can recover from, assuming the initial dose is not significant. Long-term or chronic health effects may cause cancer or other serious health effects after years of exposure. If the duration of the exposure persists, and the level of the exposure is high, the overall dose of the material that we take in will generally be higher. In these cases, the body does not have time to rid itself of the harmful materials.

Uptake Rate This term refers to the rate at which the material is taken into our bodies. It is tied in somewhat to the duration factor, but also involves not only the time, but the actual rate that we take in the materials. As examples, it can denote the volume of the air breathed during the time period or the amount of water consumed per unit time. Essentially, it is how quickly the material entered the body. Higher uptake rates generally result in higher doses within the body.

> **■ NOTE**
> Chemicals can often mix, causing effects that differ from those of either of the materials by themselves. To safely work with mixtures, it is important to know what the actual effects of the mixture will be.

Chemical Interactions It is important to note that overall chemical exposure is a function of not just the one chemical that might be present, but also the effects of multiple chemicals that may be present in a particular environment. Chemicals can often mix, causing effects that differ from those of either of the materials by themselves. To safely work with mixtures, it is important to know what the actual effects of the mixture will be. Although we have studied and quantified the effects of specific materials, we do not know in many cases what effects chemicals might have when they mix or are taken in combination with other materials. Often, chemical mixtures have a totally different effect than we might have otherwise expected.

Our exposure may be enhanced, or in some cases, reduced if the exposure occurs in the presence of other materials. These changes to our exposure patterns generally fall into four areas: additive, synergistic, potentiation, and antagonistic.

Additive chemical interactions. These additive effects occurring when chemicals mix can be summed up by the following equation: 2 + 3 = 5. In this case, the effect of mixing two materials will not result in any more of a hazard than either of the materials by themselves. The result of mixing these materials simply produces the effect as if the materials are added together. With additive effects, there is no interaction of the materials and the effects are simply the same as taking two different materials. An example of this occurs with certain prescription drugs. Taking a Tylenol tablet for a headache while also drinking alcohol may cause additive effects to our bodies because both are metabolized in our livers. If we intake both at the same time, the liver works overtime to break them down, causing an additive effect on our bodies.

Synergistic chemical interactions. These effects occur when chemicals mix and can be summed up by the following equation: 2 + 3 = 20. Synergistic effects are the most dangerous of all toxic chemical interactive effects. In many cases, we do not fully understand what happens when chemicals mix. They may produce chemical effects that are more hazardous than either of the materials alone or react more vigorously than expected. Not all synergistic effects are spectacular or attention-getting events. Some interactions are insidious and may occur without the knowledge of the person being affected.

An example of synergistic interactions occurs in people who smoke and are exposed to asbestos. The results are often far worse than would be expected from both of these two materials simply being added together. The effects of each of these are compounded and made much worse by the presence of the other. In the case of smoking and asbestos exposure, it is estimated that the effects of developing lung cancer is increased ninefold over what would be expected from either smoking or being exposed solely to asbestos. Synergistic effects should always be considered whenever multiple materials are present.

Potentiation interactions. These effects may result when chemicals mix and can be summed up by the equation of 0 + 2 = 10. In this case, one of the materials does not have any adverse effects of its own, but can potentiate the other into having far worse effects than it otherwise would. The material that is the potentiator might simply cause the other material to be taken into the body at a higher rate. Such is the case with the popular solvent dimethyl sulfoxide or DMSO. Although DMSO might not present any serious problems on its own, its interaction in a mixture often causes other materials to be rapidly absorbed by the body. If DMSO is applied to intact skin, it acts as a catalyst for transporting any other applied chemicals immediately through the skin. Essentially it opens the door for another substance to enter the body. This potentiates or enables another chemical to cause significantly higher exposures than would otherwise have occurred.

Antagonistic interactions. Antagonistic interactions are sometimes confusing but can be mathematically expressed as $4 + 6 = 8$; $4 + (-4) = 0$; or $4 + 0 = 1$. In certain rare instances, certain chemicals have an adverse effect on each other and cause harmful effects to be reduced. Such is the case of antagonism. With antagonistic chemicals, the harmful effects of one (or both) are decreased or eliminated due to their interaction with each other.

A classic example of this antagonistic effect is the drug naloxone (Narcan), which is given to combat narcotic overdoses such as heroin. The naloxone competes for the same nerve receptor site as the narcotic does and consequently lessens the bad effects of the narcotic. The naloxone is an antagonist only for narcotics and does not otherwise interact with other substances.

The You Factors

Like toxicity factors, the *You factors* are mostly fixed. Each of us is the way we are and there is little or nothing that we can do about it. We are born a particular sex, our heredity and genetic makeup largely influence what we look like, and by the time that we are in the working world, we are mostly set into our patterns of behavior. Each of us is greatly different from the other, and a number of individual differences are worth noting as we review the toxicological effects that chemicals have on us.

It is a very well-documented fact that men and women can and do respond differently to similar levels of the same chemicals. Differences in the average body fat between men and women influence how a chemical will be tolerated. A person with a greater mass may tolerate a certain dose of a chemical better than a person with a low body mass. Certain materials that only have an effect on a developing fetus would be of less concern to men than to women who are, or might become, pregnant.

Each of us responds differently once the chemical enters our body. Two points that come into play here are the **retention factors** that each of us have to various materials, along with our individual metabolism.

Retention Factors and Metabolism The amount of a substance that is taken up by our body is referred to as retention factor. Retention factor is partly a result of the material and its toxicity as well the body's ability to rid itself of the exposure. In some cases, the body is quickly able to deal with the exposure and reduce or eliminate any adverse health effects. In other cases, the body is unable to metabolize the chemical and subsequently stores the substance instead of eliminating it. A person's metabolism has a great influence on his or her retention factor.

■ **NOTE**
It is a very well-documented fact that men and women can and do respond differently to similar levels of the same chemicals.

retention factors
conditions of a chemical exposure that largely determine how the body will respond

■ **NOTE**
A person's metabolism has a great influence on his or her retention factor.

As a way of illustrating retention factor, let us consider what happens if we take a multivitamin tablet in the morning. In many cases, the body does not use all of the compounds in the tablet. Those that are not absorbed quickly pass through the digestive system and are eliminated. As a result, the body rids itself of unused substances that are of no benefit at that time. Vitamin B is a common example of a water soluble substance that the body will easily excrete. If too much is ingested, it is quickly passed through the body and eliminated in the form of bright yellow urine. Fecal material, urine, sweat, expired air, and saliva are the primary means of excretion for water-based materials. Other chemicals are not so easily traced, but make no mistake that they are either stored or excreted after an exposure.

The manner in which the body uses and ultimately disposes of chemicals is to a certain degree based on whether the material is soluble in water. If we take a chemical into our bodies that is not water soluble, our metabolism must work to convert it via biochemical processes. The chemical compound may be metabolized into a material that is water soluble and subsequently excreted. Unfortunately, many materials are not easily broken down by the body, and consequently are retained in bones, certain organs, and in the fat. Many nonwater soluble hydrocarbons such as benzene fall into this category.

> **NOTE**
> The length of time a chemical remains unchanged is the basis of the metabolism factor.

The length of time a chemical remains unchanged is the basis of the metabolism factor. Another concern involves reviewing what fraction of the total exposure is retained within us or is taken up by the target organ. Obviously, this rate varies from sex to sex, person to person, and even with race. Physical conditioning is also a factor in reviewing our overall metabolism. Again, the metabolism factor is tied closely with the retention issues as we have previously discussed.

Route of Entry

> **NOTE**
> Perhaps the most important factor from the standpoint of worker and responder exposure is that of routes of entry.

Perhaps the most important of the factors from the standpoint of worker and responder exposure is that of routes of entry. **Route of entry** refers to the pathway that materials may take to enter our bodies. In simplest terms, we can look at these routes as four pathways or highways that might be available for materials to travel into our bodies. These pathways include *inhalation, absorption, ingestion*, and *injection*.

The means by which a chemical enters our body is certainly a factor in its ability to do harm. Water is a prime example. We can drink it almost without issue, it does not permeate our skin and cause problems, but it could prove fatal if we tried to breathe it for any length of time. Certainly, this example shows that the route a material takes in entering our bodies is an important factor in the measure of its ability to do harm.

route of entry
the means by which materials may gain access to the body, for example, inhalation, ingestion, injection, and absorption

It is important to understand that the concept of route of entry is the foundation for much of our study of chemical safety, whether as a responder or as a worker at a site. In fact, knowing the route of entry that a chemical may take is important when selecting the appropriate level of personal protective equipment.

inhalation
breathing in of a substance in the form of a gas, vapor, fume, mist, or dust

Inhalation Inhalation is the route of entry that poses the highest level of concern to us in the field. Inhalation is the term used to describe chemical exposure via the respiratory system. Figure 4-1 shows that the respiratory system actually starts at our nose and mouth and continues down progressively smaller passages until it terminates in our lungs.

As you will note, our respiratory system is quite complex and encompasses a large area within our body. Ultimately the inspired air terminates in the small air sacs called **alveoli** as shown in Figure 4-2. These small air sacs number in the millions in the average healthy adult and are each surrounded by their own blood supply. The point where the alveoli and the blood supply meet is where the exchange of gases and other materials actually takes place.

Given the large numbers of individual air sacs involved, the surface area for gas exchange is incredibly large. In fact, it is reported that if we were to open up each of the hundreds of thousands of alveoli and measure their surface area, we would have approximately 700 square feet.

Additionally, a normal, healthy adult at rest breathes approximately 15 times per minute. While this rate varies with exertion and sleep, we can generally count on the fact that an adult will breathe over 21,000 times in a 24-hour period. The purpose of the respiratory system is to take in materials that our body needs to survive (oxygen) and to remove harmful products that our body produces (carbon dioxide). Based on the average respiratory rate, it is estimated that the normal, healthy adult breathes approximately 35 pounds of air per day.

■ **NOTE**
The main purpose of the respiratory system is to take in materials that our body needs to survive (oxygen) and remove waste products that our body produces (carbon dioxide).

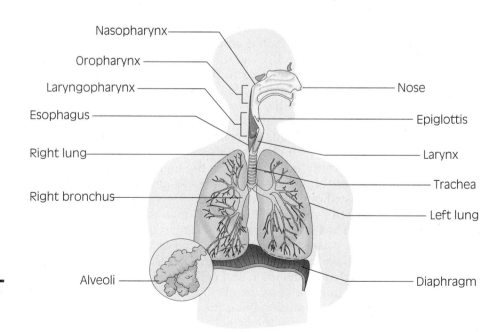

Figure 4-1 *Anatomy of the respiratory system.*

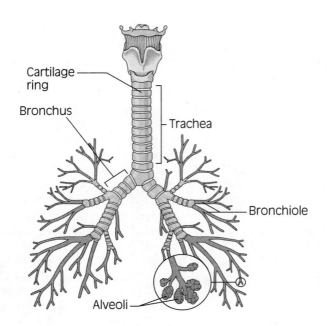

Figure 4-2 *The respiratory tree and alveoli.*

As we inhale, air is drawn into our lungs through either our nose or mouth. During the intake process, it is filtered and humidified to a certain degree by the anatomy of the nose and various other parts of the respiratory system. Refer back to Figure 4-1 for a complete diagram of the respiratory system.

Uncontaminated ambient air contains an adequate supply of oxygen, the material that each of our millions of body cells need to survive. It is the job of the lungs to allow this gas to pass directly into the bloodstream for distribution to each of the cells. This job is accomplished as the air passes into progressively smaller and smaller passages until it finally reaches the alveoli. Each of these tiny alveoli are surrounded by small blood vessels that allow the delivery of oxygen into the blood cells. This process takes place largely due to the natural process of diffusion in which materials with a higher concentration pass into areas of lesser concentrations. In this case, the higher concentrations of oxygen in the inhaled air pass through the alveolar walls and into the red blood cells where it can then be distributed via the circulatory system throughout the body. Figure 4-3 illustrates this movement of gases across the alveolar membrane.

The body also has the ability to rid itself of waste products during the exhalation phase of breathing, which is accomplished in the opposite manner in which we inhale oxygen. In this case, carbon dioxide, the primary waste product, is carried by the red blood cells to the lungs. As the red blood cells pass the thin separation between the alveoli and the capillary, another transfer of gases takes place. This transfer allows the carbon dioxide to diffuse across the barrier and into the air sac where it is expelled into the atmosphere as we exhale.

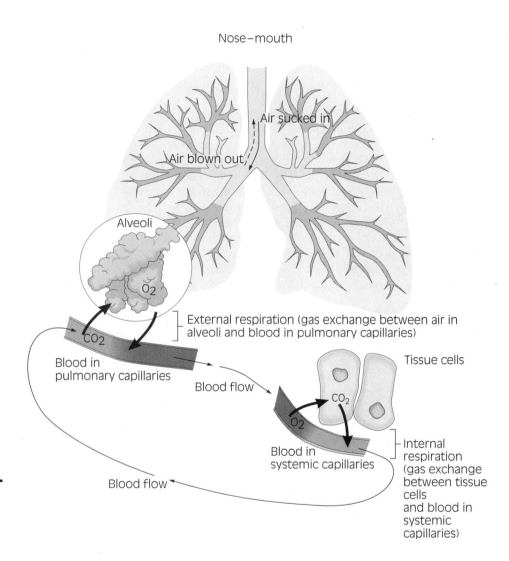

Figure 4-3 *Gas transfer occuring in the capillaries and the alveoli.*

> ⚠ **SAFETY**
> Smell should never be relied upon to determine the presence of a chemical.

The problem is that most inhaled materials cross the alveolar membrane just as oxygen does. Some chemicals cross the barrier easier than others, but the concern should be to place a high priority on respiratory protection. In field work, the concern should be to reduce the possibility of exposure through inhalation. Smell should never be relied upon to determine the presence of a chemical.

Smell can never be counted on to indicate trouble. Furthermore, the lack of an odor does not guarantee that we are not being exposed. Carbon monoxide is an excellent example to illustrate this point. It is produced from almost all types of combustion processes and is colorless, tasteless, does not cause irritation, and

ppm
parts per million; the concentration by volume, of a gas or vapor in air

> **! SAFETY**
> Carbon monoxide is one of many products that is not detected at levels lower than those that cause adverse health effects.

> **! SAFETY**
> Some chemicals do have an odor associated with them but are so toxic that by the time we detect them with our sense of smell, we are already exposed to a lethal dose.

odor threshold
the lowest concentration of a substance's vapor in air that can be smelled

> **! SAFETY**
> Inhalation of toxic materials is a significant concern when working with potentially toxic materials. Materials that are hazardous by inhalation do not necessarily have an odor and are quickly absorbed into our systems once inhaled.

is completely odorless. You *would not* be able to detect its presence by odor alone. An exposure to levels of carbon monoxide above 2,000 **ppm** (parts per million) would be fatal and may not have ever been detected by the victim. Carbon monoxide is one of many products that is not detected at levels lower than those that cause adverse health effects.

In other cases, the chemical may be detected by odor initially, but deaden the sense of smell with prolonged exposure. Such is the case with hydrogen sulfide. It has a strong odor much like that of rotten eggs at low concentrations, but in higher concentrations can actually knock out a persons ability to detect it. With hydrogen sulfide, exposures above 200 ppm will immediately deaden the sense of smell. Exposure to concentrations above approximately 900–1,000 ppm can be immediately fatal. Hydrogen sulfide is also very flammable and should also be considered a dangerous fire risk.

Another problem with detecting chemicals by their odor is that once you smell a particular substance it may be too late. Some chemicals do have an odor associated with them but are so toxic that by the time we detect them with our sense of smell, we are already exposed to a lethal dose. An example of this would be arsine gas. It is detectable at concentrations of 0.5 to 4 ppm but can be fatal at levels below the **odor threshold.** Persons exposed to levels of 500 ppm or above will most likely suffer fatal consequences. It is also a suspected carcinogen so any exposure should be considered a significant event. Arsine has an odor similar to that of garlic and is regulated as a toxic gas.

Conversely, the ability to smell a chemical does not necessarily mean that we are being exposed to dangerous levels. Ammonia vapors can be detected in concentrations of as little as 0.05 ppm but is not at dangerously high levels until it reaches a concentration of 300 ppm. It is also a combustible gas and is irritating to skin, eyes, and mucous membranes. Inhalation of toxic materials is a significant concern when working with potentially toxic materials. Materials that are hazardous by inhalation do not necessarily have an odor and are quickly absorbed into our systems once inhaled.

Absorption The process of absorption theoretically accounts for all of the exposure that we get in regard to chemicals. Chemicals technically are absorbed as we ingest them; likewise, they are also absorbed by our lungs as we breathe them. While this is the case, this is not what we mean when we use the term.

For the purpose of our study, we define **absorption** as the process by which materials enter our body by passing into or through the skin, eyes, or mucous membranes. Obviously, the manner in which this takes place means that we must be close enough to touch the materials. Generally, this contact is in the form of a liquid, however, some hazardous materials can be absorbed by us when they are in the solid or gaseous form.

As we study this area, let us first take the case of chemical exposure through the skin. The makeup of the skin is very complex. It is actually composed of

Chapter 4 Principles of Toxicology

absorption
passage of toxic materials through some body surface into body fluids and tissues

dermis
the "true skin" that covers the body; contains blood vessels and nerves that supply the skin

■ **NOTE**
It is important to understand that certain types of hazardous chemicals can and do pose a significant risk if we come in contact with them.

several layers that provide barriers of protection. Figure 4-4 shows these layers and indicates some of their basic functions.

The outermost part of the skin is a layer of mostly dead cells called the epidermis, a waxy mantel of cells that provides the first layer of defense from chemicals and other hazards.

The lowest level of our skin is the **dermis** layer, the layer that is rich in blood supply. Once chemicals enter the dermis, they can be quickly absorbed by the blood and transferred to other areas of our bodies where they can produce other harmful effects.

Generally speaking, the skin is waterproof and provides a fairly effective barrier against a wide range of materials, partly because of the makeup of the layers and the composition of the skin. Even in the case of disease-producing materials, such as germs or pathogens, intact skin is the first, and one of the most effective means, of preventing these materials from entering our bodies.

Although the skin can provide a significant barrier of protection we must also understand that there are limitations to this shield. It is important to understand that certain types of hazardous chemicals can and do pose a significant risk if we come in contact with them. Two of the major types of hazardous materials that pose this increased risk include corrosives and solvents.

Corrosives, by definition, are materials that can cause significant harm or damage to tissue upon contact. Likewise, when we studied the effects of acids and

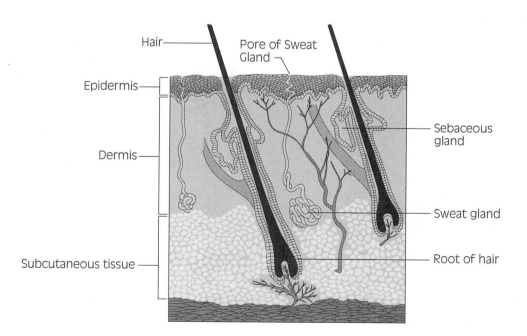

Figure 4-4 *Layers of the skin.*

> **■ NOTE**
> Corrosives and solvents are groups of chemicals that can be rapidly absorbed through our skin. Once contacted, they can allow other materials to enter our system through the openings that they produce as they damage our skin.

anhydrous
chemical compound containing no free water

bases on our skin in Chapter 3, we saw that while the effects of the two differed, both acids and bases have a profound effect on our skin. Some of the common corrosives that are damaging to the skin include sulfuric acid, hydrochloric acid, perchloric acid, sodium hydroxide, and potassium hydroxide.

Solvents produce very harmful effects if we contact them. Again, the very definition of a solvent denotes that they are capable of dissolving something. In this case, the oils and fats that help to keep us waterproof, can also be affected by the solvent. The outer waxy mantel dissolves upon contact with these materials, and once the protection afforded by the oils, fats, and epidermis are gone, the skin becomes an open sieve for the solvent. The exposure may also be magnified as other materials may also enter our bodies through the newly formed openings in the skin. Corrosives and solvents are groups of chemicals that can be rapidly absorbed through our skin. Once contacted, they can allow other materials to enter our system through the openings that they produce as they damage our skin.

In addition to the skin, our eyes are an area of potential exposure. Figure 4-5 shows the anatomy of the eye and illustrates that for the most part, the eyes are largely unprotected. The eye tissue is moist and a number of blood vessels are very close to the surface. Throughout the day, we consistently replenish this fluid through blinking. This moisture provides the vehicle that the chemicals can ride into our system, particularly in the case of **anhydrous** materials. When our eyes are exposed to these anhydrous materials, they interact with the water where they are hydrated. Sodium hydroxide pellets are a good example

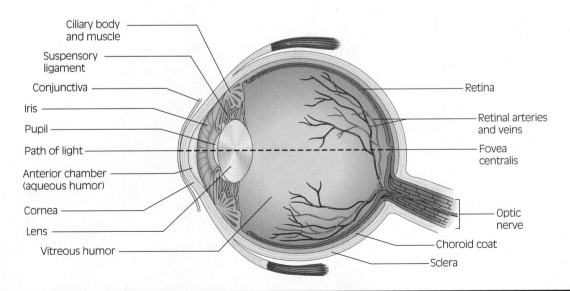

Figure 4-5 *Anatomy of the eye.*

aqueous
dissolved in water

> **■ NOTE**
> The closer that the blood system is to the surface of the membrane, the less distance that the chemicals have to travel to enter our body.

vascular
pertaining to the circulatory system

of a material that may behave in this fashion. If the dusts from these pellets become airborne, they only need a source of water to put them into an **aqueous** solution. The water supply for the creation of this solution can be found in moist skin, or in the eyes. Therefore protective efforts should be aimed at eliminating the possibility of dermal contact. The closer that the blood system is to the surface of the membrane, the less distance that the chemicals have to travel in order to enter our body.

The mucous membranes are another area of concern relative to absorption. Mucous membranes are specialized tissues that are both moist and **vascular** and are concentrated in areas around the face and groin. As with the eyes, these areas are vulnerable to exposure, making absorption a real concern.

An example of absorption would be the use of nitroglycerin tablets to treat chest pain. Patients who take nitroglycerin tablets do not swallow them. Instead, the tablets are placed under the tongue where they are rapidly absorbed into the bloodstream. This effect occurs within just a few minutes since the blood supply is concentrated and very close to the surface of the membrane. The active ingredients can rapidly gain access into the circulatory system and go to work dilating the vasculature.

Ingestion The third route of entry to our body for chemicals is through the digestive tract. This route begins at the mouth and continues some 28 to 30 feet until it passes out through the rectum or urethra. This path is essentially a long tube comprised of specially designed tissue that extracts nutrients and vitamins from the food we eat. The purpose of this type of absorption is to bring energy into the body and eliminate the waste products from the digestion process. This process takes up to 24 hours in most cases with absorption taking place at any point within the digestive tract. Figure 4-6 shows the components of the digestive tract of the average adult.

It is true that ingestion of chemicals can present a problem for those who work with, or respond to chemical incidents, but it is not considered to pose a major problem for us because we do not generally drink or eat the harmful materials that we work with. An exposure via this route is considered to be a secondary means of exposure.

A secondary exposure occurs when chemicals are taken into our bodies after they contact some other item, which is then placed into our mouths. This is the case when we work with hazardous materials which in the course of the day, contact our hands. Unless the hands are decontaminated, everything that is touched after the initial contact with the chemical is unclean. This could result in a secondary exposure when eating, smoking, or in any way touching skin, eyes, or mucous membranes.

Injection Injection is the fourth and last route that we need to consider. In many studies, this route is often discussed when covering the absorption route. While

> **■ NOTE**
> A secondary exposure occurs when chemicals are taken into our bodies after they contact some other item, which is then placed into our mouths.

injection
a route of entry that a chemical may take to enter the body

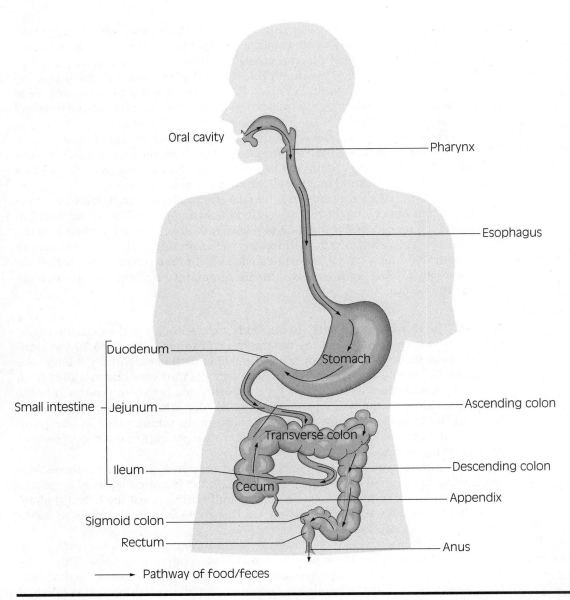

Figure 4-6 *Frontal view of the anatomy of the digestive tract.*

this may have merits, we have chosen to discuss this as a separate route of entry because it covers those exposures that occur if the skin is broken or damaged. Such is the case when we touch chemicals with our hands that have open wounds, or when our skin is punctured or damaged by something that is contaminated with a hazardous material.

The best example of this type of entry is contact with a used hypodermic needle that might be contaminated with a disease-producing **pathogen**. In this case, the skin, which as we discussed earlier, does provide an excellent barrier from many types of materials including pathogens, is no longer an effective barrier.

pathogen
a disease-causing agent

Another, and often overlooked example of the injection route, is that of compressed air forcing chemicals through the skin. Workers who inappropriately use the compressed air hose at the site to blow contaminants off their clothing and skin run a risk of exposure. By doing so, the workers may actually be injecting materials that otherwise would not penetrate their skin. Such injection can also occur whenever employees use power sprayers during painting operations.

Protection for the Routes of Entry

■ **NOTE**
Simply identifying the potential sources of chemical exposure is not enough in terms of protecting ourselves from the chemicals.

Simply identifying the potential sources of chemical exposure is not enough in terms of protecting ourselves from the chemicals. Identifying the potential route of entry for a chemical exposure is only the beginning of chemical safety. One must also develop a strategy to protect or block these routes. In virtually every case, once we know which route or routes to block, we can provide a means to do so and work safely with the chemicals involved. Be aware that in some cases, a material may be able to enter the body by more than one of the routes described. To illustrate how some materials pose a hazard through multiple routes of entry, let us look at the examples of carbon dioxide and hydrogen cyanide.

■ **NOTE**
Be aware that in some cases, a material may be able to enter the body by more than one of the routes described.

Carbon dioxide (CO_2) is a common gaseous substance. Although it is considered to be nontoxic, atmospheres with high concentrations of CO_2 (above 40,000 ppm) may pose a hazard through its ability to **asphyxiate.** Whenever work is performed in areas of high concentrations of CO_2, respiratory protection must be worn. Carbon dioxide has no dermal toxicity and is not flammable. It is not corrosive to the skin and poses no other risk to personnel than that of asphyxiation. It is not hazardous to us if we have cuts or breaks in our skin and does not pose a problem to us if we ingest it. In fact, we actually inject carbon dioxide when we drink carbonated drinks. Essentially, it is dangerous by a single route of entry and if the inhalation route is protected, the material is safely handled.

asphyxiate
to suffocate to death (from lack of oxygen)

!**SAFETY** It is important to understand all the physical hazards in order to be properly protected.

Contrast the properties of carbon dioxide to that of hydrogen cyanide. Hydrogen cyanide or hydrocyanic acid is hazardous to us through more than one route of entry. We certainly cannot breathe it for it is toxic by inhalation at approximately 50 ppm. Furthermore, if the material enters the body through injection or ingestion, it can be rapidly fatal. A person suffering a mild exposure to this chemical may exhibit such symptoms as nausea, vomiting, headache, and irregular respirations. As a gas or even dissolved in liquid form, hydrogen cyanide can permeate the skin and cause death. Unfortunately, this material poses equally high risks through all four routes of entry. It is therefore critical to identify all the hazards and wear the appropriate personal protective equipment. It is important to understand all the physical hazards in order to be properly protected.

EFFECTS OF CHEMICAL EXPOSURE

It is important to understand the effects that chemicals may have, should we allow them to enter our body for two reasons. First, the regulations require that we discuss the effects of chemical exposure as part of our program, and second, it only makes sense that we understand the consequences of chemical exposure.

Modes of Action

In reviewing the effects that hazardous materials have on us it is necessary to look at the manner in which they exert their effects. These chemical effects are accomplished through **acute** and **chronic** exposure. Both of these terms represent exposure modes, but they are not the same and occur at two distinctly different levels.

Acute exposure is likely to be the exposure best known to most field or response personnel. Acute exposure refers to the ability of a chemical substance to do damage to us as a result of a one time or relatively brief exposure of short duration. There are a number of variations of this definition, but it is clear in all of them that an acute exposure is one that is short in duration and that is related to a single exposure of the material. In general, the acute exposure, in order to exhibit toxic effects, must be of a higher concentration than we are physically able to deal with at that time.

Examples of acute exposures include the immediate skin destruction that occurs secondary to a dermal exposure of concentrated sulfuric acid or death that occurs from a relatively brief exposure to a highly toxic gas such as hydrogen cyanide. Acute exposures can occur whether we work as a waste site worker or as an emergency responder. It is not the work function that determines the exposure, it is the chemical itself.

Chronic effects occur when we are exposed over a long time to relatively low levels of a particular chemical or group of chemicals. Again, depending on the definition, this time is typically considered to be more than a few months and can be as long as several years.

Examples of chronic exposure effects are those that can occur when someone smokes cigarettes for a long time and develops cancer or heart disease or workers exposed to asbestos in low levels, for extended periods who later develop lung disease. Chronic exposure effects are more pronounced in those occupations that place workers in contact with the materials on a daily basis, rather than those that involve emergency response.

acute exposure
exposure of short duration, usually to relatively high concentrations or amounts of material

chronic exposure
long-term contact with a substance

■ NOTE
It is not the work function that determines the exposure, it is the chemical itself.

A material can have both acute and chronic effects. Many materials can exhibit effects from a single exposure as well as from repeated doses of smaller amounts. A good example of this concept can be found in some alcohol compounds. Ethyl alcohol is the form of alcohol found in liquor. Although it can produce effects from a single and relatively large exposure, it can also produce adverse health effects if ingested for long periods of time. Intoxication can occur on an acute level, and liver disease is common in those who ingest alcoholic beverages on a chronic basis.

Chemical Effects

In addition to the differences between acute and chronic exposure, it is also worthwhile to note that chemical exposure can manifest itself in two distinctly different forms: **local** and **systemic.**

local
an effect of a chemical exposure limited to the area of contact

systemic
health effects occurring in body systems such as gastrointestinal, reproductive, respiratory, or cardiovascular

Local effects of chemical exposure are those that have an effect on the area that contacts the hazardous material. An example of this would be a chemical burn following contact with an acid. Local exposures most often can be identified early and acted upon. If the effect is local, the material is generally still in contact at that point and therefore can be washed off or removed in some other way.

Systemic effects of chemical exposure are different than local and can be very serious. Because the area of the body affected is not necessarily the area of the body that contacts the material, the symptoms might be dismissed or attributed to something else. A classic example is the onset of nausea and vomiting following exposure to any chemical that is absorbed through the skin. Many times the onset of such symptoms might not be associated with the exposure of the chemical.

■ **NOTE**
The central nervous system is often affected by chemical exposure.

It is important to realize that the central nervous system is the body system most often affected by chemical exposure. Many times, the initial results of exposure of our central nervous system (CNS) to chemicals is that of confusion. With confusion, there could come additional exposure because the victim may not know that he is being exposed. A classic example of this type of exposure occurs with our exposure to carbon monoxide. In many cases, the exposed individual is confused and does not believe that he is experiencing any adverse effects. This may prolong the exposure to more carbon monoxide, which could lead to total incapacitation or death.

Chemical Exposure Time Frames

Local and systemic effects from either acute or chronic exposures do not necessarily occur as soon as we come into contact with them. In fact, a person may not suffer the effects of chemical exposure for some period of time following the initial exposure. The effects of chemical exposure time frames fall into three distinct areas. While determining these time frames is not an exact science, it does provide us with a basis from which to learn. These three areas include immediate, delayed, and long-term effects.

Immediate Effects Some materials that we might contact have immediate effects while others may not. Immediate effects are those that occur within minutes of our exposure to the chemical. These immediate effects can take many forms but always occur within seconds or minutes of the exposure. The effects could range from minor irritation to significant injury or death.

From a safety perspective, the effects of the exposure occur immediately. Consequently, the effects usually provide a warning or notice to the person involved of the potential hazardous effects that might be occurring. Because of these warning properties, immediate chemical effects can be treated or responded to in an effort to reduce the overall exposure.

Delayed The delayed effects of exposure to a chemical are not seen for up to several hours after the initial event. Delayed health effects are often mistaken for other problems and often are not associated with chemical exposure. A second issue with delayed chemical exposure is that even if the effects are noted, it is not possible for the victim to stop the exposure because the damage is already well underway.

Examples of this type of exposure include delayed pain and damage resulting from exposure to hydrofluoric acid, which attacks the bones following dermal contact. Also, the slow onset of disease after exposure to **biohazards** or **pulmonary edema** resulting from an exposure to chlorine gas illustrate this point. In each of these cases, by the time the symptoms are noted, the exposure has stopped and therefore cannot be reduced.

Long Term This type of chemical effect occurs long after the initial exposure. The symptoms may not surface for months or even years after the exposure actually happened. Again, as with delayed effects, by the time the actual symptoms are noted, the exposure may be long over. Once again, there may be little or nothing that can be done to stop the health effects from happening.

Examples of this type of exposure include the formation of cancer that typically occurs years after the original exposure, the development of organ damage as a result of prolonged exposure to various materials (such as asbestosis), or the birth defects that might occur following exposure to teratogenic materials. In each of these cases, the exposed person is not necessarily aware of the effects of the exposure because the damage will take months to years to manifest itself.

Delayed and long-term effects of exposure to chemicals present unique hazards in that once the effects of exposures are noted, it is too late to reduce the exposure. Because of this, these exposures should not be taken lightly.

■ **NOTE**
Immediate effects are those that occur within minutes of our exposure to the chemical.

! **SAFETY**
Delayed health effects are often mistaken for other problems and often are not associated with chemical exposure.

biohazards
a disease-causing agent similar in concept to a pathogen

pulmonary edema
fluid in the lungs

EXPOSURE LIMITS

For those who work with, or plan to work with, hazardous waste it should be reassuring that a number of research groups (**OSHA, NIOSH, ACGIH**) are involved in establishing chemical exposure levels. These **exposure limits** as they are called, are designed to provide guidance for employers and employees alike in the areas of chemical safety. It is the working supposition that if workers stay below these predetermined chemical exposure levels, they will suffer no harmful effects and will be able to work comfortably. Bolstered by volumes of laboratory research, these limits are established to provide for safe working conditions.

A word of caution regarding published exposure limits. Keep in mind that a number of individual factors must be considered when discussing chemical exposure. Although the limits listed by each of the organizations are conservative in their approach, they cannot possibly take into account all of the individual differences that come into play. They are based on the average worker who can be from 18 to 65 years old, male or female, of various ethnic backgrounds, and with a whole range of health conditions. In no way can the values take all of these into account, nor can they be based on the most cautious approach.

Another point to consider regarding these values is that they are difficult to quantify. Unlike some other studies where rats or monkeys may be used to determine toxicity, it is relatively difficult to get a rat to work in a factory for 8 hours a day, 40 hours a week, for 30 years. Partly because of this, the levels are often an educated "best guess" of those who have considerable knowledge of the materials in question. As more data and information about the materials becomes available, the levels are adjusted accordingly.

A last important point relative to exposure limits is that they are primarily intended to protect us from airborne exposure to the materials listed. The inhalation route should be of the utmost concern for all of us, and the formation of airborne exposure limits reinforces this point. The following terms are presented to denote airborne exposure and are expressed in parts per million (ppm) or mg/m^3 of a substance in air.

The two groups actively involved in establishing safe exposure levels are the Occupational Safety and Health Administration (OSHA) and the American Conference of Governmental Industrial Hygienists (ACGIH). Although OSHA and the ACGIH have their own particular domain, there are some commonalties between the two in their involvement with setting standards for workers to follow. Essentially, the exposure limits listed below are stopgap measures for chemical exposure. They are designed to give workers an idea of how dangerous a substance is and what doses may be fatal. The following definitions should assist you in deciphering the technical information found on MSDSs and other reference material. It is important to understand these terms as they are commonly used in all disciplines of hazardous materials work.

OSHA
Occupational Safety and Health Administration

NIOSH
National Institute for Occupational Safety and Health

ACGIH
American Conference of Governmental Industrial Hygienists

exposure limits
values established by scientific testing that denote safe levels that enable workers to maintain a margin of safety when functioning in contaminated atmospheres

■ **NOTE**
Delayed and long-term effects of exposure to chemicals present unique hazards in that once the effects are noted, it is too late to reduce the exposure. Because of this, these exposures should not be taken lightly.

■ **NOTE**
Although the limits listed by each of the organizations are conservative in their approach, they cannot possibly take into account all of the individual differences that come into play.

> **■ NOTE**
> The two groups actively involved in establishing safe exposure levels are the Occupational Safety and Health Administration (OSHA) and the American Conference of Governmental Industrial Hygienists (ACGIH).

TLV/C

The TLV/C stands for the Threshold Limit Value—Ceiling level. As the term denotes, this is the upper limit of exposure for anyone in the workplace. It is a level that could produce harm or at least enough irritation that workers might suffer some type of health effect or at least would not be able to function in the working environment. This level should not be exceeded even for an instant. You may correctly assume that this value is to protect us from the acute exposures that might occur in our workplace. This value is established by the ACGIH.

STEL

STEL stands for the Short-Term Exposure Limit. First we need to understand just how long a "short term" is. In the case of this value, a short-term exposure is defined as one that occurs for up to 15 minutes. It actually is a value that reflects the average exposure that occurs over the 15-minute period, because continuous air monitoring is not always feasible. This value also is established by the ACGIH.

Like the TLV/C, the STEL reflects our protection from very short exposures, or as we discussed, those which are acute. Because the exposure is relatively brief, and because it is relatively low, workers are allowed to reenter an area where the level of contamination is at the STEL more than once in a workday. In fact, workers are allowed to work at the STEL level up to four times per day assuming they have at least a 60-minute break between these exposures.

TLV-TWA

The TLV-TWA is the Threshold Limit Value-Time Weighted Average. It is the maximum value, averaged over a time period. For this term, the time frame is a full 8-hour workday, and it assumes that workers will be exposed to this level for 8 hours per day, 5 days per week, for a full working career. Because this value is used to denote our safe exposure during our normal work day, the term presumes that if we stay below this level, we will not suffer any effects from that exposure. Additionally, it is also presumed that if we stay below this level, we will not need to wear additional respiratory protection or chemical protective equipment.

This term is used exclusively by the ACGIH and published in their book *Threshold Limit Values and Biological Exposure Indices*, which is updated annually. Because the ACGIH book is a standard, published by a nongovernmental agency, it is a recognized guideline and not a law. It is a simply a recommendation based on ACGIH studies and does not have the weight of law as OSHA does.

> **■ NOTE**
> Because the ACGIH book is a standard published by a nongovernmental agency, it is a recognized guideline and not a law.

PEL

PEL stands for the Permissible Exposure Limit of a substance. Like the TLV-TWA used by the ACGIH, this term is usually used to denote the 8-hour, time-

Chapter 4 Principles of Toxicology

> **NOTE**
> The term *PEL* is used to denote levels that are established as the law and used by OSHA.

weighted average exposure that can be used for a 40-hour workweek over a working career of exposure. The term PEL is used to denote levels that are established as the law and used by OSHA.

Although not an occupational limit, the term Immediately Dangerous to Life or Health (IDLH) is very important to understand. IDLH is a level that denotes that one of three things will happen if exposure occurs:

1. There is an immediate threat to life.
2. If exposure occurs there would be a significant risk of an irreversible adverse health effect.
3. At this level, the individual's ability to escape is significantly impaired.

As workers, we should never work in IDLH atmospheres unless steps have been taken to block the route of entry for the material(s) in question. IDLH levels are usually present in emergency situations where chemical concentrations tend to be extreme.

TOXICOLOGY TERMINOLOGY

Let us now direct our attention to some other terms that relate to chemical exposure.

Carcinogen

This term is very important as it is the basis for a great deal of hysteria relating to chemical exposure. Simply put, a carcinogen is a material that is known or suspected to cause cancer. This being the case, workers should try to eliminate any exposure to these types of materials. Because cancer has a high mortality rate, exposures to carcinogens should not be taken lightly.

However, a number of facts need to be discussed relative to cancer and our exposure to carcinogens.

- Cancer is the second leading cause of death within the United States currently (second to cardiovascular disease).
- The statistical odds of getting cancer is one in four.
- The odds are one in five that each of us may die from cancer.
- Our occupational risk for contracting cancer (statistically) is low.
- Most chemical substances that have been tested on rats have produced cancer when the rat is exposed to high doses. Yet many of these substances have not been shown to cause cancer in humans. (*A review of the chemicals known to cause cancer in humans can be found in appendix B.*)
- Several agencies study chemicals for their ability to produce cancer. These include OSHA, the International Agency For Research on Cancer (IARC), the National Toxicology Program (NTP), and the ACGIH. These

> **■ NOTE**
> As we work with materials in the workplace, we should study whether they are known or suspected carcinogens.

teratogen
a substance or agent that can cause malformations in the fetus of a pregnant female

> **■ NOTE**
> Pregnant women in the first trimester are especially susceptible to exposures from teratogenic chemicals.

> **■ NOTE**
> Examples of teratogens include cigarette smoke and alcoholic beverages.

LD_{50}
a single ingested or dermally absorbed dose of a material expected to kill 50% of a group of test animals

LC_{50}
the concentration of a material in air that is expected to kill 50% of a group of test animals with a single exposure (usually 1 to 4 hours)

agencies generally break down the carcinogenic effects into several major categories including known human carcinogens, suspected human carcinogens, animal carcinogens.

- Sometimes, just the irritation effects of a substance are enough to cause cancer. Asbestos and wood dust may not have any chemical that causes cancer in humans, but chronic irritation by them may contribute to the cancer.
- A relatively small number of materials that are known to cause cancer in humans, but it is always prudent to limit our exposure to materials that are suspected of causing cancer. As we work with materials in the workplace, we should study whether they are known or suspected carcinogens.

Teratogen

The word **teratogen** is derived from a Greek word meaning monster or monster baby, because the Greeks mistakenly believed that if a couple had a child who was deformed, it was a result of angering the gods. We know that in some cases, birth defects can be traced to a chemical exposure of a teratogenic material. If a pregnant woman is exposed to a teratogen, she risks delivering a malformed child. Pregnant women in the first trimester are especially susceptible to exposures from teratogenic chemicals. The mother seldom suffers any effect from the chemical exposure and does not run the risk of problems with subsequent pregnancies. The drug thalidomide was a major cause of birth defects in the 1950s. Pregnant women took this drug for morning sickness associated with pregnancy. Unfortunately, many birth defects were attributed to this drug and it was later removed from the market in the United States. Examples of teratogens include cigarette smoke and alcoholic beverages.

LD_{50}/LC_{50}

The terms LD_{50} and LC_{50} stand for lethal dose 50% and lethal concentration 50%. These terms are used to denote levels of chemical exposure where lethality or death will occur in 50% of the test cases. Occasionally, other terms are used to denote similar levels such as **LDlo, LDhi, LClo,** and **LChi**. These stand for lethal dose lowest level, lethal dose highest level, lethal concentration lowest level, and lethal concentration highest level. These terms denote the exposure levels that will be fatal to a specified number of animals in a testing study.

In our example, we expose test animals, often rats, to high levels of a material. We monitor the dose taken into their bodies, or a dermal exposure, and note the point that the first animal dies. We then know that the material is lethal at this low level and we have subsequently reached the LDlo. We keep increasing the dose and eventually kill off one-half of the population, or 50% and now know the LD_{50}. As we increase the dose even higher, we will eventually kill off the entire

Chapter 4 Principles of Toxicology

LDlo
lowest administered dose of a material capable of killing a specified test species

LDhi
the concentration of a chemical by dermal contact or absorption that will be fatal to 100% of a test population

LClo
lowest concentration of a gas or vapor as measured in the air which is capable of killing a specified species over a specified time

LChi
the airborne concentration of a gas or vapor that will be fatal to 100% of a test population

■ **NOTE**
The D in LDlo, LD_{50}, and LDhi stands for *dose*: This means an absorbed or dermal dose or can denote an oral dose.

■ **NOTE**
The C in LClo, LC_{50}, and LChi stands for *concentration* This means airborne exposure.

population of our study group and develop our LDhi. The D in LDlo, LD_{50}, and LDhi stands for *dose*: This means an absorbed or dermal dose or can denote an oral dose.

We can do the same thing with lethal concentration. In this case, the exposure is an airborne material that must be inhaled by the test population. Unlike the lethal dose, which is measured in milligrams of the material found in each kilogram of the animal (mg/kg), lethal concentration is expressed in terms of parts per million in the air. The C in LClo, LC_{50}, and LChi stands for *concentration*: This means airborne exposure. Again, lethal dose refers to exposure via the absorption and ingestion routes, whereas lethal concentration refers to exposure that is inhaled or airborne.

With the differentiation between dose and concentration now clear, let us now direct our attention to seeing what differences there are between the *lo*, *50*, and *hi* levels. The *lo* level denotes the lowest exposure level where death occurs due to the exposure. If we think about it, this low level is not very useful because it reflects the death of the weakest and most vulnerable of the test population. On the other hand, the *hi* level is also not very useful because it reflects the level of exposure needed to kill the strongest of our rat population. The level that is used to describe the exposure level needed to kill the average member of the population is the *50%* level. For this reason, it is the best indicator of lethal effects and is one used in the following definitions.

So, we now come back to the premise that all materials are poisonous to some degree and can kill you if the dose is correct. Do not forget this important concept regarding toxicology—*the dose makes the poison*. Virtually all chemicals can be classified as poisons and hurt us if taken in large enough doses.

Summary

- Toxicology is a science devoted to the study of poisons. This study includes identifying the chemical's mode of action and health effects as well as other facets of exposure mechanisms.
- The branches of the science of toxicology include: clinical toxicology, descriptive toxicology, mechanistic toxicology, forensic toxicology, environmental toxicology, regulatory toxicology, and industrial toxicology.
- The term *toxic* refers to the ability of a material to cause adverse effects to living tissues. These adverse effects can be transmitted when the living organism is exposed through inhalation, ingestion, injection, or absorption.
- A poisonous agent is one that is harmful to living organisms. The dose of a chemical largely determines the severity of a poisonous exposure.
- An acute exposure will cause a person to exhibit symptoms within a relatively short exposure time.
- Chronic exposures occur over very long periods of time, from a period of months to several years. Many carcinogens are harmful due to prolonged exposures over long periods of time.
- Local effects of chemical exposure are those that occur at or near the site of exposure. Strong corrosives such as sulfuric acid cause localized skin damage at the site of contact only.
- Systemic health effects are those that affect entire body systems after an exposure. Carbon monoxide is a systemic poison that may be brought into the body through inhalation but ultimately affects vital organs and the circulatory system.
- Inhalation is a major route of entry to the body for chemicals because the lungs provide a very direct path of access for chemicals. Inhalation hazards pose the highest threat to workers and responders.
- Ingestion is a route of entry whereby chemicals are brought into the body by contaminated food, hands, or used personal protective equipment.
- Injection is a direct route for chemicals to enter the body. If a person has cuts or other open wounds that are unprotected, a chemical can gain immediate access to the circulatory system. Compressed air lines can also inject chemicals through the skin when they are used as a means of decontamination.
- Absorption refers to the route of entry chemicals take when they come into contact with skin, eyes, and mucous membranes.
- Exposures can occur over three distinct time frames: immediate, delayed, and long term. Immediate health effects are those that are felt by a person from seconds to minutes after exposure. For example, most corrosives cause pain upon contact. Delayed health effects are those that occur up to several hours after

exposure. Hydrofluoric acid, for example, may not alert a person to the fact that they have been exposed. Long-term health effects occur after years of exposure. Cancer is a common long term health effect that may result from years of exposure to substances such as benzene.

- Toxicity factors are those physical characteristics of a substance that make them harmful.
- LD_{50} stands for lethal dose, 50% mortality rate to an animal test population. This is an expression of a dermal or ingested substance.
- LDlo is the point at which the first subject in the animal test population dies as a result of a dermal or ingested test chemical.
- LDhi is an expression of a chemical exposure strong enough to kill 100% of the animal test population.
- The LC_{50}, LClo, and LChi follow the same logic as LD_{50}, LDlo, and LDhi. The difference between the two values is that *LC* illustrates an airborne concentration.
- The odor threshold of a chemical is the lowest point at which it can be detected by smell. Some materials such as ammonia have a very low odor threshold.
- PEL stands for permissible exposure limit. It is an OSHA term that denotes a chemical exposure level that should not be exceeded during an individual's work day.
- A TLV-TWA is a term used by the ACGIH to denote an occupational exposure limit. This exposure limit is the expression of a chemical exposure based on an 8-hour workday, 5 days a week for an entire working career. If a worker is exposed to chemicals below these levels, he/she should suffer no ill effects as a result of the exposure. These values are often expressed as ppm, ppb, or mg/m^3.
- The short-term exposure limit or STEL is a term used by the ACGIH to denote occupational exposure limits. Essentially, the STEL is an expression of an exposure that is higher than the TLV-TWA, but below the PEL, and is of a short duration. A STEL is limited to 15 minutes with at least an hour break from any contamination afterward. These levels allow for short excursions into chemical atmospheres, but are based on short exposure times.
- IDLH describes the upper end of chemical exposure limits for workers. The term stands for immediately dangerous to life and health and indicates atmospheric levels of corrosive, toxic, or asphyxiating chemicals. Exposures above the IDLH value could result in death, irreversible health effects, or the inability of a worker to escape the area of contamination. This exposure is based on a 30 minute time frame exposure.
- A simple asphyxiant is a material that could cause suffocation if atmospheric levels are high enough to displace oxygen. Nitrogen gas is a simple asphyxiant and would drop oxygen concentrations to harmful levels if released in suffi-

cient quantities without adequate ventilation. Simple asphyxiants differ from cellular asphyxiants in their mechanism of action. Cellular asphyxiants act at the cell and tissue level and most often interact with the body's effort to use and transport oxygen. Carbon monoxide and many cyanide compounds are example of tissue asphyxiants.

- Carcinogens are those materials which are known or suspected to cause cancer in humans.
- Teratogens are those chemicals which may cause birth defects in a developing fetus.
- Additive effects occur when two substances combine but do not result in any more of a hazard than each one would pose alone.
- Synergistic interactions are those that occur when two agents are combined with results that are far worse than either substance would pose individually.
- Antagonistic reactions are those which are reduced in severity by the combination of two or more agents. Essentially, one agent reduces or offsets the effects of another.
- Potentiation reactions occur when a material of low hazard interacts with another and is made worse or more dangerous as a result of the reaction.

Review Questions

1. Chemicals can enter the body through which of the following routes?
   A. Inhalation
   B. Skin and eye contact
   C. Ingestion
   D. All of the above

2. Of the routes of exposure, which one poses the highest hazard potential to field personnel?
   A. Inhalation
   B. Ingestion
   C. Conduction
   D. Absorption

3. Which of the following is not a recommended method to protect yourself from inhalation hazards?
   A. Ventilation systems
   B. Smell of the chemical
   C. Air purifying respirators
   D. Air supplying respirators

4. True or False? The health effects of chemicals are always immediate, letting you know that they are hurting you and providing you with the opportunity to reduce the exposure.

5. The best example of an immediate health effect of a chemical exposure is:

A. Cancer
B. Burning or irritation
C. Lung disease
D. Organ damage

6. True or False? Acute and chronic exposures are the same.

7. If a chemical contacts the skin or eyes, it should be washed with water for a minimum of
    A. 30 seconds
    B. 15 minutes
    C. 5 minutes
    D. 30 minutes

8. Explain the difference between the terms *toxic* and *poison*.

9. True or False? Chemicals that mix might create products that are more hazardous than either of the chemicals by themselves.

10. Which of the following most closely defines LD_{50}?
    A. Lethal dose at 50 ppm
    B. Lethal dose that eliminates 50% of an animal test population.
    C. Lethal concentration at 50 ppm
    D. Lethal concentration that eliminates 50% of an animal test population.

11. Generally speaking, would a systemic toxic exposure be more or less problematic than local exposures? Why?

12. Which of the following statements is most correct?
    A. All man-made chemicals are poisonous, natural materials are generally non-toxic.
    B. All chemicals are potentially poisonous—the difference is in the dose.
    C. Most chemicals that are flammable do not pose a toxic hazard.
    D. All flammable and combustible materials are toxic.

13. Define carcinogen and teratogen.

14. Briefly describe the four routes of entry.

15. Occupational exposure to chemicals results in about what percent of the overall cancer rate in the United States?

16. Explain what is meant by the term *odor threshold*.

17. What does an asphyxiant do?

18. Define IDLH.

19. Explain the differences between PEL and STEL.

20. What source of information provides some of the most complete information of the health and toxicological effects of a chemical?

Chapter 5

Hazardous Materials Identification Systems

Learning Objectives

Upon completion of this chapter, you should be able to:

- Understand the basic use of and need for identification systems.
- Explain why identification systems are in place.
- Understand what identification systems can tell a responder or waste site worker.
- List the advantages and disadvantages of identification systems.
- Understand, describe, and interpret the DOT system of labels and placards.
- Describe the conditions and situations when the DOT Identification System is used.
- Understand, describe, and interpret the NFPA 704 system.
- Describe the conditions and situations when the NFPA 704 is used.
- Understand the Hazardous Materials Information System (HMIS).
- Identify the uses of shipping papers.
- Understand the uses of Uniform Hazardous Waste Manifests.
- Demonstrate the ability to use the DOT *North American Emergency Response Guidebook*.
- Be able to differentiate between steel, polyethylene, and fiber drums.

INTRODUCTION

The amount and type of hazardous substances that are transported, stored, and used in the United States is staggering. Since the end of the Second World War, the amount and variety of chemicals used in industry has risen dramatically. Daily production and consumption requires products as diverse as TNT, pesticides, inert gases, and gasoline. All of these, by one means or another have to be transported throughout the country and indeed the world. To assist responders and waste workers in basic identification of such hazardous materials, two systems have come into common usage. First, the Federal Department of Transportation (DOT) has adopted and enforces a United Nations standard of labels and placards for hazardous materials being transported. This identification system is mandated and required by law to be used as per 49 CFR 172. Second, the National Fire Protection Association (NFPA) has established a system that identifies chemicals stored at a fixed site. The NFPA is a recommending authority only and has no lawful enforcement, but this identification system can be enforced by the Uniform Fire Code and is usually adopted by a local jurisdictional authority such as county or city government.

The DOT and NFPA standards are only two of a wide array of identification systems in use. Although many types of identification systems are in use, the two most common are the DOT system and the NFPA 704 system. Private agencies may have a system of identifying what they use and how it is stored. Hospitals require specific labeling of gases and piping as does the refinery industry. However, all of these systems have one thing in common: Identification systems must be used properly and understood by all.

■ **NOTE**
Although many types of identification systems are in use, the two most common are the DOT system and the NFPA 704 system.

Identification systems are used in every aspect of modern society. In some situations, they are hardly recognized in day-to-day use. For example, when pouring a cup of coffee at a convenience store, we choose regular or decaffeinated coffee by the color (brown or orange) of the carafe handle. If a coffee drinker does not know that each color means a different type of coffee, then he or she cannot accurately choose one or the other. The odds are 50/50 of getting the desired cup of coffee. Are these good odds in choosing the right course of action with hazardous materials? These odds also signify that identification systems are only as good as the people who use them.

■ **NOTE**
With few exceptions in the realm of hazardous materials, identification systems do not portray exactly what the material is, but primarily its physical properties and how it may act if released.

Keep in mind, these systems are for initial hazard recognition only and do not provide in-depth information. With few exceptions in the realm of hazardous materials, identification systems do not portray exactly what the material is, but primarily its physical properties and how it may act if released. Identification systems are designed to inform personnel working in the hazardous materials field how to react in case of a discharge or other accident. The systems are designed to give this basic information from a distance. This is to your advantage because product contact and spread of the product can be reduced. Imagine walking into a visible cloud without respiratory protection to read a small label that says "Poison by inhalation." Or incorrectly identifying a burning container to the fire department because the orange placard (explosives) is on an orange background. It is much better to be able to recognize and identify hazards correctly the first time from a safe, upwind, and uphill location.

DOT AND NFPA IDENTIFICATION SYSTEMS

In 1974, the Hazardous Materials Transportation Act was enacted by the DOT to impose regulation on identifying hazardous materials in shipment. This was to be accomplished with the use of shipping papers per each mode of transport and the proper use of labels and placards. According to the DOT, the term *hazardous materials* means any material that may present a danger during shipment by truck, rail, air, or water. However, be aware that modes of transportation may not show a placard, but a hazardous material could be present. Why? It all depends on how much is being shipped. For example, DOT hazard class labels will be present on any package of hazardous materials with a capacity of 110 gallons or less. Additionally, small hazardous loads that total less than 1,000 pounds are not required to be placarded unless any amount of the materials displayed in Table 5-1 are being carried in the vehicle.

As a responder, it is not as necessary to understand how exactly labeling and placarding applies, but primarily how to interpret it. Be aware of where to find specific information regarding placards in 49 CFR 172.101. As seen in Table 5-2, this section of the table gives you proper shipping name, hazard class, packaging, labeling, and specifics in air and vessel shipment.

The DOT, under 49 CFR 172 deals with chemicals that are considered pure and for a specific use. These are called *hazardous materials*. The Environmental

> ■ **NOTE**
> According to the DOT, the term *hazardous materials* means any material that may present a danger during shipment by truck, rail, air, or water.

Table 5-1 *Materials in any quantity requiring placards.*

Category of Material	Type of Placard
1.1	Explosives 1.1
1.2	Explosives 1.2
1.3	Explosives 1.3
2.3	Poison Gas
4.3	Dangerous When Wet
6.1 (Packing Group I, inhalation hazard only)	Poison
7 (Radioactive Yellow III only)	Radioactive

Table 5-2 49 CFR 172.101 hazardous materials table.

Symbols	Hazardous Materials Description and Proper Shipping Names	Hazard Class or Division	Identification Numbers	Packing Group	Label(s) Required (if not excepted)	Special Provisions
(1)	(2)	(3)	(4)	(5)	(6)	(7)
D	Aldrin, *liquid*	6.1	NA2762	II	POISON	
D	Aldrin, *solid*	6.1	NA2761	II	POISON	
	Alkali metal alcoholates, self-heating, corrosive, n.o.s.*	4.2	UN3206	II	SPONTANEOUSLY COMBUSTIBLE, CORROSIVE.	
				III	SPONTANEOUSLY COMBUSTIBLE, CORROSIVE.	
	Alkali metal alloys, liquid, n.o.s.*	4.3	UN1421	I	DANGEROUS WHEN WET.	A2, A3, B48, N34
	Alkali metal amalgams.	4.3	UN1389	I	DANGEROUS WHEN WET.	A2, A3, N34
	Alkali metal amides.	4.3	UN1390	II	DANGEROUS WHEN WET.	A6, A7, A8, A9, A20, B106
	Alkali metal dispersions, *or* Alkaline earth metal dispersions.	4.3	UN1391	I	DANGEROUS WHEN WET.	A2, A3
	Alkaline corrosive liquids, n.o.s.,* see Caustic alkali liquid, n.o.s.*					
	Alkaline earth metal alcoholates, n.o.s.*	4.2	UN3205	II	SPONTANEOUSLY COMBUSTIBLE.	
				III	SPONTANEOUSLY COMBUSTIBLE.	
	Alkaline earth metal alloys, n.o.s.*	4.3	UN1393	II	DANGEROUS WHEN WET.	A19, B101, B106
	Alkaline earth metal amalgams,	4.3	UN1392	I	DANGEROUS WHEN WET.	A19, B101, B106, N34, N40
	Alkaloids, liquid, n.o.s.,* *or* Alkaloid salts, liquid, n.o.s.*	6.1	UN3140	I	POISON	A4, T42
				II	POISON	T14

*n.o.s. indicates "not otherwise specified."

Chapter 5 Hazardous Materials Identification Systems

	(8) Packaging Authorizations (§173.***)			(9) Quantity Limitations		(10) Vessel Stowage Requirements	
Exceptions	Nonbulk Packaging	Bulk Packaging		Passenger Aircraft or Railcar	Cargo Aircraft Only	Vessel Stowage	Other Stowage Provisions
(8A)	(8B)	(8C)		(9A)	(9B)	(10A)	(10B)
None	202	243		5 L	60 L	B	
None	212	242		25 kg	100 kg	A	40
None	212	242		15 kg	50 kg	B	
None	213	242		25 kg	100 kg	B	
None	201	244		Forbidden	1 L	D	
None	201	244		Forbidden	1 L	D	
None	212	241		15 kg	50 kg	E	40
None	201	244		Forbidden	1 L	D	
None	212	241		15 kg	50 kg	B	
None	213	241		15 kg	100 kg	B	
None	212	241		15 kg	50 kg	E	
None	211	242		Forbidden	15 kg	D	
None	201	243		1 L	30 L	A	
None	202	243		5 L	60 L	A	

> **NOTE**
> According to the EPA, hazardous wastes are substances that have no commercial value and cannot be sold or recycled.

Protection Agency (EPA) has regulations dealing with substances that have no commercial value such as contaminates from a cleanup operation. These are called *hazardous wastes*. According to the EPA, hazardous wastes are substances that have no commercial value and cannot be sold or recycled. The EPA regulates hazardous wastes by the Resource Conservation and Recovery Act (RCRA) and under 40 CFR. However, the EPA shares the listing of hazardous wastes with the DOT under 49 CFR 172.101. Thus, personnel involved in cleaning up a chemical, whether in basic or waste form, can find the reportable quantity (RQ) in Appendix A to 49 CFR 172.101. This RQ is an important factor to know. If a discharge is above the RQ, then the responsible party is required to make notification to the U. S. Coast Guard National Response Center (NRC) in Washington, DC (800-424-8802).

DOT Labels and Placards

Once it is determined that a hazardous material being transported requires identification, the appropriate label or placard must be properly affixed to the shipping container. Whether this involves a 40-ton container or a cardboard box, placarding or labeling must be done correctly. Placards are usually $10\frac{3}{4}''$ square and placed on large transport containers such as rail cars, and truck trailers. Labels are usually four inches square and designed for smaller portable containers with a capacity less than 110 gallons such as cardboard boxes, gas cylinders, and plastic drums. For example, intermodal box containers, as transported on ships, trains, and trucks, must have the appropriate placard on both ends and on both sides and not be obstructed by any other markings. Further requirements state that placards and labels be worded in English and be durable. They will be placed right side up and should be in contrast to the background color of the container itself. There is more variety in labels than placards. For example, there is a label for etiologic (disease-causing) agents and rectangular labels in place to identify chemicals on aircraft.

No matter what the exact requirements of labeling and placarding, the DOT system has some disadvantages. First, a placard or label does not tell a responder or witness exactly what material is in the container. For example, the placard with the identification number 1203 on a bulk liquid tank truck means to most emergency response people a shipment of unleaded gasoline. However, if identification number 1203 is referred to in the Department of Transportation's (DOT) *North American Emergency Response Guidebook*, the material may also be gasohol or motor spirits. Chances are the liquid is indeed unleaded gasoline, but the 1203 placard does not allow for an exact identification. If the material is not being shipped in bulk form, a plain language word describes the class only. An example of this is the placard for nonflammable gases, which could be nitrogen, argon, neon, or even ammonia (a toxic, corrosive, and potentially explosive gas). Second, any label or placard can be obscured by physical factors. A placard or label in conditions of darkness, fog, heavy rain, smoke, fire, distance, or even being

Chapter 5 Hazardous Materials Identification Systems

covered by graffiti, can give the responder trouble in making a positive identification. Proper recognition can also be hampered if the container does not have a label or placard on both sides and it has rolled or turned over on the side that has the identification. Third, if the placarded material in shipment has multiple hazards, the DOT requires identification of only what it considers the prime hazard. Only one hazard may be noted by the placard and other hazards not identified may pose a higher hazard to a responder.

Even with such disadvantages listed in the previous paragraph, the DOT system is still a functional and useful tool. It is designed to be simple, to allow persons with little information to make an initial size-up and evaluation. What good would a placard be if it were hard to find, difficult to read, had small print, and confusing? Remember, personnel involved in making the initial investigation are not always qualified waste workers or responders. They need to make a quick identification and be able to pass on correct information. To identify the hazard stated by the DOT label or placard, personnel can note the following four characteristics:

1. The hazard class number (UN number) of the label or placard
2. The color of the placard or label
3. The plain language name or identification number (if a bulk shipment)
4. The symbol of the placard or label

■ **NOTE**
If the placarded material in shipment has multiple hazards, the DOT requires identification of only what it considers the prime hazard. Only one hazard may be noted by the placard and other hazards not identified may pose a higher hazard to a responder.

■ **NOTE**
With the DOT system of labels and placards, the lower the hazard class number, the higher the hazard. For example, Class 1 explosives are considered the most hazardous in transportation.

The DOT has determined a specific classification order in the shipment of hazardous materials. With the DOT system of labels and placards, the lower the hazard class number, the higher the hazard. For example, Class 1 explosives are considered the most hazardous in transportation. All hazardous materials are categorized in UN hazard classes from 1 to 9 with class 1 being the most potentially dangerous. In other words, a shipment of explosives (class 1) is considered by the DOT to be more hazardous than a shipment of radioactive material (class 7). The UN number is found in the lower corner of both labels and placards. There may also be subclass numbers directly above the class number. Such is the case with explosives, which may have a subclass number from 1.1 (detonating agents) to 1.6 (extremely insensitive articles). The only problem with making an error in class identification is if involved personnel do not read English or are illiterate. Following is a list of all the DOT classes:

Class 1	Explosives
Class 2	Gases
Class 3	Flammable Liquids
Class 4	Flammable Solids
	Spontaneously Combustible
	Dangerous When Wet
Class 5	Oxidizers
	Organic Peroxides

Class 6	Poisonous
	Etiologic Agents
Class 7	Radioactive
Class 8	Corrosives
Class 9	Miscellaneous

> **■ NOTE**
> The colors and/or color patterns are designed to stand out. Each color or color combination on a label or placard symbolizes a specific hazard.

The color of the placard is the most readily observable factor of the DOT system. The colors and/or color patterns are designed to stand out. Each color or color combination on a label or placard symbolizes a specific hazard. For example, an all orange placard denotes explosives. An all green placard indicates a nonflammable gas. If a label is seen that is all red, then a flammable or combustible gas or liquid is present. Some placards have two color standards but still stand out due to the specific design. Red and white, whether a split field or in stripes means a flammable solid. Half yellow and half white means a radioactive material. However there are some disadvantages with colors. A witness to a release may be colorblind, or the color can be difficult to determine under adverse light conditions. Two examples of this are direct sunlight glaring off a metal placard and looking at orange and white placards under sodium vapor lights at night. If either of these are the situation, then moving around to change perspective may help.

The third way to identify the hazard is by reading the plain language name (such as Oxidizer) or by reading and interpreting the identification number if it is a bulk shipment. In the case of materials with class titles such as explosive, corrosive, or radioactive, the hazard is clear to most people. However, not everyone who works with hazardous materials realizes the inherent dangers of substances such as oxidizers or nonflammable gases. Classes such as these represent their own special dangers, but people may not react with as much caution as perhaps they would with radioactive or poisonous materials. A second disadvantage is hazard class numbers. Personnel may have trouble identifying the name due to illiteracy or not understanding English. If a bulk shipment is involved, then an identification number (usually four characters) appears instead of the plain language name. This number must be properly interpreted to determine what is in the container. This can be done by using a reference guide such as the appropriate table in 49 CFR 172.101, the DOT *Emergency Response Guidebook* or the *International Maritime Dangerous Goods* (IMDG) book. In most cases, the *Emergency Response Guidebook* is the best book to use because of its ease of use. Without such a guide, determination of a hazard class by number alone is difficult. With the guide a responder can determine the exact name of the involved contaminant. For example, 1017 is chlorine and the type of packaging (gas cylinder or drum), along with class number (2 for gas, 5 for oxidizer) determines whether a gas or a solid of chlorine is involved. If the number 1956 is seen, then the material may be a poisonous liquid, a compressed gas, or a liquefied gas among others. In situations like the latter, not enough information is given by the identification number alone to determine what exact material is involved.

The final way to determine the hazard is by the symbol. Symbols work very well in many cases in describing the hazard by using a descriptive picture as shown in Table 5-3. The skull and crossbones symbol means a toxic or poisonous substance to almost every society on earth. Most people in this post-Cold War era will also readily recognize the trefoil or spinning propeller of radioactive material. However, like the plain language name, some may not appreciate the dangers of hazardous material classes such as oxidizers (a burning "O") or the outline of a cylinder (nonflammable gases). Others are self-descriptive and will give a good idea of the hazard posed even to the novice. A prime example is the symbol for corrosives, which displays two test tubes dripping a liquid onto a bar of metal and a human hand with both metal and hand being eaten away: Little is left to the imagination.

It can be seen in Table 5-3 that all four ways of determining a hazard can offset any problems involved personnel may have in doing so. All four ways of showing a class hazard can be used together or separately to determine what hazard is present. For those with reading disabilities or trouble with English, color and symbol can assist in determining the hazard. For others with problems with color blindness, class number, plain language name, and/or identification number

■ **NOTE**
All four ways of showing a class hazard can be used together or separately to determine what hazard is present.

Table 5-3 *DOT symbols.*

Hazard Class	Color	Symbol
Class 1 Explosives	Orange	Exploding device
Class 2 Gases	Yellow	Burning "O"
	Red	Flame
	White	Skull and crossbones
	Green	Cylinder
Class 3 Flammable liquids	Red	Flame
Class 4 Flammable solids	Red/white stripes	Flame
	Red/white field	Flame
	Blue	Flame
Class 5 Oxidizers/organic peroxides	Yellow	Burning "O"
Class 6 Poisons/etiologic agents	White	Skull and crossbones
	White	Sheaf of wheat with cross
	White	Broken circles
Class 7 Radioactive	Yellow/white field	Trefoil/spinning propeller
Class 8 Corrosives	Black/white field	Melting metal bar and hand
Class 9 Miscellaneous	Black stripes, white field	Black and white stripes

will be of greater benefit. In most situations an identification number allows the responder to make a more accurate identification of the hazardous material, but a reference guide is needed to do so. Never assume you know all the numbers. Always verify by using the appropriate guide.

NFPA 704 System

As discussed, the DOT identification system relates to the shipment or transport of hazardous materials and is required by federal law. Once hazardous materials have reached their destination, that is, being stored or used at a fixed facility, a new system of identification, the NFPA 704 placard, is used. As seen in Figure 5-1 it is a simple system based on four colors and a number system from 0 to 4. Like the DOT system it is designed to be easy to use and simple in identification. Unlike the DOT system, it is a recommending guideline and neither the NFPA nor the federal government can enforce its use. However, in some parts of the country its use is mandated by the Uniform Fire Code. As such, it is usually adopted by the local government authority such as a county or city government and enforced at that level. Thus, in some areas the NFPA 704 placard can be found everywhere it may be required, whereas in other areas its use may be severely curtailed or not even used at all. This does not imply that some local governments or agencies are not concerned about hazardous materials safety. They may choose to utilize other identification or protection systems previously in place.

The NFPA 704 system may be found in the form of a label or a placard, depending on the size of the container. In many situations, a fixed facility such as a small factory will have one NFPA diamond for the entire building. It should be placed where emergency responders are likely to see it when they enter by the main route of approach. In other words, the diamond should not be put in a

Figure 5.1 *NFPA 704 diamond.*

remote, hard to see location. If a larger building or facility is involved, then it can be placed at specific sites to identify the hazards in that one area. The NFPA symbol or the applied hazard numbers may also be found on material safety data sheets (MSDS); usually in the first section near the top as seen in Figure 5-2. In either situation it is designed for quick determination of the hazards present.

Each section of the NFPA symbol represents a type of hazard. On the far left is a blue diamond for health hazards. On top dead center is a red diamond for flammability hazards. To the far right is a yellow diamond for reactivity hazards. On the bottom is a white diamond for other specific hazards. In the blue, red, and yellow diamonds there should be a number from 0 to 4. With this system, the higher the number, the higher the hazard. If no number appears in a diamond (due to changing stock or amounts) the emergency responder should treat the situation as if a 4 appeared in the diamond. Some criticism does arise because the NFPA system relates a higher number to a high hazard whereas the DOT system relates a lower number to a higher hazard. When looking at an NFPA placard, use NFPA guidelines. When looking at a DOT placard, use DOT guidelines. There is one similarity between the two in that red means a flammable hazard, but all other colors and numbers mean different hazards. For example, blue in the NFPA system means a health hazard. In the DOT system blue means a Class 4 flammable solid that is dangerous when wet. Do not confuse the two!

The lower white diamond of the NFPA 704 system may have a symbol or lettering that relates a hazard that is not exactly covered by the other three diamonds. An example of this is radioactive materials, which can be identified by a trefoil or spinning propeller. Other examples include Oxy which denotes a oxidizer, Cor which denotes corrosives, and especially a W with a slash through it which means the chemical reacts with water.

> ■ NOTE
> With the NFPA 704 system, the higher the number, the higher the hazard. If no number appears in a diamond, the emergency responder shall treat the situation as if a 4 appeared in the diamond.

Disadvantages of the NFPA 704 System

In many cases, a fixed facility has more than one chemical present. Each chemical may have an NFPA label or placard on its individual container, but the facility has a placard on the front that lists the highest hazard of each categorized chemical. For example, a building that is used to store chlorine, ammonia, methyl ethyl ketone and acetylene has:

- Chlorine—Rated individually as 4 (health), 0 (flammability), and 0 (reactivity)
- Ammonia (anhydrous)—Rated individually as 3 (health), 1 (flammability), and 0 (reactivity)
- Methyl ethyl ketone—Rated individually as 1 (health), 3 (flammability), and 0 (reactivity)
- Acetylene—Rated individually as 1 (health), 4 (flammability), and 2 (reactivity)

In this case, the NFPA placard on the front door would read (4 health), 4

Genium Publishing Corporation
One Genium Plaza
Schenectady, NY 12304-4690 USA
(518) 377-8854

Material Safety Data Sheets Collection:

Sheet No. 300
Acetone

Issued: 11/77   Revision: F, 9/92

Section 1. Material Identification		39
Acetone (CH_3COCH_3) Description: Derived by the dehydrogenation or oxidation of isopropyl alcohol with a metallic catalyst, the oxidation of cumene, the vapor phase oxidation of butane; and as a by-product of synthetic glycerol production. Used as a solvent for paint, varnish, lacquer, fat, oil, wax, resin, rubber, plastic, and rubber cement; to clean and dry parts of precision equipment; in the manufacture of chemicals (methyl isobutyl ketone, methyl isobutyl carbinol, methyl methacrylate, bisphenol-A, acetic acid (ketene process), mesityl oxide, diacetone alcohol, chloroform, iodoform, bromoform), explosives, aeroplane dopes, rayon, photographic films, isoprene; acetylene gas storage cylinders; in purifying paraffin; in nail polish remover; in the extraction of various principles from animal and plant substances; in hardening and dehydrating tissues; in cellulose acetate (especially as spinning solvent); as a solvent for potassium iodide and permanganate; as a delusterant for cellulose acetate fibers; in the specification testing of vulcanized rubber products. **Other Designations:** CAS No. 67-64-1, AI3-01238, Chevron acetone, dimethylformaldehyde, dimethylketal, dimethyl ketone, β-ketopropane, methyl ketone, propanone, 2-propanone, pyroacetic acid, pyroacetic ether. **Manufacturer:** Contact your supplier or distributor. Consult latest *Chemical Week Buyers' Guide*[73] for a suppliers list. **Cautions:** Acetone vapor is a dangerous fire and explosion hazard. High vapor concentrations may produce narcosis (unconsciousness). Prolonged or repeated skin contact causes dryness, irritation, and mild dermatitis.	R 1 I 1 S 1* K 3 * Slight skin absorption HMIS H 1 F 3 R 0 PPE* * Sec. 8	NFPA 1 3 0

Section 2. Ingredients and Occupational Exposure Limits

Acetone, 99.5% plus 0.5% water

1991 OSHA PELs *
8-hr TWA: 750 ppm (1800 mg/m³)
15-min STEL: 1000 ppm (2400 mg/m³)
1990 IDLH Level
20,000 ppm
1990 NIOSH REL
TWA: 250 ppm (590 mg/m³)

1992-93 ACGIH TLVs
TWA: 750 ppm (1780 mg/m³)
STEL: 1000 ppm (2380 mg/m³)
1990 DFG (Germany) MAK
1000 ppm (2400 mg/m³)
Category IV: Substances eliciting very weak effects (MAK >500 mL/m³)
Peak: 2000 ppm, 60 min, momentary value†, 3 peaks/shift

1985-86 Toxicity Data ‡
Human, eye: 500 ppm
Human, inhalation, TC_{Lo}: 500 ppm produced olfaction effects, conjunctival irritation, and other changes involving the lungs, thorax, or respiration.
Rat, oral, LD_{50}: 5800 mg/kg altered sleep time and produced tremors.
Mammal, inhalation, TC_{Lo}: 31500 μg/m³/24 hr administered to pregnant female from the 1st to 13th day of gestation produced effects on fertility (post-implantation mortality).

* In the cellulose acetate fiber industry, enforcement of the OSHA TWA for "doffers" was stayed on 9/5/89 until 9/1/90; the OSHA STEL *does not* apply to that industry.
† Momentary value is a level which the concentration should never exceed.
‡ See NIOSH, *RTECS* (AL3150000), for additional irritation, mutation, reproductive, and toxicity data.

Section 3. Physical Data

Boiling Point: 133.2 °F (56.2 °C) at 760 mm Hg
Freezing Point: -139.6 °F (-95.35 °C)
Vapor Pressure: 180 mm Hg at 68 °F (20 °C), 400 mm Hg at 103.1 °F (39.5 °C)
Saturated Vapor Density (Air = 1.2 kg/m³, 0.075 lb/ft³): 1.48 kg/m³, .093 lb/ft³
Refractive Index: 1.3588 at 20 °C
Appearance and Odor: Colorless, highly volatile liquid; sweetish odor.
* Odor thresholds recorded as a range from the lowest to the highest concentration.

Molecular Weight: 58.08
Specific Gravity: 0.7899 at 20 °C/4 °C
Water Solubility: Soluble
Other Solubilities: Alcohol, benzene, dimethyl formamide, chloroform, ether, and most oils.
Odor Threshold: 47.5 mg/m³ (low), 1613.9 mg/m³ (high)*

Section 4. Fire and Explosion Data

Flash Point: 0 °F (-18 °C), CC	**Autoignition Temperature:** 869 °F (465 °C)	**LEL:** 2.6% v/v	**UEL:** 12.8% v/v

Extinguishing Media: *Do not* extinguish fire unless flow can be stopped. For small fires, use dry chemical, carbon dioxide (CO_2), water spray or alcohol-resistant foam. For large fires, use water spray, fog, or alcohol-resistant foam. Use water in flooding quantities as fog because solid streams may be ineffective. **Unusual Fire or Explosion Hazards:** Acetone is a dangerous fire and explosion hazard; it is a Class IB flammable liquid. Vapors may travel to a source of ignition and flash back, fire-exposed containers may explode, and a vapor explosion hazard may exist indoors, outdoors, or in sewers. **Special Fire-fighting Procedures:** Because fire may produce toxic thermal decomposition products, wear a self-contained breathing apparatus (SCBA) with a full facepiece operated in pressure-demand or positive-pressure mode. Structural firefighters' protective clothing provides limited protection. If feasible, remove all fire-exposed containers. Otherwise, apply cooling water to sides of containers until well after fire is extinguished. If the fire becomes uncontrollable or container is exposed to direct flame, consider evacuation of a one-third mile radius. In case of rising sound from venting safety device or any discoloration of tank during fire, withdraw immediately. For massive cargo fires, use unmanned hose holder or monitor nozzles. Do not release runoff from fire control methods to sewers or waterways.

Section 5. Reactivity Data

Stability/Polymerization: Acetone is stable at room temperature in closed containers under normal storage and handling conditions. Hazardous polymerization cannot occur. **Chemical Incompatibilities:** Acetone may form explosive mixtures with hydrogen peroxide, acetic acid, nitric acid, nitric acid + sulfuric acid, chromic anhydride, chromyl chloride, nitrosyl chloride, hexachloromelamine, nitrosyl perchlorate, nitryl perchlorate, permonosulfuric acid, thiodiglycol + hydrogen peroxide. Acetone reacts vigorously with oxidizing materials and ignites on contact with activated carbon, chromium trioxide, dioxygen difluoride + carbon dioxide, and potassium-*tert*-butoxide. Other incompatibles include air, bromoform, bromine, chloroform + alkalies, trichloromelamine, and sulfur dichloride. **Conditions to Avoid:** Keep acetone away from plastic eyeglass frames, jewelry, pens, pencils, and rayon garments. **Hazardous Products of Decomposition:** Thermal oxidative decomposition of acetone can produce CO_2 and carbon monoxide (CO).

Section 6. Health Hazard Data

Carcinogenicity: The IARC,[164] NTP,[169] and OSHA[164] do not list acetone as a carcinogen. **Summary of Risks:** Acetone has been placed among solvents of comparatively low acute and chronic toxicities. In industry, the most common effects reported are headache from prolonged vapor inhalation and skin irritation resulting from its defatting action. Exposures to less than 1000 ppm acetone vapor produces only slight eye, nose, and throat irritation. Acetone does not have sufficient warning properties to prevent repeated exposures. It is a narcotic at high concentrations, i.e., above 2000 ppm. Concentrations above 12000 ppm cause loss of consciousness.

Continue on next page

Copyright © 1992 Genium Publishing Corporation. Any commercial use or reproduction without the publisher's permission is prohibited.

Figure 5.2 *Portion of MSDS showing NFPA diamond.*

(flammability), and 2 (reactivity). The numbers are not added up, but the worst one (the higher number) of each hazard is used. Although this is not a distinct disadvantage, it can lead to confusion. An actual chemical that is categorized 4,4,2 is not even present. For the reporting witness or emergency response team such as a fire department it is not possible to determine what exact chemical(s) are present, how many and where they are exactly located by this system. However, imagine the situation if no placard existed. Normal actions of investigation and firefighting could have catastrophic results. Like the DOT system, the placards can be obscured by physical factors such as smoke, fog, darkness, and distance. In many cases the lettering of the NFPA 704 placards are black which can be especially difficult to read against the blue (health) background.

NFPA 704 Health Hazards

The NFPA system relates health hazards to acute, short-term exposures because this is the type of exposure that is likely to occur with involved personnel in case of a release. Some emergency responders use the health hazard to determine the level of initial response as far as level of personal protective equipment, but this is considered rudimentary by others. The following is a brief description of the hazard for each number and an example of a chemical:

- **0** Materials whose exposure would offer no hazard beyond that of ordinary combustible materials. They are not considered to be acutely toxic. Examples include magnesium and oxygen gas.
- **1** Materials only slightly hazardous to health. Once again, these materials are not considered acutely hazardous. Examples include ammonium perchlorate, red phosphorus, and propylene.
- **2** Materials that are hazardous to health but which you can be exposed to with appropriate eye and respiratory protection. Examples include benzene, hydrogen peroxide, methyl chloride, and polychlorinated biphenyls (PCBs).
- **3** Materials that are extremely hazardous to health but which workers can be exposed to with self-contained breathing apparatus and firefighter's protective clothing including coat, pants, helmet, boots, gloves, and bands around the arms, legs, and waist. No skin area should be exposed. Examples include liquefied natural gas (methane), methyl bromide, and phenol (carbolic acid).
- **4** Materials too dangerous to health to expose workers to. A few whiffs of the material could cause death or the vapors could penetrate clothing, including that worn by the average firefighter. The normal clothing and breathing apparatus worn by firefighters does not provide protection from these materials. Examples include the gases germane, fluorine, and the poison hydrogen cyanide.

NFPA 704 Flammability

The NFPA system rates the assigning degrees within this category on how easily a fire would occur. This often involves a review of the flash point of the materials coupled with the method of combating a fire and is influenced by the degree of hazard assigned. The following is a brief description of the hazard for each number and an example of a chemical:

- **0** Materials that will not burn under ordinary conditions. Examples include ammonium nitrate, chlorine, and bromine.
- **1** Materials that must have significant preheating before they can burn. Generally speaking, these materials have a flash point greater that 200°F. If materials in this range are involved in fire, water may be effective in extinguishing them because it can cool the material below its flash point. Examples include methyl bromide, calcium metal, and oxalic acid.
- **2** Materials that must be moderately heated before ignition can occur. They have a flash point between 100°F and 200°F. Water might be effective in extinguishing these materials because it can cool the materials below their flash point, however it should be used with extreme caution. Examples include peracetic acid, formic acid, and naphthalene.
- **3** Materials that can be ignited under almost all normal temperatures. Flash points for materials in this category are below 100°F. Water is generally ineffective on these materials because it cannot cool the material below its flash point. Examples include cumene, propyl nitrate, and styrene.
- **4** Very flammable or volatile materials such as flammable gases and very volatile flammable liquids. Extinguishing fires involving these materials should be directed toward shutting off the flow of the material and protecting other exposed areas. Examples include acetylene, carbon monoxide, diborane, and ethylene.

NFPA 704 Reactivity Hazards

The NFPA system assigns the degree of reactivity upon the susceptibility of the materials to release energy either by themselves or in combination with water. The following is a brief description of the hazard for each number and an example of a chemical:

- **0** Materials that (by themselves) are normally stable even under fire exposure conditions and that do not react with water. Examples include carbon monoxide, chloroform, and hydrogen sulfide.
- **1** Materials that (by themselves) are normally stable but that might become unstable at elevated temperatures and pressures or that might

react with water with some release of energy. Generally, these are minor reactions posing relatively minor hazards. Examples include phosgene gas, hydrogen peroxide, and magnesium.

2. Materials that (by themselves) are normally unstable and readily undergo violent chemical change but do not detonate. This class includes materials that can undergo chemical change at elevated temperatures and pressures. This class also includes materials that may react violently with water or may form potentially explosive mixtures with water. Examples include hydrogen cyanide, methyl isocyanate, and sodium.

3. Materials that (by themselves) are capable of detonation, explosive decomposition, or explosive reaction but which usually require a strong initiator or being heated under confinement before initiation. This class includes materials that are sensitive to temperatures and pressures or that react explosively with water without requiring heat or confinement. Fire fighting activity in this class should be done from an explosion-resistant location. Examples include ethylene oxide (ETO), nickel carbonyl, and silane gas.

4. Materials that (by themselves) are readily capable of detonation or of explosive decomposition or explosive reaction at normal temperatures and pressures. This class includes materials sensitive to mechanical or thermal shock. In the case of a major fire involving these materials, the area should be evacuated. Examples include fluorine gas, nitromethane, and picric acid.

HAZARDOUS MATERIALS IDENTIFICATION SYSTEMS

Another identification system used on chemical containers is known as the hazardous materials identification system (HMIS). As seen in Figure 5-3, this system is similar to the NFPA 704 placards in that the blue, red, and yellow colors mean

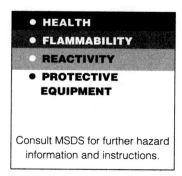

Figure 5.3 *HMIS label for containers.*

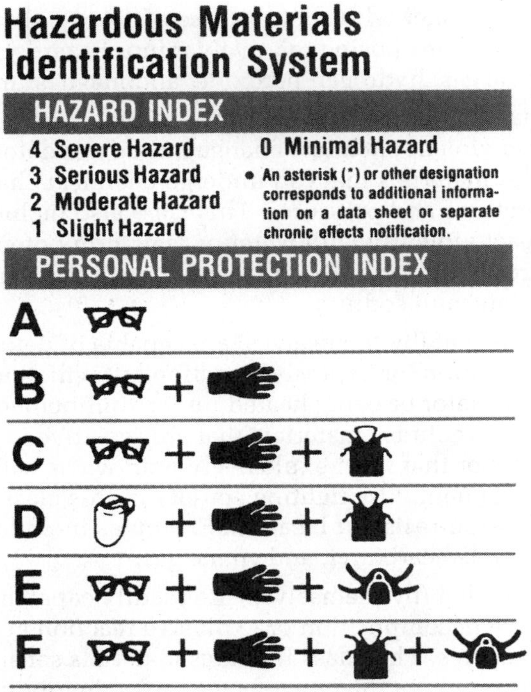

Figure 5.4 *HMIS letters and PPE.*

■ **NOTE**
The HMIS letter A through D system conflicts with the NFPA standards of chemical protective clothing.

the same along with the same numbering system (0–4). However, Figure 5-4 shows the difference between the NFPA 704 system and the HMIS system. With the HMIS system, the color white describes types of personal protective equipment (PPE). The designation letters in the white box run from A to K and then X. As progression is made toward K, a higher level of personal protective equipment is worn. The disadvantage is that, the HMIS letter A through D system conflicts with the NFPA standards of chemical protective clothing. With the HMIS system, the letter A in the white box denotes a pair of safety glasses should be worn. With the NFPA standards, the letter A stands for a totally encapsulating gas protective chemical protective suit with supplied air respiratory protection. There is a significant difference between the two types of protection!

SHIPPING PAPERS AND HAZARDOUS WASTE MANIFESTS

When hazardous materials are in transit or being loaded or offloaded they are required to be accompanied by shipping papers and/or a uniform hazardous waste manifest. Figure 5-5 shows what is probably the most common type of shipping paper, a bill of lading used in trucking. It is usually found (as seen in Figure 5-6) in the map pocket of the truck door. Shipping papers inform the investigator

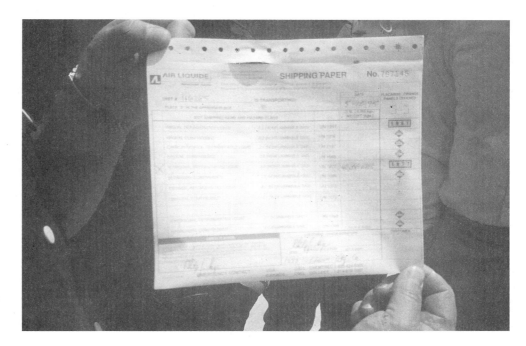

Figure 5.5 *Shipping papers.*

Figure 5.6 *Location of shipping papers in truck door.*

State of California—Environmental Protection Agency
Form Approved OMB No. 2050–0039 (Expires 9-30-96)
Please print or type. *Form designed for use on elite (12-pitch) typewriter.*

See Instructions on back of page 6.

Department of Toxic Substances Control
Sacramento, California

IN CASE OF EMERGENCY OR SPILL, CALL THE NATIONAL RESPONSE CENTER 1-800-424-8802; WITHIN CALIFORNIA, CALL 1-800-852-7550

GENERATOR

UNIFORM HAZARDOUS WASTE MANIFEST

1. Generator's US EPA ID No. | Manifest Document No. | 2. Page 1 of | Information in the shaded areas is not required by Federal law.

3. Generator's Name and Mailing Address

A. State Manifest Document Number 96124011

B. State Generator's ID

4. Generator's Phone ( )

5. Transporter 1 Company Name | 6. US EPA ID Number

C. State Transporter's ID

D. Transporter's Phone

7. Transporter 2 Company Name | 8. US EPA ID Number

E. State Transporter's ID

F. Transporter's Phone

9. Designated Facility Name and Site Address | 10. US EPA ID Number

G. State Facility's ID

H. Facility's Phone

11. US DOT Description (including Proper Shipping Name, Hazard Class, and ID Number)	12. Containers No.	12. Containers Type	13. Total Quantity	14. Unit Wt/Vol	I. Waste Number
a.					State EPA/Other
b.					State EPA/Other
c.					State EPA/Other
d.					State EPA/Other

J. Additional Descriptions for Materials Listed Above

K. Handling Codes for Wastes Listed Above
a. b.
c. d.

15. Special Handling Instructions and Additional Information

16. **GENERATOR'S CERTIFICATION:** I hereby declare that the contents of this consignment are fully and accurately described above by proper shipping name and are classified, packed, marked, and labeled, and are in all respects in proper condition for transport by highway according to applicable international and national government regulations.

If I am a large quantity generator, I certify that I have a program in place to reduce the volume and toxicity of waste generated to the degree I have determined to be economically practicable and that I have selected the practicable method of treatment, storage, or disposal currently available to me which minimizes the present and future threat to human health and the environment; OR, if I am a small quantity generator, I have made a good faith effort to minimize my waste generation and select the best waste management method that is available to me and that I can afford.

Printed/Typed Name | Signature | Month Day Year

TRANSPORTER

17. Transporter 1 Acknowledgement of Receipt of Materials
Printed/Typed Name | Signature | Month Day Year

18. Transporter 2 Acknowledgement of Receipt of Materials
Printed/Typed Name | Signature | Month Day Year

FACILITY

19. Discrepancy Indication Space

20. Facility Owner or Operator Certification of receipt of hazardous materials covered by this manifest except as noted in Item 19.
Printed/Typed Name | Signature | Month Day Year

DO NOT WRITE BELOW THIS LINE.

DTSC 8022A (1/95)
EPA 8700—22

White: TSDF SENDS THIS COPY TO DTSC WITHIN 30 DAYS.
To: P.O. Box 3000, Sacramento, CA 95812

Figure 5.7 *Uniform hazardous waste manifest.*

or emergency responder what exact materials are present, how much, and who to contact for further information.

In addition, if a material is being shipped as an EPA regulated hazardous waste, it must be accompanied by a uniform hazardous waste manifest. This document will give you information on the shipping name, the hazard class, and the exact total quantity. The form should be readily accessible, but separate or distinguishable from other shipping papers. Figure 5-7 displays a blank uniform hazardous waste manifest.

The type and location of the shipping papers involved depends on the mode of transportation and are given in Table 5-4.

Shipping papers are required to be in the locations listed in Table 5-4. A common problem is that the set of shipping papers available may not exactly agree with what is actually present due to constant loading and offloading of material from the mode of transport. Although shipping papers must be current by law, it is sometimes very difficult to accurately list real time locations of every hazardous materials. For example, a train in a marshaling yard is being put together or taken apart, or a large container ship at a dock is constantly loading and discharging cargo. These situations are not usually the case with trucks or trains in transit so a higher level of accuracy may be expected. In addition, every shipper must make previous arrangements for technical assistance with an emergency agency that is manned 24 hours a day, such as the Chemical Emergency Transportation Center (CHEMTREC). The phone number for CHEMTREC is 1-800-424-9300. A phone number such as this must appear on the shipping papers along with the following information:

- Material description (technical name)
- Immediate health hazards
- Fire or explosion risk

■ **NOTE**
Shipping papers are required to be in the locations listed in Table 5-4.

■ **NOTE**
Although shipping papers must be current by law, it is sometimes very difficult to accurately list real time locations of every hazardous material.

Table 5-4 *Mode of transport and shipping papers required.*

Mode of transport	Name of document	Location
Truck	Bill of lading	Driver's side door map pocket or on driver's seat if unoccupied.
Rail	Way bill Consist bill (Wheel Report)	Engine cab manned by engineer
Airplane	Airbill	Cockpit or by main passenger door to airplane
Ship	Dangerous cargo manifest Vessel stow plan	Wheelhouse, ship's office, or in possession of Chief Mate

- Immediate response to be taken in the event of an accident
- Immediate methods for handling fires
- Initial actions for handling spills or leaks
- Preliminary first aid in case of exposure

Some shippers attach a copy of the appropriate MSDS onto the shipping paper to avoid reprinting all of the required information. There is nothing wrong with this practice as long as it is easy to determine the information.

PENALTIES

The requirements for labels and placards under 49 CFR 172 are quite detailed and honestly, not the easiest to read and understand. However, if hazardous materials or hazardous waste are shipped improperly, either by mistake or intentionally, penalties can be assessed. In the case of transportation, the regulations fall under federal jurisdiction, so federal agencies enforce the rules. For example, a mislabeled container on a freight ship would be cited by the U. S. Coast Guard. The citation would go to a Hearing Officer who would assess a penalty based on the seriousness of the situation and the past record of the responsible parties. What is more common is local law agencies enforcing state or county health and safety regulations for the incorrect use of labels and placards. For example, in California, section 27903 of the California Vehicle Code can be cited if a party fails to display appropriate placards and markings of most hazardous materials. In terms of monetary assessment, penalties again depend on the seriousness of the situation. The penalty for the first offense (a misdemeanor) is $1,000. The penalty for a second offense (felony) can range from $5,000 to $25,000.

DOT NORTH AMERICAN EMERGENCY RESPONSE GUIDEBOOK

No discussion of DOT labeling and placarding would be complete without an understanding of the DOT *North American Emergency Response Guidebook* (see Figure 5-8). This publication is updated approximately every three years and is a useful resource when referring to DOT labels and placards.

The book is designed for the first responder for initial response action only and should be used in conjunction with at least two other reference sources. In the words of the book, "*it is primarily a guide to aid first responders in quickly identifying the specific or generic hazards of the material(s) involved in the incident and protecting themselves and the general public during the initial response phase of the incident.*" The user of the book should take these words to heart and be aware that it is only one guide for quick assessment only. However, in most situations it performs adequately to make an initial identification of a hazardous material. The guidebook is simple to use and designed to provide basic information. It is divided into the following color-coded sections:

■ **NOTE**
The DOT guidebook is designed for the first responder for initial response action only and should be used in conjunction with at least two other reference sources.

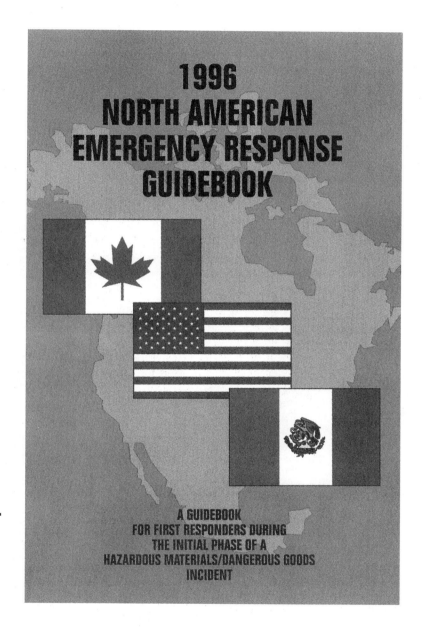

Figure 5.8 *Cover of DOT* North American Emergency Response Guidebook.

- Yellow Section (see Figure 5-9)—Lists materials in order by their identification number and refers to a guide number.

ID No.	Guide No.	Name of Material
1829	137	Sulfur trioxide, uninhibited
1829	137	Sulphur trioxide
1829	137	Sulphur trioxide, inhibited
1829	137	Sulphur trioxide, stabilized
1829	137	Sulphur trioxide, uninhibited
1830	137	Sulfuric acid
1830	137	Sulfuric acid, with more than 51% acid
1830	137	Sulphuric acid
1830	137	Sulphuric acid, with more than 51% acid
1831	137	Oleum
1831	137	Oleum, with less than 30% free Sulfur trioxide
1831	137	Oleum, with less than 30% free Sulphur trioxide
1831	137	Oleum, with not less than 30% free Sulfur trioxide
1831	137	Oleum, with not less than 30% free Sulphur trioxide
1831	137	Sulfuric acid, fuming
1831	137	Sulfuric acid, fuming, with less than 30% free Sulfur trioxide
1831	137	Sulfuric acid, fuming, with not less than 30% free Sulfur trioxide
1831	137	Sulphuric acid, fuming
1831	137	Sulphuric acid, fuming, with less than 30% free Sulphur trioxide
1831	137	Sulphuric acid, fuming, with not less than 30% free Sulphur trioxide
1832	137	Sulfuric acid, spent
1832	137	Sulphuric acid, spent
1833	154	Sulfurous acid
1833	154	Sulphurous acid
1834	137	Sulfuryl chloride
1834	137	Sulphuryl chloride
1835	153	Tetramethylammonium hydroxide
1836	137	Thionyl chloride
1837	157	Thiophosphoryl chloride
1838	137	Titanium tetrachloride
1839	153	Trichloroacetic acid
1840	154	Zinc chloride, solution
1841	171	Acetaldehyde ammonia
1843	141	Ammonium dinitro-o-cresolate
1845	120	Carbon dioxide, solid
1845	120	Dry ice
1846	151	Carbon tetrachloride
1847	153	Potassium sulfide, hydrated, with not less than 30% water of crystallization
1847	153	Potassium sulfide, hydrated, with not less than 30% water of hydration
1847	153	Potassium sulphide, hydrated, with not less than 30% water of crystallization
1847	153	Potassium sulphide, hydrated, with not less than 30% water of hydration
1848	132	Propionic acid
1849	153	Sodium sulfide, hydrated, with not less than 30% water
1849	153	Sodium sulphide, hydrated, with not less than 30% water
1851	151	Medicine, liquid, poisonous, n.o.s.
1851	151	Medicine, liquid, toxic, n.o.s.
1854	135	Barium alloys, pyrophoric
1855	135	Calcium, metal and alloys, pyrophoric

Figure 5.9 *Example of DOT ERG Yellow Section (also see color insert).*

Chapter 5 Hazardous Materials Identification Systems

- Blue Section (see Figure 5-10)—Lists materials by name in alphabetic order and refers to a guide number.

Name of Material	Guide No.	ID No.	Name of Material	Guide No.	ID No.
Compressed gas, toxic, oxidizing, n.o.s. (Inhalation Hazard Zone A)	124	3303	Corrosive liquid, acidic, organic, n.o.s.	153	3265
Compressed gas, toxic, oxidizing, n.o.s. (Inhalation Hazard Zone B)	124	3303	Corrosive liquid, basic, inorganic, n.o.s.	154	3266
			Corrosive liquid, basic, organic, n.o.s.	153	3267
Compressed gas, toxic, oxidizing, n.o.s. (Inhalation Hazard Zone C)	124	3303	Corrosive liquid, flammable, n.o.s.	132	2920
			Corrosive liquid, n.o.s.	154	1760
Compressed gas, toxic, oxidizing, n.o.s. (Inhalation Hazard Zone D)	124	3303	Corrosive liquid, oxidizing, n.o.s.	140	3093
Consumer commodity	171	8000	Corrosive liquid, poisonous, n.o.s.	154	2922
Copper acetoarsenite	151	1585	Corrosive liquid, self-heating, n.o.s.	136	3301
Copper arsenite	151	1586			
Copper based pesticide, liquid, flammable, poisonous	131	2776	Corrosive liquid, toxic, n.o.s.	154	2922
Copper based pesticide, liquid, flammable, toxic	131	2776	Corrosive liquid, water-reactive, n.o.s.	138	3094
Copper based pesticide, liquid, poisonous	151	3010	Corrosive liquid, which in contact with water emits flammable gases, n.o.s.	138	3094
Copper based pesticide, liquid, poisonous, flammable	131	3009	Corrosive solid, acidic, inorganic, n.o.s.	154	3260
Copper based pesticide, liquid, toxic	151	3010	Corrosive solid, acidic, organic, n.o.s.	154	3261
Copper based pesticide, liquid, toxic, flammable	131	3009	Corrosive solid, basic, inorganic, n.o.s.	154	3262
Copper based pesticide, solid, poisonous	151	2775	Corrosive solid, basic, organic, n.o.s.	154	3263
Copper based pesticide, solid, toxic	151	2775	Corrosive solid, flammable, n.o.s.	134	2921
Copper chlorate	141	2721	Corrosive solid, n.o.s.	154	1759
Copper chloride	154	2802	Corrosive solid, oxidizing, n.o.s.	140	3084
Copper cyanide	151	1587	Corrosive solid, poisonous, n.o.s.	154	2923
Copra	135	1363			
Corrosive liquid, acidic, inorganic, n.o.s.	154	3264	Corrosive solid, self-heating, n.o.s.	136	3095

Page 112

Figure 5.10 *Example of DOT ERG Blue Section (also see color insert).*

- Orange Section (see Figure 5-11)—Each guide number is here and describes basic action to be taken.
- Green Section (see Figure 5-12)—A table of Initial Isolation and Protective Action Distances if the material is highlighted in the yellow or blue section.

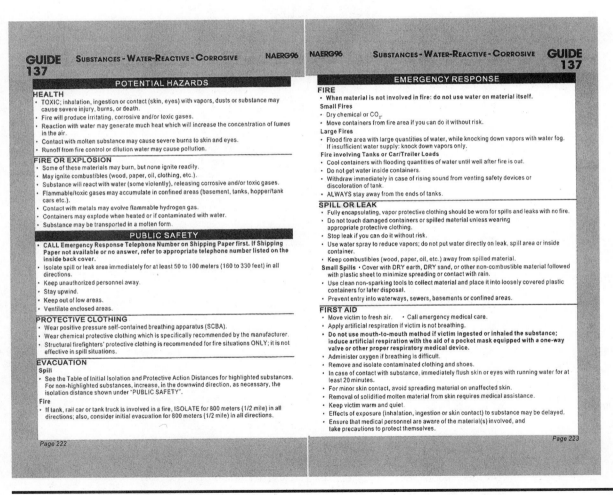

Figure 5.11 *Example of DOT ERG Orange Section (also see color insert).*

Chapter 5   Hazardous Materials Identification Systems

TABLE OF INITIAL ISOLATION AND PROTECTIVE ACTION DISTANCES

ID No.	NAME OF MATERIAL	SMALL SPILLS (From a small package or small leak from a large package)				LARGE SPILLS (From a large package or from many small packages)			
		First ISOLATE in all Directions		Then PROTECT persons Downwind during-		First ISOLATE in all Directions		Then PROTECT persons Downwind during-	
				DAY	NIGHT			DAY	NIGHT
		Meters	(Feet)	Kilometers (Miles)	Kilometers (Miles)	Meters	(Feet)	Kilometers (Miles)	Kilometers (Miles)
1829	Sulfur trioxide	60 m	(200 ft)	0.2 km (0.1 mi)	0.8 km (0.5 mi)	185 m	(600 ft)	0.6 km (0.4 mi)	2.9 km (1.8 mi)
1829	Sulfur trioxide, inhibited								
1829	Sulfur trioxide, stabilized								
1829	Sulfur trioxide, uninhibited								
1829	Sulphur trioxide								
1829	Sulphur trioxide, inhibited								
1829	Sulphur trioxide, stabilized								
1829	Sulphur trioxide, uninhibited								
1831	Oleum	60 m	(200 ft)	0.2 km (0.1 mi)	0.8 km (0.5 mi)	185 m	(600 ft)	0.6 km (0.4 mi)	2.9 km (1.8 mi)
1831	Oleum, with not less than 30% free Sulfur trioxide								
1831	Oleum, with not less than 30% free Sulphur trioxide								
1831	Sulfuric acid, fuming								
1831	Sulfuric acid, fuming, with not less than 30% free Sulfur trioxide								
1831	Sulphuric acid, fuming								
1831	Sulphuric acid, fuming, with not less than 30% free Sulphur trioxide								
1834	Sulfuryl chloride	95 m	(300 ft)	0.3 km (0.2 mi)	1.1 km (0.7 mi)	215 m	(700 ft)	1.0 km (0.6 mi)	3.9 km (2.4 mi)
1834	Sulphuryl chloride								
1836	Thionyl chloride	DANGEROUS:	When spilled in water, see list at the end of			this table.			
1838	Titanium tetrachloride	60 m	(200 ft)	0.2 km (0.1 mi)	0.6 km (0.4 mi)	155 m	(500 ft)	0.5 km (0.3 mi)	2.1 km (1.3 mi)
1859	Silicon tetrafluoride	60 m	(200 ft)	0.2 km (0.1 mi)	0.6 km (0.4 mi)	155 m	(500 ft)	0.5 km (0.3 mi)	1.9 km (1.2 mi)
1859	Silicon tetrafluoride, compressed								
1892	Ethyldichloroarsine	95 m	(300 ft)	0.5 km (0.3 mi)	1.6 km (1.0 mi)	275 m	(900 ft)	1.4 km (0.9 mi)	6.1 km (3.8 mi)

Figure 5.12 *Example of DOT ERG Green Section (also see color insert).*

Procedures for Using the Guide

■ **NOTE**
The DOT *North American Emergency Response Guidebook* lists chemicals by identification number (yellow section, name (blue section), basic action to take in case of spill (orange section), and initial isolation/protective action distances (green section).

In order to understand this guide properly it is important to remember what each colored section means. The DOT *North American Emergency Response Guidebook* lists chemicals by Identification number (yellow section), name (blue section), basic action to take in case of spill (orange section), and initial isolation/protective action distances (green section). To use the guide properly, take the following steps. First, refer to the yellow or blue section to identify the product present by either the identification number (yellow section) or by plain language name (blue section). If explosives are involved, use the table on page 1 and go directly to the appropriate orange guide. If only a placard is seen (with no identification number or name) then pages 14 and 15 that display the placards shall be used as displayed by Figure 5-13.

Second, the yellow and blue sections provide a guide number for obtaining fundamental emergency response information. This response information can be found in the orange section and will furnish basic information to be taken by first responders. The first part of this orange section describes potential hazards such as fire and health effects. The second part of the orange section suggests public safety measures such as notification, isolation and ventilation considerations, protective clothing, and evacuation. Finally, the orange section lists considerations for emergency response such as fire, spills or leaks, and first aid. The user shall then turn to the green section for evacuation distances if the identification number or name is highlighted in the yellow or blue section. This section identifies

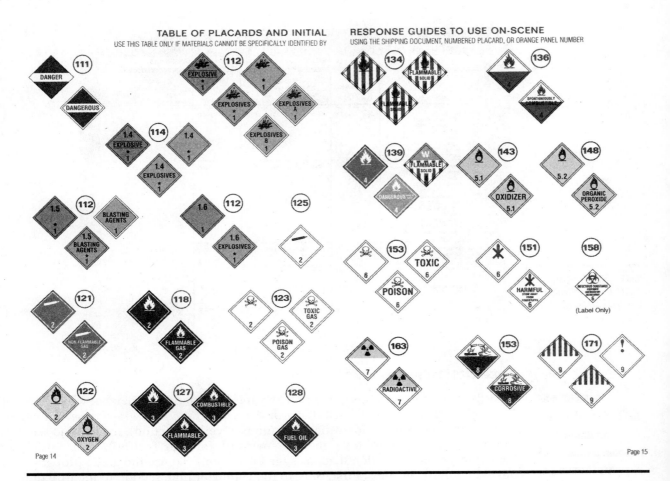

Figure 5.13 *Placards and initial response guides to use on scene.*

isolation distances and protective action distances according to small spills (less than 55 gallons from one or more containers) and large spills (greater than 55 gallons from one or a combination of containers). One disadvantage to the guide is that these distances are only good for about 30 minutes after a spill. After this period, other references and outside resources should be available. Another disadvantage is that the guide does not go into specific detail as does an MSDS. To offset this, there is some further information in the back of the book regarding incident considerations for the user in terms of protective clothing and fire and spill control, but it is still limited.

For example, if informed of a small leak from a container labeled only 1829 (see color insert), an initial responder could take the following steps using the DOT *North American Emergency Response Guidebook*.

- Research 1829 in the yellow section. It is sulfur trioxide, is highlighted, and refers the user to Guide number 137.
- Turn to the orange section to read Guide 137. This guide is for water reactive or corrosive substances. According to the Public Safety section, isolation should be 160 to 330 feet in all directions.
- Next, reference the green section because the material was highlighted in the yellow section. Here, isolation and protection distance are better defined for this specific material. Because the discharge can be considered small, isolation shall be 200 feet in all directions.

DRUM PROFILES

As previously stated, you must always be on the lookout for visible labels or other markings that could indicate what the released chemical is. Get a 360° look at your problem when possible and locate as many clues as you can. In many cases a label that offers some basic information can be located. Simply identifying the type of container the product is leaking from may offer good information. Drums are an example of this concept and are a commonly encountered storage vessel for all kinds of chemicals. Abandoned drums, leaking drums, or bulging drums, offer a constant source of headache for the typical waste worker or responder. Truly seeing the drum can at least give a starting point for what type of material may be inside. The following points give some basic information on the various types of drums and what may be found inside. Learn as much as possible about the typical storage vessels found in your jurisdiction.

Closed Head Steel Drum

Construction Features: These drums are constructed of low carbon steel with a mechanically seamed side. There are typically two rolling rings along the vertical body with the top and bottom being mechanically sealed. The top of a drum is referred to as the head. The top and bottom "lip" are called chimes, and are sometimes used as an anchor to pick up a drum with a lifting device, or to roll the drum along flat, horizontal surfaces. The head of the drum typically has two holes or "bungs" from which product is extracted. See Figure 5-14 for an example of the drum profile and bung arrangement.

The larger bung is usually 2 inches in diameter, has a screw cap, and is used as a port for pumping or otherwise removing or inserting non**viscous** materials. The smaller bung is generally 3/4 of an inch in diameter and functions in many cases as a vent hole. Both can be easily removed using a special bung wrench, which is shown in Figures 5-15 and 5-16. Some bung wrenches are made of

■ **NOTE**
Be on the lookout for visible labels or other markings that could indicate what the released chemical is.

Safety
● Learn as much as possible about the typical storage vessels found in your jurisdiction.

■ **NOTE**
The top of the drum is referred to as the head.

viscous
an expression of the resistance of a fluid to flow

brass
an alloy consisting of copper and zinc

Figure 5-14 *Closed head steel drums in storage.*

lab packing
a packing procedure performed when several smaller containers, such as 1-gallon glass bottles or cans, need to be stabilized for transport inside a larger container; in many cases, a 55-gallon drum is used for overpacking

brass so as not to create sparks when being used around flammable liquids. These drums may be painted with an array of colors but the most common are blue, red, or black.

What's Usually Inside: Hydrocarbon-based materials, which are easily pumped or transferred. Examples include acetone, methyl ethyl ketone, ethyl alcohol, gear oil, diesel fuel, or other similar substances.

Open Head Steel Drum

Construction Features: These drums are constructed of low carbon steel with a mechanically seamed side. There are typically two rolling rings along the vertical body with the bottom only being mechanically sealed. The top is a one-piece lid with no bungs that is fastened to the drum with a removable ring. The ring is secured with a nut and bolt once the lid is properly placed. The ring functions much like a pipe clamp in that it closely fits the circumference and is subsequently tightened down to ensure closure. Figure 5-17 shows the features of a typical open head steel drum.

What's Usually Inside: Hydrocarbon-based materials that are highly viscous. Examples include lubricating greases, extremely heavy gear oils, and sludge. **Lab packing** is also done in this type of container. To accomplish a lab pack, place a several inch layer of loose absorbent in the bottom of the drum. The smaller containers are placed on that base, and surrounded by more loose absorbent which creates a base for the next layer. More containers are placed atop the new layer and subsequently surrounded by more loose absorbent. This procedure is repeated until the open head drum is full. In essence, several strata of containers are created which are stabilized by the loose absorbent.

These drums are also used as a typical over pack for leaking or unsound drums. Figure 5-18 shows students placing a closed head steel drum into an open head steel drum used as an overpack.

Chapter 5 Hazardous Materials Identification Systems

Figure 5-15 *Typical brass bung wrench used to open/close the 2" cap.*

Figure 5-16 *Same brass wrench as seen in Figure 7-2 and used to operate the 3/4" vent cap.*

polyethylene
a thermoset plastic exhibiting a high degree of chemical resistance, temperature resistance, and tensile strength

Closed Head Polyethylene Drum

Construction Features: These drums are fashioned with **polyethylene** plastic as the overall construction material. This plastic is rigid and provides a very strong containment system. A good example of the inherent strength of these vessels is demonstrated by the fact that sulfuric acid is commonly found in black polyethylene drums. Concentrated sulfuric weighs approximately 15 pounds per gallon. Contrast this with water which weighs approximately 8 pounds per gallon. This means that a 55-gallon drum of sulfuric would weigh around 825 pounds. This weight is far more than a similar drum of gasoline would weigh so the strength

Figure 5-17 *Steel open head drum.*

Figure 5-18 *Students practicing the "v-roll" technique of overpacking.*

of plastic drums should never be an issue in the field. Bung arrangements are the same as for closed head steel drums and the same bung wrench usually works for either type of drum. The containers also have two rolling rings on the vertical plane but are not as easy to roll as steel drums. These drums come in many colors but the most common are **translucent** white, black, blue, or yellow.

What's Usually Inside: These containers are the classic vessel for acids and bases. At times they may contain a mild solvent but as a general rule you will not find strong solvents in these drums. Aggressive solvents break down the plastic and cause a leak. Typical chemicals found in these drums include sulfuric acid, hydrochloric acid, mixtures such as **Skasol**, sodium hydroxide, potassium hydroxide, cleaning fluids, and soaps. Figure 5-19 shows a polyethylene drum used to store product at a fixed facility.

Open Head Fiber Drum

Construction Features: These drums are somewhat shorter and may have a larger circumference than the drums previously listed. They are essentially made of cardboard much like a typical cardboard shipping box. The construction is a little denser and may be wax covered or treated to repel liquids. Most have a plastic bag for a liner and should always have an open head. The method of closure is the same as its open head steel drum counterpart.

What's Usually Inside: These drums should always contain some kind of solid. Typical materials include sodium hydroxide pellets, potassium hydroxide pellets, soap flakes, or any other nonmetallic solid.

Closed Head Stainless Steel Drum

Construction Features: These drums should immediately alert you that there is a bad chemical inside. These drums are very expensive and facilities do not store garden variety chemicals inside. These drums are usually gray and look a lot like an enormous keg of beer! Do not tap these drums however as a very strange brew may be unexpectedly served up. At times they are lined with a type of wax to help reduce the possibility of any reaction with the metal inside. They have two rolling rings on the vertical plane of the drum and appear much more pronounced than the rolling rings on a typical steel drum. They have the same bung arrangement as a closed head steel drum and require a bung wrench to gain access. If you have an incident involving one of these drums approach it with a high level of concern as the material inside will most likely be very dangerous.

What's Usually Inside: These drums almost always carry strong oxidizers like >72% nitric acid. Most exotic oxyacids are transported in these vessels and it is necessary to identify the shipper (if possible) if it is found on the roadway or as an abandoned drum. Odd-looking drums usually contain odd chemicals—pay attention.

---

translucent
quality of a material that allows light to pass but does not allow it to be seen through clearly

■ **NOTE**
Aggressive solvents break down the plastic and cause a leak.

Skasol
trade name for a corrosive etch, typically hydrochloric acid in a 17% concentration

Safety
Odd-looking drums usually contain odd chemicals—pay attention.

Figure 5-19 *Black polyethylene drum. Common storage vessel for corrosives.*

Once again, this is merely an illustration of the detail you can find if you just pay attention to the subtle stuff on the scene. Make it a point to augment your classroom information with the real world details of the local storage and transportation vessels. Most local laboratories, manufacturing facilities, computer fabrication sites, petrochemical facilities, and such are happy to discuss the specifics of the containers unique to their operations. Hands-on education is far and away better than reading about them in a textbook.

Summary

- Although many types of identification systems are in use, the two most common are the DOT system and the NFPA 704 system. Never confuse the markings of one system with another.
- Most identification systems do not exactly tell what chemical is present, only the potential hazards posed by the chemical itself.
- Any identification system in use is only effective if it is implemented properly and understood by those who use it.
- On a DOT label or placard, there are four factors to be used in determining the hazard: Color, UN number, name or identification number, and symbol. All four factors can be used together or separately to determine what material is present.
- Proper labeling and placarding requirements for hazardous materials and hazardous wastes can be found in 49 CFR 172.101.
- The DOT system is used for hazardous materials being transported and the NFPA system is used when hazardous materials are being stored at a fixed facility.
- The DOT system is mandated by federal law and the NFPA 704 is a recommended guideline to be enforced by local government if so desired.
- The NFPA 704 system has the advantage of being easy to use, but can be limited in some situations due to not informing responders how much, how many, and where chemicals are located.
- Shipping papers and uniform hazardous waste manifests shall accompany hazardous materials during shipment and must be accurate and readily accessible.
- The DOT *North American Emergency Response Guidebook* is for initial action only and is only one of many available reference resources. Isolation and protective action distances are effective for only about 30 minutes.
- The DOT *North American Emergency Response Guidebook* lists chemicals by identification number (yellow section), name (blue section), basic action to take in case of spill (orange section), and initial isolation/protective action distances (green section).
- The hazardous materials information system (HMIS) is similar to the NFPA 704 diamond, but the white section denotes levels of personal protective equipment that may conflict with other standards.

Review Questions

1. What are the two most common identifications systems and where are they used?
2. What other identification systems do you use or have come across?
3. List three disadvantages of the DOT identification system.
4. Describe the difference between a hazardous material and a hazardous waste and which agency regulates each one.
5. List the United Nations hazard classes.
6. If you are shipping 10 pounds of 1.1 explosives, is labeling or placarding required? If so, why and what other materials require placards in any amount?
7. According to the DOT *North American Emergency Response Guidebook,* what is the difference between a large and a small spill?

Scenario

Objectives:

1. To allow an opportunity to work with the DOT *North American Emergency Response Guidebook* in the identification of hazardous materials.
2. To show the limitations of the DOT *North American Emergency Response Guidebook.*

Use the information given to you in Chapter 5 (along with all figures and tables) to answer the following questions.

As part of an Emergency Response Team at a large factory, a worker has reported to you several containers knocked over by a forklift on the loading dock. He reports he saw a label with the number 1831 on it. Based on the information found in the appropriate response guide, what are its health effects and what is the isolation distance with a 75-gallon spill? Another container had an all yellow diamond label on the side. What is probably in that container? A third container lying next to the above container had a red diamond with a flame on it? What could be in that container?

Chapter 6

Respiratory Protection

Learning Objectives

Upon completion of this chapter, you should be able to:

- Identify the five types of respiratory hazards likely to be found in situations involving hazardous materials and wastes.
- Name the elements of a respiratory protection program required by federal OSHA regulations.
- Identify the two classes of respiratory protective equipment.
- List the advantages and disadvantages of each class of respiratory protective equipment.
- Select the appropriate type of respiratory protective equipment for specific situations likely to be encountered in the hazardous waste field.
- List the maintenance and inspection procedures for each type of respiratory protective equipment.
- Describe the steps necessary to effectively don and doff an air purifying respirator.
- Identify a supplied air (umbilical) system.
- Describe the steps necessary to effectively don and doff a self-contained breathing apparatus (SCBA).

INTRODUCTION

As we have studied previously, inhalation hazards are a commonly encountered problem when working with hazardous chemicals. In Chapter 4, we learned that our respiratory system is perhaps the most vulnerable of our body systems to the harmful effects of chemicals. Therefore it is vital that both the individual who works with chemicals in the workplace and those who respond to chemical emergencies understand the need for proper respiratory protection. It also stands to reason that proper selection and use of respiratory protection be an important element of our training program.

> **Safety**
> The respiratory system is extremely vulnerable to the effects of hazardous materials.

RESPIRATORY HAZARDS

Respiratory hazards are posed by chemicals that can be present in the air or atmosphere. Airborne contaminants can be divided into five major groups:

1. Particulates including dusts, fumes, and mists
2. Gases and vapors
3. Combination of particulates and gases
4. Oxygen-deficient atmospheres
5. High or low temperature areas

The first group of respiratory hazards is actually comprised of several components that ultimately end up as larger, and often visible particulates, in the atmosphere. Dusts are formed when solid materials are broken down by operations such as drilling, grinding, demolition, or sanding. Generally speaking, the smaller the dust, the longer it hovers in the air. Smaller dusts may also be easier to inhale. Also, smaller dusts may not be as visible and therefore more difficult to detect. The smaller dusts are more respirable and may find their way deeper into our respiratory systems.

> **Safety**
> The smaller dusts are more respirable and may find their way deeper into our respiratory system.

Fumes are produced when metal or plastic solids are heated and then cooled quickly, creating very fine particles that drift in the air. As an example, fumes are generated during welding operations and potentially whenever high temperatures are present.

Mists are tiny liquid droplets usually created during operations involving pressure or **atomizing** of the liquids. Examples include spraying operations and processes involving aerosols or high pressure.

Gases are substances that are airborne at ambient temperature and pressure. They are materials that can travel far and fast. Because they are very mobile, gases can pass immediately through your airway to your lungs where they are absorbed into the bloodstream. Unlike particulate materials, gases tend to be difficult to see and usually require detection devices to identify their presence.

atomizing
the process of rapidly compressing a liquid into a mist or vapor

evaporation
a physical change of a solid or liquid into its gaseous or vapor state

> **Safety**
> • In addition to confined spaces, fire conditions, and chemical releases can produce oxygen-deficient conditions.

> **Safety**
> • An oxygen-deficient atmosphere may not have any characteristic odor to alert those involved that a serious condition is present.

> **Safety**
> • Remember! Be aware of vapors and gases, particulates, low oxygen conditions, and possible thermal conditions.

Vapors are the gaseous form of a solid or liquid that is created by the process of **evaporation**. Like gases, vaporous materials are mobile and can enter the respiratory system very quickly. They are often invisible to the naked eye and may be difficult to detect without proper air monitoring procedures.

The next major inhalation hazard is oxygen deficiency, which occurs when the level of oxygen within the environment drops to dangerously low levels. An oxygen-deficient atmosphere is defined by the Occupational Safety and Health Administration (OSHA) as an atmosphere where the percentage of oxygen in the air drops below 19.5% by volume. In addition to confined spaces, fire conditions, and chemical releases can produce oxygen-deficient conditions.

The effects of oxygen deficiency on the body are numerous and quite profound. A summary of these effects is found in Table 6-1. As you see, this condition initially disrupts the central nervous system with effects such as confusion and dizziness. This is often followed by an increase in the breathing rate and heart rates leading to a higher demand for the limited supply of oxygen. Eventually, this condition leads to death if steps are not taken to remedy the situation.

An additional problem regarding oxygen deficiency is that it is a condition not easily recognized by the persons involved. An oxygen-deficient atmosphere may not have any characteristic odor to alert those involved that a serious condition is present. This condition, coupled with the problems associated with the confusion produced, make oxygen deficiency one of the most serious issues confronting protection of our respiratory system. Only by using special detection devices can we know that oxygen levels may be dropping.

Additionally, work sites or emergency scenes may be complicated by thermal conditions. Very hot or cold air can damage tissue in your nose, mouth, trachea, and lungs, interfering with normal breathing. Obviously, if these conditions are found in the atmosphere that we breathe, steps need to be taken to remove the condition or otherwise protect our respiratory system. Remember! Be aware of vapors and gases, particulates, low oxygen conditions, and possible thermal conditions.

Table 6-1 *Effects of low oxygen on the human body.*

O_2 Percentage	Effects on Body
21	Normal atmospheric air; normal respirations, no problems
19.5	OSHA O_2 deficiency
17	Beginning of confusion, increased respiratory rate, slight loss of coordination
12	Fatigue (extreme in some cases), decreased level of consciousness, rapid respiratory rate
9	Coma or death

RESPIRATORY PROTECTION FUNDAMENTALS

A number of OSHA regulations deal with the use of respiratory protection. The key regulation at the federal level, which is also referenced in the hazardous waste operations and emergency response (HAZWOPER) regulation, is found in 29 CFR, part 1910.134. This standard is now under revision and when issued, may have additional or different information than what is presented here.

In reviewing this standard, we find that protection of the respiratory system is a priority in the workplace. Recall that the exposure limits discussed in Chapter 4 (PEL, TLV, etc.) were largely developed around airborne contaminants. Because of this, OSHA mandates that employers who have levels of contaminants that could pose a hazard to workers via the respiratory system, develop a respiratory protection program. These high contamination levels occur in some instances in the workplace, but are even more common in emergency response. Any workplace that has the need for respiratory protection must have a written program.

As with other programs mandated by OSHA, a respiratory protection program is required to be in writing and available for employees to review as part of their worker protection system. Because airborne contaminants are a significant hazard for those who work with hazardous wastes or deal with chemical emergencies, we will spend some time reviewing the principles of respiratory protection as mandated by OSHA.

> **Safety**
> Any workplace that has the need for respiratory protection must have a written program.

Written Respiratory Protection Programs

OSHA requires that employers who have respiratory hazards in the workplace develop a program to mitigate them. When asked how one protects the respiratory system, the answer usually involves the concept of a some type of respirator or self-contained breathing apparatus. Although respirators are a very effective means of protecting the respiratory system, they may not always be the best choice. Prior to wearing respiratory protection, OSHA mandates that engineering controls be used as the first means of mitigation. In fact, the regulation as found in 29 CFR, part 1910.134 (a)(1) reads as follows:

> In the control of those occupational diseases caused by breathing air contaminated with harmful dusts, fogs, fumes, mists, gases, smokes, sprays, or vapors, the primary objective shall be to prevent atmospheric contamination. This shall be accomplished as far as feasible by accepted engineering control measures (for example, enclosure or confinement of the operation, general and local ventilation, and substitution of less toxic materials). When effective engineering controls are not feasible, or while they are being instituted, appropriate respirators shall be used.

Engineering controls might easily provide the respiratory protection for you!

Even at waste sites or in some emergency situations, it may be possible to mechanically remove contaminants in advance of the requirement for respiratory

> **Safety**
> Engineering controls might easily provide the respiratory protection for you!

INCIDENT MANAGEMENT

Elements of Successful Incident Management . . 2
Emergency Response Guidebook Explained . . . 6
Chemical Resistance Guide 8

Elements of Successful Incident Management

Proper notification is an important facet of getting the action started off properly. Note prevailing weather conditions, time of day, and preplans for a facility if the plans exist.

Response to the incident must be accomplished in a safe manner. Approaching from an uphill, upwind direction should be the first protective action responders take.

A proper size up must be accomplished once the responders arrive. This process involves observing the area for visible vapors, puddles, container damage, or other clues to identify the nature of the emergency.

Command should be established early in the incident. Establish a command post in a safe and convenient location.

The highest hazard area should be identified and isolated. Entry into the restricted zones must be controlled by the response personnel. This action prevents responders and civilians from wandering into any contaminated area.

Identification of the released product must be accomplished. This phase includes looking for labels, placards, or other means to positively identify the chemical hazard.

Once the product is identified, reference sources must be consulted in order to understand the properties of the released substance.

The IC and/or other key staff should weigh the risks of intervention against the probability of success. This is the point at which a decision to take definitive action is reached.

3 INCIDENT MANAGEMENT

If offensive action is to be taken, a proper decon corridor must be established.

The IC, Safety Officer, and any other key staff should be involved in the development of an action plan. This plan should include incident objectives, safety items, required tools and tactics, necessary PPE, type of decon, and any other safety information pertinent to the response.

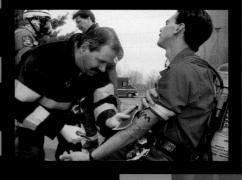

Entry and decon personnel should be medically monitored prior to working in PPE.

The final dress out should be accomplished with acute attention to detail.

INCIDENT MANAGEMENT 4

Implement the plan and evaluate the progress. Change tactics if the current course of action does not appear to be working. It is always acceptable to move back to a safe location and change the mitigation approach.

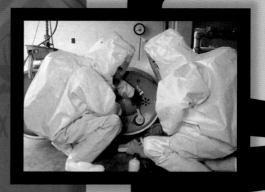

Provide for a complete and systematic decon, appropriate for the expected chemical hazard.

Once the incident has been stabilized, it is time to move into the clean up phase. This part of the effort may be delegated to clean up contractors or accomplished by the original responders.

It is wise to conduct a post incident analysis after any significant incident. This phase is important in order to debrief all involved, and to share any lessons learned during the course of the response.

Emergency Response Guidebook Explained

The white section has general information about the guide and a visual summary of the placards.

This section identifies each chemical by its placard number. Any highlighted chemical will also be found in the green section, "Table for Initial Isolation Distances", at the end of the book. These are highly hazardous chemicals.

INCIDENT MANAGEMENT 6

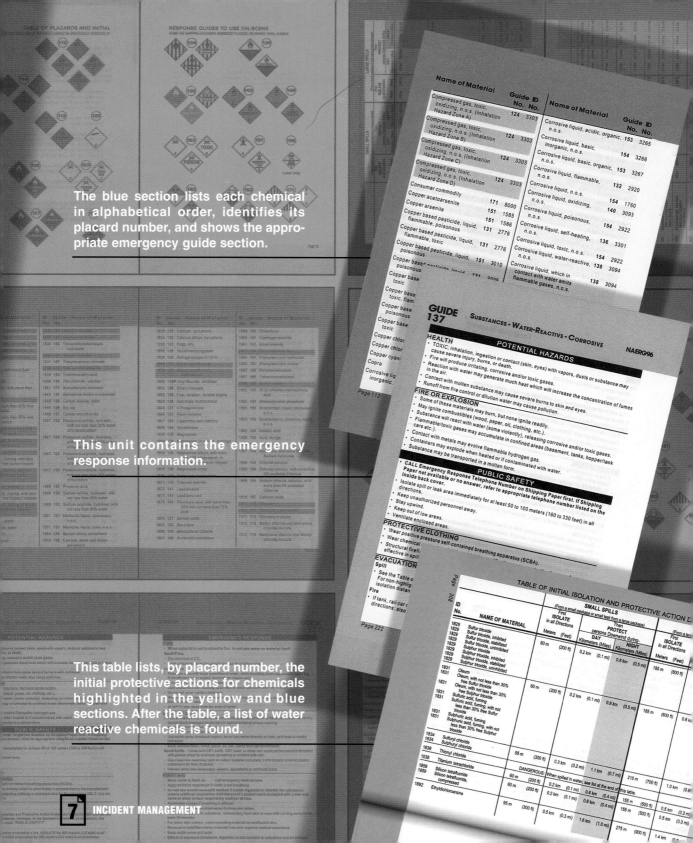

The blue section lists each chemical in alphabetical order, identifies its placard number, and shows the appropriate emergency guide section.

This unit contains the emergency response information.

This table lists, by placard number, the initial protective actions for chemicals highlighted in the yellow and blue sections. After the table, a list of water reactive chemicals is found.

7 INCIDENT MANAGEMENT

North Chemical Resistance Guide

This Chemical Resistance Guide incorporates three types of information. First, the guide indicates breakthrough time and permeation rate. Breakthrough time is defined as the elapsed time between initial contact of the liquid chemical with the outside surface of the glove and the time at which the chemical can be detected at the inside surface of the glove by means of analytical equipment. *When breakthrough occurs, the glove is no longer providing total protection.* Permeation rate, measured in milligrams per square meter per second ($mg/m^2/sec$) is the measured steady state flow of the permeating chemical through the glove elastometer. Glove thickness plays an important role in resistance to permeation.

Secondly, this guide indicates degradation resistance. Degradation is a deleterious change in one or more of the glove's physical properties. The most obvious forms of degradation are the loss of the glove's strength and excessive swelling. Several published degradation lists (primarily "The General Chemical Resistance of Various Elastomers" by the Los Angles Rubber Group, Inc.) were used to determine degradation.

The third item of information provided by this chart is the color code that gives the user a general recommendation on which gloves should be evaluated and tested first. The color code incorporates data from other sources.

The permeation data in this guide are based on permeation tests performed in accordance with ASTM Standard F 739 under laboratory conditions by North Safety Products or an independent AIHA accredited laboratory. Neither North Safety Products nor the independent laboratory assumes any responsibility for the suitability of an end user's selection of gloves based on this guide.

Data on chemicals not listed here can be obtained by calling the North Hand Protective Division Customer Service Department at (800) 456-8315 or your North Hand Protection Territory manager.

USER PRECAUTIONS

Protective gloves and other protective apparel selection must be based on the user's assessment of the workplace hazards. Apparel materials do not provide unlimited protection against all chemicals, and the user must determine before use that the apparel will resist permeation and degradation by the chemicals (including chemical mixtures) in the environment of intended use.

To obtain maximum life, protective glove and other protective apparel should have chemicals removed from the surface by washing or other appropriate methods after each use. The apparel should be stored away from the contaminating atmosphere.

Punctured, torn or otherwise ruptured apparel must be removed from service; unserviceable apparel may be disposed of only in accordance with applicable waste disposal regulations.

KEY TO DEGRADATION AND PERMEATION RATINGS

E	Excellent	Fluid has no effect
G	Good	Fluid has minor effect
F	Fair	Fluid has moderate effect
P	Poor	Fluid has severe effect, ranging from moderate to complete destruction
ND		None detected
ID		Insufficient data, data not available or conflicting data

PERMEATION CHART COLOR KEY

- Good for Total Immersion
- Good For Accidental Splash/ Intermittent Exposure
- Not Recommended

PHYSICAL PERFORMANCE CHART

Physical Characteristics	Silver Shield®	Viton*	Butyl	Nitrile
Abrasion Resistance	F	G	G	E
Cut Resistance	P	G	G	E
Puncture (Snag) Resistance	F	G	G	E
Flexibility	E	G	G	E
Heat Resistance	F	G	G	G
Ozone Resistance	E	E	G	F
Tensile Strength	E	G	G	E
Low Gas Permeability	E	E	E	P

Note: Products in these categories vary in capabilities, Laboratory tests are necessary for specific recommendations.

* Viton is a Registered Trademark of the Du Pont Co.

Chemical		Silver Shield (4 Mil)			Viton (9 Mil)			Butyl (17 MIl)			Nitrile Latex (11 Mil)	
	D	BT	PR	D	BT	PR	D	BT	PR	D	BT	PR
Acetaldehyde	E	> 6 hrs	ND	P	0 min	281.9	E	9.6 hrs	0.066	F	4 min	161
Acetone	E	> 6 hrs	ND	P	ID	ID	E	> 17 hrs	ND	P	5 min	172
Acetonitrile	E	> 8 hrs	ND	ID	ID	ID	E	> 8 hrs	ND	ID	ID	ID
Acrylic Acid	ID	ID	ID	G	5.9 hrs	0.23	E	> 8 hrs	ND	F	ID	ID
Acrylonitrile	E	ID	ID	F	1 min	176	G	3.1 hrs	0.000048	P	3 min	176
Aldehyde	E	> 6 hrs	ND	P	0 min	281.9	E	9.5 hrs	0.0665	P	4 min	161
Aniline	E	> 8 hrs	ND	G	10 min	18.7	F	>8 hrs	ND	P	1.1 HRS	45.0
Benzaldehyde	ID	ID	ID	F	9.9hrs	4.0	E	9 hrs	ND	P	ID	ID
Benzene	E	> 8 hrs	ND	G	6 hrs	0.012	P	31min	32.3	P	ID	ID
Benzoyl Chloride	ID	ID	ID	E	> 8 hrs	ND	F	6.2 hrs	16.6	P	ID	ID
Bromobenzene	E	ID	ID	E	8 hrs	ND	P	32 min	39.8	P	13 min	9.1
Butyl Acetate	E	> 6 hrs	ND	P	ID	ID	G	1.9 hrs	7.61	P	29 min	54.4
p-t Butyltoluene	E	> 8 hrs	ND	P	> 8 hrs	ND	G	1.7 hrs	8.0	P	ID	ID
Butyraldehyde	E	ID	ID	P	54 min	9.0	E	> 15 hrs	ND	P	ID	ID
Carbon Disulfide	G	> 8 hrs	ND	E	> 8 hrs	ND	P	7 min	98.0	P	1 min	51
Carbon Tetrachloride	E	> 6 hrs	ND	E	> 13 hrs	ND	P	ID	ID	G	3.4 hrs	5
Cellosolve	G	> 6 hrs	ND	F	ID	ID	G	ID	ID	P	ID	ID
Chlorobenzene	E	ID	ID	E	> 8 hrs	ND	P	35 min	308	P	ID	ID
Chloroform	E	> 6 hrs	ND	E	9.5 hrs	0.46	P	ID	id	P	4 min	352
Chloronaphthalene	E	> 8 hrs	ND	E	> 16 hrs	ND	P	ID	id	P	2.9 hrs	> 1.32
Chloroprene	ID	ID	ID	ID	> 8 hrs	ND	P	28 min	18	ID	ID	ID
Cyclohexane	E	> 6 hrs	ND	E	> 7 hrs	ND	P	1.1 hrs	20.3	P	ID	ID
Cyclohexanol	E	> 6 hrs	ND	E	> 8 hrs	ND	E	> 11 hrs	ND	E	> 16 hrs	ND
Cyclohexanone	E	> 6 hrs	ND	P	29 min	86.3	E	> 16 hrs	ND	P	ID	ID
Dibutylphthalate	E	> 6 hrs	ND	E	> 8 hrs	ND	E	> 16 hrs	ND	E	>16 hrs	ND

protection. When this is not possible, or not feasible, the use of an appropriate respirator shall be required. Further, a written respiratory protection program should be enacted to cover the following areas. Although the listed items are not to the letter of the written OSHA standard, they do include the salient points.

- Procedures for the proper selection and use of respirators including specific requirements regarding the types required for the hazard or combination of hazards likely to be encountered. Only certain types of approved respirators are allowed to be used.
- Training program requirements for those who will wear respiratory protection. The number of hours required for this training program is not established in the regulation, however employees must receive training in the elements of the respiratory protection program in place at the site. Clearly, training in respiratory protection can be included as part of HAZWOPER training, whether for emergency response or for waste operations.
- Inspection, maintenance, and sanitation procedures should be specified. Regular inspections should be conducted and maintenance performed in accordance with the manufacturers requirements.
- Storage procedures for each piece of respiratory protection equipment must be specified in order to maintain them in proper working order.
- Medical surveillance practices need to be outlined to include a physical examination by a physician to determine whether the person is capable of wearing respiratory protective equipment. Respirators can place demands on our bodies above those normally encountered. Additionally, the types of operations requiring the use of respirators can cause harm to some individuals. For this reason, OSHA mandates that anyone who wears a respirator be cleared by a physician prior to use. Follow-up medical examinations are required on a regular basis. Medical exams are generally performed annually.
- Monitoring of the program and surveillance of the work area condition with special attention to the potential exposures of the worker is a required part of the respiratory protection program. If monitoring shows deficiencies, appropriate action is indicated.
- Record keeping requirements for all components of the program is mandated. This includes employee exposure and medical screening records, maintenance records, air quality records, and training programs.
- Fit testing of the specific respirators used in the work area by the employees must be provided for. Fit testing is done to ensure that the respirators selected fit the individual faces of those who use them.

If respiratory protection is to be worn as part of employment, the employer is required to develop a written respiratory protection program in accordance with 29 CFR, part 1910.134.

> **Safety**
> Medical exams are generally performed annually.

> **Safety**
> If respiratory protection is to be worn as part of employment, the employer is required to develop a written respiratory protection program in accordance with 29 CFR, part 1910.134.

RESPIRATORY PROTECTION EQUIPMENT

If it is not possible to mitigate the five types of respiratory hazards by engineering controls, the use of respiratory protective equipment is necessary. For this reason, training on the proper use, selection, and maintenance of respiratory protective equipment is an important element of a HAZWOPER training program. Respiratory protective equipment that is used in emergency response or in hazardous waste cleanup activities is divided into two main categories: air purifying systems and air supplying systems.

Air Purifying Systems

An air purifying respirator (APR) is a piece of equipment that has special filters designed to remove particulates (dust, mists, fumes) from the air or that has canisters or chemical cartridges designed to protect the wearer from gases and vapors. The system utilizes either a full-face, or half-face mask, which fits onto the face of the wearer and is connected to a cartridge or cartridges that filter out the airborne contaminants. Inspired air is drawn through a series of filters or cartridges that *purify* the air. An example of a full-face air purifying respirator can be seen in Figure 6-1.

Figure 6.1 *Worker making repairs using a full-face air purifying respirator (APR). Photo courtesy of Mine Safety Appliances*

The mechanism for filtering contaminants varies. In the case of solid particulates, the inspired air is drawn through paper or other materials that act as a barrier to block particles of various size from entering the mask. As the size of the particles decreases, the filtering capability of the material must be effective enough to handle the tiniest microscopic contamination. All air purifying respirator cartridges are not created equal. You must fully understand the limitations of each type of filtering system that you may be using.

Like any other type of filter, as these air purifying units are used, they become clogged, causing a decrease in air flow through the unit, which is recognized by the user as difficulty in breathing. The user works harder to breathe through the clogged filter thus causing rapid fatigue. Once this occurs, the user should exit the contaminated area and change the filter.

Another type of filtering system may be used for airborne contaminants such as vapors or gases. These contaminants are filtered from the atmosphere with substances such as activated charcoal. This kind of filter **adsorbs** or **absorbs** the contaminants and traps them before they are inspired. This filtering process is a result of a chemical reaction that occurs due to an electrical attraction. The activated charcoal carries a positive or negative electrical charge depending on the filter type. The activated charcoal in acid gas filters carries a positive charge in order to attract the negatively charged ions that exist in acidic vapors. Because of the age-old concept that opposites attract, the charged charcoal particles attract and hold the contaminants that are drawn past it. The down side of this type of filtration is that it is only effective when the chemical substance is known. Based on this, it is necessary to have a complete knowledge of the airborne hazard if we are to select the correct filter. If the hazard is not fully known an air purifying respirator cannot be used! If the hazard is known and an approved cartridge is available, we may be able to select a filter to remove harmful dusts, mists, gases, or vapors from the air. This cartridge selection, however, is not solely based on knowing the chemical. In order to safely select the proper filter the wearer must also ensure that the following conditions are met.

1. The atmospheric oxygen concentration must be above 19.5%.
2. The wearer must not be working in a superheated or extremely cold atmosphere.
3. The airborne contamination levels must not be above immediately dangerous to life and health (IDLH) conditions.
4. The material for which the cartridge is used against must have good warning properties.

Table 6-2 lists available cartridges typically offered by most manufacturers. As you review the list of cartridges, notice that many provide protection from a combination of materials.

All chemical cartridges become saturated over time. The amount of time depends on a number of factors including the type of filter and the amount of con-

---

Safety
● All air purifying respirator cartridges are not created equal.

adsorb
a process or reaction that causes molecules to adhere to the surface of another substance

absorb
a process whereby one substance penetrates into the structure of another

Safety
● If the hazard is not fully known an air purifying respirator cannot be used!

Safety
● At no time can an air purifying system be used in the presence of an unknown material.

Table 6-2 *Guidelines for selecting APR cartridges.*

Type of Cartridge	Color	Typical Protection Against
Organic vapor	Black	Organic vapors.
Acid gas	White	Chlorine, hydrogen chloride, sulfur dioxide, or formaldehyde.
Organic vapors/acid gases	Yellow	Organic vapors, chlorine, hydrogen chloride, sulfur dioxide, or chlorine dioxide, hydrogen fluoride, or hydrogen sulfide (escape only).
High efficiency particulate air (HEPA) dust, mist, fumes	Purple	Dusts, fumes, and mists having a time-weighted average (TWA) less than 0.05 mg/m^3 including asbestos, radon daughters attached to these dusts, and radionuclides.
Organic vapors, HEPA	Black/purple	Organic vapors and dusts, fumes and mists with a TWA less than 0.05 mg/m^3, asbestos fibers, radon daughters attached to these dusts, particulate radionuclides, and pesticides.
Organic vapors/acid gases, HEPA	Yellow/purple	Organic vapors, chlorine, hydrogen chloride, sulfur dioxide, or chlorine dioxide, hydrogen fluoride or hydrogen sulfide (escape only); and dusts, fumes, and mists having a TWA less than 0.05 mg/m^3 including asbestos, radon daughters attached to these dusts, radionuclides, and pesticides.
Acid gases/HEPA	White	Chlorine, hydrogen chloride, sulfur dioxide, and dusts, fumes, and mists having a TWA less than 0.05 mg/m^3 including asbestos, radon daughters attached to these dusts, and radionuclides.

!**Safety**
Do not store respirator cartridges out of their original packages. Doing so reduces the overall life expectancy of the unit.

breakthrough
the passage of a substance from one side of a barrier to the other

tamination in the air. The time it takes for a cartridge to become saturated is largely unpredictable, and the process of adsorbing or absorbing airborne particles begins once it is exposed to the air. Unused cartridges come in tightly sealed, airtight containers designed to protect the filters prior to use. Once open, the cartridge begins the process of chemically treating the air to which it becomes exposed. APR cartridges have a limited shelf life, and due to packing imperfections, may even be working when contained within their sealed package. Do not store respirator cartridges out of their original packages. Doing so reduces the overall life expectancy of the unit. Once used, many systems require that the cartridge be discarded because it might not be effective at the next use.

Once the cartridge reaches its maximum usage, and **breakthrough** occurs, the user must leave the contaminated atmosphere as soon as possible. For this reason, one of the key conditions that must be in place prior to using an air purifying respirator, is to ensure that the material being filtered has **warning properties.** Warning properties are the physical characteristics of a chemical that would alert the wearer of their presence in the absence of respiratory protection. These properties

Chapter 6 Respiratory Protection

warning property
a physical characteristic of a chemical compound that can be detected in some fashion by humans

> **⚠ Safety**
> Warning properties are the physical characteristics of a chemical that would alert the wearer of their presence in the absence of respiratory protection.

> **⚠ Safety**
> If a tight seal is not maintained, the negative pressure within the mask will allow contaminants to enter through the gaps rather than through the filter system.

> **⚠ Safety**
> It's worth repeating: A knowledge of the warning properties of the hazards in the area is vital if we are to ensure that our mask is protecting us and that no exposure is occurring.

typically include such things as odor, or taste of the material, which would provide the necessary warning that the substance is present in the air. In some cases, skin or eye irritation may warn of breakthrough. Therefore respirators for carbon monoxide or other colorless, odorless, tasteless gases are not indicated.

Ammonia is a good example of a substance that has very distinctive warning properties. When exposed to levels as low as 0.05 ppm a person will note a distinctive odor or suffer minor eye irritation. When working in an air purifying respirator in an atmosphere containing ammonia, the wearer would immediately leave the area if they detect any irritation or odor within the mask. Such warning properties indicate that exposure to the material is occurring or that the mask is not properly sealed to the face.

In addition to the possibility that breakthrough is occurring, the detection of some of the material's warning properties could indicate another problem with the APR. If we think about the conditions within the APR as we inhale, it is easy to understand that a momentary negative pressure is created in the mask. This negative pressure is expressed relative to the ambient air pressure in the outside atmosphere. The presence of this negative pressure within the mask makes it very important to ensure that a proper seal occurs between the mask and the wearer. If a tight seal is not maintained, the negative pressure within the mask will allow contaminants to enter through the gaps rather than through the filter system.

Other problems associated with equipment malfunction can lead to the failure of the respirator. Examples of some of these malfunctions include problems associated with exhalation valves, facepiece seals, or ineffective straps for securing the APR to the face. In all cases, detection of the warning properties would require the wearer to immediately leave the area and correct the cause. It's worth repeating: A knowledge of the warning properties of the hazards in the area is vital if we are to ensure that our mask is protecting us and that no exposure is occurring.

With so many possibilities for failure, the use of an air purifying system must only be done in those instances where the level of atmospheric contamination is minimal. OSHA has established values that are referred to as *protection factors,* which denote the acceptable level of the airborne contaminants for which various types of respiratory protective equipment may be used. Such protection factors used by OSHA are based on the permissible exposure limit (PEL) of the material and are found in Table 6-3. Note that the accepted level of contamination that can be worked in with an APR can be many times greater the PEL. It may sound complicated at first but essentially protection factors allow you to establish another limit besides IDLH for a condition of use. For example, a half-face dust mask will allow you to work in a level 10 times the PEL for the material present in the air. As you see, a protection of 10× is illustrated beside "half-face, dust" under the heading of "particulate removing" type of respiratory protection. In order for protection factors to be useful, the contaminant must be identified, the work atmosphere fully monitored, and the PEL for the specific substance must be identified. Once those items are complete, the protection factors can be applied as explained above. At no time can this level ever exceed the IDLH of the material.

Table 6-3 *OSHA protection factors.*

Air Purifying
    Particulate removing
        Single-use, dust 5×
        Quarter face, dust 5×
        Half face, dust 10×
    Gas and vapor removing
        Half-face 10×
        Full-face 50×

Atmosphere supplying
    Supplied air
        Half face, demand 10×
        Full face, demand 50×
        Half face, continuous 1,000×
        Full face, continuous 2,000×
    Self-contained breathing apparatus:
        Pressure demand 10,000×

> **Safety**
> ● At no time can this level ever exceed the IDLH of the material.

Another class of respiratory contaminant is that of a lack of oxygen. Because APRs do not supply any additional oxygen, there must be sufficient amounts in the atmosphere for the wearer to breathe. Table 6-1 shows us the ill effects of low oxygen levels in the body. Note that normal air contains approximately 21% oxygen. At this level, no harmful effects on the body are present, and normal respiratory effort and function is occurring.

As the level of oxygen drops, physiological changes start to happen. The first of the effects noted begins when the oxygen level drops below 17%. At this point we see that some effects involving the brain begin to occur. Below 17%, our brain, which needs more oxygen than any other major organ, begins to have problems. These initial effects are ominous in that they involve the loss of some of our reasoning power and are expressed as a state of mental confusion. Because of this, OSHA specifies that there be a minimum of 19.5% oxygen in the atmosphere, and that air purifying respirators not be used when this level cannot be maintained. To wear an APR in an oxygen deficient atmosphere would be like trying to breathe with your head in a plastic bag. Therefore it becomes important to monitor the oxygen levels within the atmosphere unless it can be clearly determined that oxygen deficiency is not possible. Examples of conditions where such monitoring may not be needed include outdoor areas or other places that are known to be well ventilated and open.

> **Safety**
> ● To wear an APR in an oxygen-deficient atmosphere would be like trying to breathe with your head in a plastic bag.

Chapter 6 Respiratory Protection

The last class of respiratory hazard is that of heat and cold. Since the air drawn into the mask of an air purifying system is only filtering the specific chemical hazard, air purifying systems do little to provide thermal protection. In those instances where extremely hot or cold temperatures are likely to be encountered, air purifying systems cannot be used. APRs can provide some protection against specific types of respiratory hazards. They are not effective in areas of high heat, low oxygen, or whenever the material is not known.

While they do have serious restrictions limiting their use, it should be noted that such systems are often the best to use when it can be done safely and the conditions noted previously are met. Some of the advantages of air purifying respirators include the following:

> **Safety**
> APRs can provide some protection against specific types of respiratory hazards. They are not effective in areas of high heat, low oxygen, or whenever the material is not known.

- Air purifying systems are small, lightweight, and allow freedom of movement far above that of other types of respirators. Their use does not limit the wearer to enter tight areas nor do they result in a significant burden on the wearer due to weight or changes in the wearer's center of gravity.
- Far less training in the use of an air purifying respirator is needed over other types of respirators.
- Maintenance on air purifying systems is considerably less than with other types of respirators. Often the maintenance involves simple cleaning and replacement of the cartridge.
- The cost of an air purifying system is significantly less than air supplying systems. This cost includes both initial purchase and ongoing maintenance.

TYPES OF AIR PURIFYING SYSTEMS

Half-Face Air Purifying Respirators

Although all air purifying respirators are similar in the manner in which they work, there are two major types, each having their own advantages and disadvantages. The first type of air purifying respirator is the half-face respirator. An example of one of these units is shown in Figure 6-2. As can be seen, a half-face respirator provides a mask that covers the mouth and nose only. As we will discuss later, this feature has both good and bad points. Like the other types of respiratory protection, a half-face respirator is attached to the face using straps that go around the head. Unlike full-face respirators, the straps of the half-face respirator are limited and do not ensure as snug a fit. As we will discuss later, this is also both a good and bad feature of such units.

While it does have some limitations, the half-face respirator offers a number of advantages over the full-face type. These include the following:

- Half-face respirators are quite easy to use and require even less training than other respirators.

- Half-face units are some of the most comfortable to wear because they are not as tight on the face, do not significantly limit visibility, and are usually softer and more pliable than other types of units. Given this, it is more likely that people may wear them.
- Additionally, because half-face units do not cover the eyes, the wearer does not have any restriction to the peripheral vision or have a fogging problem. Additionally, the individual who wears eye glasses may wear them in conjunction with safety glasses.

Obviously, the half-face systems are not without problems. These problems are as follows:

- Half-face respirators may not provide 100% protection for the eyes. This limitation often requires the wearing of some other form of eye protection because many types of respiratory hazards are also hazardous to the eyes and mucous membranes.
- The straps on a half-face respirator do not allow the respirator to be drawn as tightly to the face of the wearer as is possible with full-face units therefore the protection that a half-face respirator provides is less than its full-face counterpart, even when using the exact same cartridge. This is evident as we review the protection factors noted in Table 6-3.
- Because the half-face respirator is an air purifying type, it has the same limitations relative to use as with other air purifying types.

A recommended donning procedure for a half-face respirator is shown in Figure 6-3.

Figure 6.2 *A half-face respirator properly worn.*

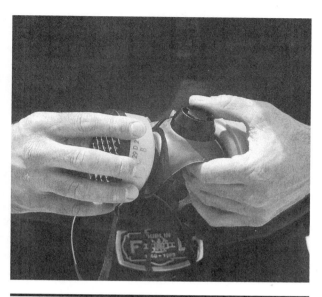

Figure 6.3 *Procedure for donning a half-face respirator. (a) Inspect cartridge for NIOSH and MSHA stamp of approval and ensure the appropriate cartridge is selected.*

(b) Firmly attach the cartridge to the unit. Do not over tighten, but make sure the fit is secure.

(c) Place the unit firmly over the mouth and nose and stretch the straps over the top of the head.

(d) Pull the straps tight until the unit is securely fastened to the face. Rotate the head in several directions to assist in attaining the final fit.

Full-Face Air Purifying Respirators

A full-face APR attaches to the head and face in a more secure manner than the half-face system. It generally utilizes either four or five straps which attach to the mask from the top of the head to the chin. With this arrangement of straps, the full face unit fits tightly on the face of the wearer. A review of Figure 6-1 will reacquaint you with the general look of a full-face APR. Some of the full-face units have an additional nose cup feature within the mask. The addition of the nose cup provides a second seal within the mask and reduces the potential for fogging of the mask.

Advantages of the full-face air purifying masks over the half-face models include the following:

- A tighter seal is possible with the full-face unit over the half-face model. Additionally, the eyes are protected within the perimeter of the mask.
- The use of a full-face system provides full protection for the face and eyes, eliminating the need for other types of facial protection.
- Some manufacturers have conversion kits to adapt the full-face air purifying units to their air supplying units. Such conversion kits reduce the need to have two types of masks and to perform fit testing with two different masks.

Disadvantages of the full-face units are primarily related to the problems associated with air purifying systems in general. Other disadvantages noted between the full-face and half-face units include the following:

- Because of the configuration of the mask, full-face units are often prone to fogging. Once fogged there is no means of defogging an air purifying mask other than to remove it and clean it manually. To prevent fogging, many units are equipped with nose cups. This helps to channel the high-moisture-content expired air into the exhalation valve.
- Full-face masks cause some loss of peripheral vision. This loss varies with the design of the masks but can be significant, resulting in the loss of contact with parts of our environment.
- Full-face units are more expensive than comparable half-face models.

Full-face APRs provide a high level of facial protection in addition to providing the user with a limited degree of respiratory protection.

The donning procedure for a full-face respirator would be the same as for a half-face unit. For a review, see Figure 6-3.

Respirator Selection

It should be obvious by now that the type of atmosphere or contaminant you will encounter dictates the type of respiratory protection you must wear. Although we have highlighted some of the conditions for using an APR, the following criteria

> **!Safety**
> Full-face APRs provide a high level of facial protection in addition to providing the user with a limited degree of respiratory protection.

should reinforce the concepts. Whether you are dealing with a hazardous material spilled on the roadway or a waste site requiring clean up, the rules are the same. Using the wrong gear can be just as harmful (sometimes even more harmful) as using no protection at all.

To avoid this problem, many of us believe that we should simply wear the most protective units available and not worry about it. This philosophy is flawed in that more is not always better. For example, when we use a self-contained breathing apparatus (SCBA) in a situation where no respirator is necessary, or only a half-face APR would provide adequate protection, we give away a degree of safety, comfort, and cost, that may not be necessary. SCBA units restrict our mobility, upset our center of gravity, require significant care, maintenance, and training. Remember that the safest level is not always the most protective level. With this in mind, we should choose the lowest level of respiratory protection that is appropriate for the hazard. We are aiming for the lowest level that can be used to safely to enter the site and do our work. Keep in mind we are not trying to "get away with" a lower level of protection, we are simply attempting to use the most appropriate equipment.

Following is a discussion on the thought process that can be used for selecting an air-purifying respirator. While it is not a perfect system, it provides us with a good starting point from which we can begin our process of respiratory protection selection. While written in a somewhat simplified format, the process must be approached with the utmost care. If there is ever a doubt regarding the answer to any of these questions, err on the side of being more conservative. It is your life, or the life of other co workers, that rests on the proper decision regarding this subject. Here are the six questions that will help us in our selection process.

1. Is there any type of respiratory hazard present in the work area?

First we need to determine the presence or absence of airborne contamination. As we ask this question, we not only need to determine the presence of airborne contamination, but also whether the materials are hazardous via the inhalation route. Recall that all materials do not necessarily pose a respiratory hazard.

Hopefully, what we see from outside the contaminated area may key us in on the presence of a respiratory hazard. Visible clouds or dusts are dead giveaways that some type of respiratory protection may be required. A gaseous release inside a building versus those occurring outside can vary our selection of respiratory protection. An indoor release with poor ventilation can be highly hazardous due to the accumulation of vapors in dead air space. Just because we can smell the material does not mean that we are exposed to harmful levels. In our previous discussions, we have made the point that odor threshold may be lower than the PEL for the material present. Thorough air monitoring may be required to accurately determine the levels of airborne contamination.

If we find that there are no respiratory hazards present, and that our work will not create unsafe levels, it is appropriate to consider that the most appro-

> **Safety**
> Remember that the safest level is not always the most protective level.

> **Safety**
> Just because we can smell the material does not mean that we are exposed to harmful levels.

priate level of protection may in fact be no respirator at all. This still may require that we wear other types of personal protective equipment, but if there are no respiratory hazards present, no respiratory protection is necessary.

If the answer is "yes," "unknown," or that during the course of work a respiratory hazard is created, we must then proceed with our decision matrix to determine the proper type of respirator. The first question to ask in determining the level of respiratory protection is whether there are unsafe levels of airborne contaminants present at the site.

> **!Safety**
> The first question to ask in determining the level of respiratory protection is whether there are unsafe levels of airborne contaminants present at the site.

2. Do we know what specific materials are present in the area to be entered?

The second question to be addressed is whether specific materials are present at the work site. Keys to making this determination include many of the items in Chapter 5, which discussed identification systems. Other things that we should take into account in emergency response include shipping papers, manifests, the type of containers, and information obtained from witnesses or workers at the site.

In fixed facilities releases, we can generally gather a great deal of information by talking with the people who currently work in the area. Often, the persons who work in the area will have specific knowledge regarding the chemicals in the area at the time of the release. In other instances, we can obtain information about the materials as we study the type of processes that ordinarily take place in the area, or when we review the records of the previous uses of the site. This is particularly helpful in the case of the cleanup of an abandoned waste site. A review of the records often reveals the processes or materials used.

If in our review of the chemical we can specifically identify the substance, we can then proceed with the rest of the decision process, and still include the use of an air purifying respirator. If, on the other hand, we cannot positively identify the material present, we must use a level of respiratory protection that is not product specific.

3. Is there a suitable air purifying respirator cartridge for the expected contaminant?

We must ascertain that there is an appropriate cartridge available for use with the material identified in question 2. In order to check this, we must find out from the manufacturer if it makes a cartridge for the identified hazard. All of the cartridges are listed and approved by the National Institute for Occupational Safety and Health (NIOSH) and **MSHA** (Mine Safety and Health Administration) and can be employed if the factors for use are favorable.

If, in answering this question, we find that there are no approved respirator cartridges available, or that all the conditions for APR use are not met, we can effectively exclude the use of an air purifying respirator. On the other hand, if a suitable cartridge is available and all conditions for use are met, we can proceed with the rest of our decision matrix.

> **MSHA**
> Mine Safety and Health Administration

4. What is the specific level of airborne contamination present at the site?

Once we have determined that there are airborne hazards we must now accomplish the following actions:

- Confirm that we have identified the material found in the area.
- Establish that the airborne contamination levels are above the PEL.
- Secure an approved respirator cartridge.

Next, it is necessary to determine the specific level of contamination that may be present in the atmosphere. This determination may seem similar to the first of our questions, but it is much more specific and must often be determined through the use of specialized instruments.

In the first question, we simply found out that a chemical hazard was present above acceptable levels. In determining this, we found that some type of respirator must be worn. As we answer this fourth question, it is necessary to determine the specific levels of contamination present so that we can more appropriately define the type of respirator needed.

To make this determination, it may be necessary to monitor the area to be entered, especially when the area is enclosed, if the amount of contamination is expected to be high, or if the material has a high **vapor density**. In such cases, the material is likely to be found in concentrations that may exceed our protection factors, or worse, to exceed the IDLH of the substance.

If it turns out that the level of airborne contaminants is above the IDLH for the material, or even above our protection factors for the type of respirator worn, we must go to a higher level of respiratory protection. In all cases, levels above the IDLH value necessitate the use of an air supplied unit. If, however, the levels are below the IDLH and our protection factors, we then can still consider the use of the air purifying system.

5. Is there any type of thermal extreme that would prohibit the use of an air purifying system?

It is important to determine the possibility of thermal extremes in the area to be entered. Remember that while air purifying systems are suitable to remove many contaminants from the atmosphere, they do little or nothing to remove the hazards associated with high or low temperatures.

Air supplying systems, because their source of air is independent of the atmosphere, are adequate to protect the wearer from such temperature extremes. It is partly for this reason that firefighters wear SCBA units during fire ground operations. These air supplying systems provide breathing air at moderate temperatures while protecting the respiratory tract from atmospheres that can exceed 1,000°Fahrenheit.

6. Are adequate oxygen levels present in the area to be entered?

---

vapor density
the weight of a vapor or gas compared to the weight of an equal volume of air

It would be wise for us to review the earlier information regarding oxygen levels. As we found, the oxygen level in a normal, uncontaminated sea level atmosphere is about 20.9%. Small reductions in this level may go undetected and do little or nothing to affect us. Levels below 19.5% are determined to be oxygen deficient by OSHA standards and require the use of air supplying respirators. As the oxygen level drops to 17% or below, we begin to experience physiological effects that can eventually lead to serious injury or death. Air purifying respirators do not add oxygen to the inspired air and cannot be used in oxygen deficient atmospheres. In such cases, an air supplying system must be selected. The following section explains those air supplying units and identifies appropriate conditions for use.

> **Safety**
> Levels below 19.5% are determined to be oxygen deficient by OSHA standards and require the use of air supplying respirators.

AIR SUPPLYING RESPIRATORS

Air supplying respirators are the most complete means of providing respiratory protection whenever there are inhalation hazards present. Air supplying units derive their name from their function. These units *supply* air to the wearer by one of two primary means.

Figures 6-4 and 6-5 show examples of each of the two major types of supplied air systems. Figure 6-4 illustrates the self-contained breathing apparatus (SCBA) while Figure 6-5 shows an *umbilical* type of system. The umbilical system derives its name from the long hose or umbilical cord that supplies air to the worker. As we can see, both types provide air to the mask from a tank and hose configuration, air compressor, or a combination of the two. Because each of these units provides clean air directly to the user, they do not have the same limitations as the air purifying systems. In either case, the air that is breathed by the wearer is independent of the atmospheric conditions of the environment.

> **NOTE**
> The umbilical system derives its name from the long hose or umbilical cord that supplies air to the worker.

For obvious reasons, OSHA mandates that the air used within any air supplying system meet certain minimal specifications. Air supplied to the user directly or indirectly comes from an air compressor. The air must be free from harmful levels of contaminants that exist when compressors are used to supply air either directly to the wearer or into a tank. Compressed air, compressed oxygen, liquid air, and liquid oxygen used for respiration should be of very high purity since the life of the wearer depends on it. OSHA mandates that the quality of the breathing air shall meet at least the requirements of the specifications for Grade D breathing air as described in Compressed Gas Association Commodity Specification G-7.1, 1966. In order to verify the quality of air as Grade D, the air source should be tested regularly. National Fire Protection Association (NFPA) Standard 1404 states that the air quality of all air supplying systems shall be tested at least every 3 months by a qualified laboratory. Examples of potential contaminants include carbon monoxide and vaporized oils and lubricants that are liberated from the operating compressor. In other words, exhausts and other sources of contamination should not have an opportunity to be part of the make up air for supplied air systems. Image the catastrophe that would occur if large amounts of carbon monoxide were unknowingly allowed into a SCBA tank.

> **Safety**
> NFPA Standard 1404 states that the air quality of all breathing systems shall be tested at least every 3 months by a qualified laboratory.

Chapter 6 Respiratory Protection

Figure 6.4 *A properly worn SCBA before putting on the mask.*

Figure 6.5 *Worker illustrating a supplied air umbilical system.*

> **Safety**
> Air supplying systems provide the highest degree of thermal protection available because the air breathed in the system is carried with the user or originates outside the area.

A properly functioning air supplying system offers huge advantages over the air purifying type of respirator. These advantages include:

- Air supplying systems provide protection from all types of dusts, fumes, mists, gases, vapors, and particulates. They are not product specific and the hazards within the area do not need to be known.
- The concentration of the contaminant in the air does not need to be known. Contamination levels above IDLH are not an issue for a properly fitted and functioning air supplying respirator system.
- The oxygen level within the area does not exclude the use of an air supplying respirator.

Air supplying systems provide the highest degree of thermal protection available because the air breathed in the system is carried with the user or originates outside the area.

positive pressure
a characteristic of supplied air systems to maintain a few psi of air pressure inside the face piece even after the wearer has inhaled

- The temperature of the air within the hazard zone is of little concern to the user of an air supplying system.
- Air supplying respirators have the potential to provide **positive pressure** within the mask. Such positive pressure further increases the protection afforded to the user because there is less chance that hazardous contaminants can be drawn into the mask even when a leak is present.

Air supplying systems provide the maximum level of respiratory protection against the five types of respiratory hazards.

Although well suited for almost all types of hazardous atmospheres, the use of an air supplying respirator is not automatic. These units have some disadvantages that limit their application. As we discovered in our discussion on respirator selection, one should always try to use the lowest level of respiratory protection needed to safely enter an area. Like so many issues in emergency response and hazardous waste operations, there are always some down sides to any decision. In this case, a higher level of protection is not necessarily the safer level of protection. This will become even more obvious in Chapter 7 when we discuss the levels of personal protective equipment. Examples of some of the limiting factors with air supplying systems are as follows:

> **Safety**
> Air supplying systems provide the maximum level of respiratory protection against the five types of respiratory hazards.

- All types of air supplying respirators are awkward to wear and limit the mobility of the user. This can restrict access to some areas and limit the ability to escape in the event of an emergency.
- Air supplying respirators add considerable weight to the user. The additional weight can cause fatigue and further reduce the effectiveness of the worker. For example, SCBA units are heavy with some weighing up to 35 pounds. This weight can take a toll on the wearer especially if he or she is also using other protective equipment.
- Air supplying systems can be very expensive. A single SCBA unit may cost up to $3,000. Air line systems can cost even more.
- Because of their complexity, they have higher maintenance requirements than air purifying systems. This increased maintenance results in increased down time, higher costs, and increased potential for failure of the units.
- Almost all types of air supplying units have a limited air use duration. Most SCBA units are rated to provide between 30 and 60 minutes of air, although some higher time units are available. Although rated to provide this level of use, practical experience shows that the actual use time is between 20 and 40 minutes.

Self-Contained Breathing Apparatus

There are two major types of air supplying respirators. In the first type, the air is stored in a tank that the worker wears into the hazardous atmosphere. The tank is carried on the back and held in place with a system of straps. The tank supplies air through hoses and pressure regulators to the mask of the user. All of the air breathed comes directly from the tank with no appreciable air coming into the mask from the outside atmosphere. Everything needed to ensure that the air is safe to breathe is contained on the wearer. For this reason, these units are known as self-contained breathing apparatus (SCBA).

Although there are some major differences in the types of systems available, all SCBA units share one common feature: the presence of a low pressure warning device that is used to alert the user of the limited amount of air left in the system. In most cases, this time is set by the manufacturer at approximately 5 minutes. With 5 minutes of air remaining in the system, the warning device activates and alerts the user to leave the contaminated area. Warning devices can either be bells, whistles, or even a constant vibration indicating that the system is low on air.

> **Safety**
> Warning devices can either be bells, whistles, or even a constant vibration indicating that the system is low on air.

SCBA units come in a variety of shapes and sizes and are also divided into two major subgroups. The two groups are differentiated by the manner in which the air is maintained within the unit. In the most common type, compressed air ranging from 2,216 pounds per square inch (psi) up to 4,500 psi is put into a tank, which is then placed in a backpack system to hold it on the user. The size of the tank varies depending on the amount of air required for the time duration of the pack. A standard size tank with 2,216 psi contains approximately 47 cubic feet of air. This provides the user with around 30 minutes of work time although actual use is often far less than the rating.

If we double the pressure within the tank, we find that we can get the same volume of air into smaller cylinder (remember Boyle's gas law?). The smaller tanks are considerably lighter and allow access into more confined areas. Because the time rating of the tank is dependent on the amount of air within the tank expressed in cubic feet, the smaller tanks can have the same rating as larger ones. The advantage is that the cylinders are smaller and lighter, which results in less fatigue for the worker.

Whether high pressure or low, the basic unit provides air to a pressure regulator or series of regulators that provide the necessary air flow to the mask. Regulated air enters the mask via a low pressure hose and leaves through an exhalation valve, in essence, creating an open system. This characteristic of allowing the air to be exhaled into the atmosphere is the basis for the name of this type of system—the *open circuit* type of SCBA.

The second major type of unit which has come into and out of favor is the *closed circuit* type. Closed circuit, or rebreathing systems, are composed of a backpack and tank, and employ a mask similar to that of the open circuit type. This is where the similarities between the two units end. The basic operational principle of the closed circuit units is that they operate by recycling the air breathed by the wearer. The operational principles of a closed circuit unit is to capture and recycle the expired air within the "closed" nature of the system. As air leaves the mask, it is collected in the backpack where it is filtered to remove

> **■ NOTE**
> The basic operational principle of the closed circuit units is that they recycle the air breathed by the wearer.

carbon dioxide. After it is filtered, a small amount of oxygen is added from an attached tank. This tank is an integral part of the system and is included in the backpack assembly. This system is far different than the open circuit unit which only uses compressed air and no additional oxygen.

The primary advantage of the closed circuit system is that time ratings of up to 4 hours are possible, although 1-hour and 2-hour units are more popular. This increase in time is largely due to the recycling of expired air.

Such advantages would seem to be significant, but the closed circuit units have largely disappeared from the hazardous materials industry. The closed circuit units are costly to purchase, require more maintenance, and create a concern due to the introduction of a strong oxidizer (oxygen) into hazardous atmospheres. The reality is that a cylinder change often provides a much-needed break for the wearer. Although such units have the potential to become the operational standard for prolonged work periods, they are not without downsides. One of these downsides revolves around that fact that few personnel can stand to be wrapped up in personal protective equipment (PPE) for hours on end.

Whether open or closed circuit, SCBA units are the leading type of air supplying respirators. Such units provide the following advantages over other types of air supplying equipment:

- SCBA units provide maximum flexibility when entering areas because all of the breathing air is contained on the back of the user.
- SCBA units are one of the most commonly used types of systems, thus providing us with the opportunity to standardize training and maintenance.

Disadvantages of SCBA units that are not present with air line units include:

- SCBA units are most limited in the amount of time that they provide. Most units are limited to a maximum of 30 minutes.
- SCBA units are some of the most heaviest of respiratory protection equipment.
- Because of their weight, SCBA units cause the user to have a higher center of gravity leading to loss of balance and added wear and tear on the user.

Donning Procedures There are several interpretations and methods accepted for donning a SCBA. Your local custom or type of unit may require a deviation from the following illustration. Regardless of the variations in SCBA units or donning procedures, it is important to conduct a quick fit check to ensure that the unit is properly sealed to your face.

To conduct a positive fit check, take in a deep breath, then place your hand over the area of the exhalation valve. In some cases, the intake for the air supply may also have to be sealed. While placing your hand over the valve, exhale with enough force to slightly push the mask away from your face. Once the mask is lifting, stop blowing and see if the unit remains away from your face or whether it

> **Safety**
> A cylinder change often provides a much-needed break for the wearer.

returns to its natural position. If it returns, it could indicate a leak in the seal. If this were to occur, it is important to reposition and/or tighten the straps to ensure that a proper seal is in place. Once the mask and/or straps are repositioned, repeat the test.

A negative pressure test is much like the positive test in that once the unit is in place, the wearer simply creates a negative pressure within the mask by sealing the inlets for air to enter the mask and forcefully inhaling. In the case of an APR, the inlets are where the cartridges are attached. With an SCBA unit, the air intake line will need to be covered. Some types of SCBA units do not allow you to conduct this test in this fashion. In such cases, the manufacturer should be consulted for proper procedures to be followed. At any rate, it is important to ensure that once a negative pressure is created in the mask, it maintains an adequate seal around the face.

A generally accepted sequence of events for donning an SCBA is shown in Figure 6-6.

■ **NOTE**
In the case of an APR, the inlets are where the cartridges are attached.

Figure 6.6 *Procedure for donning an SCBA. (a) Inspect all parts of the unit and mask for defects or extreme signs of wear.*

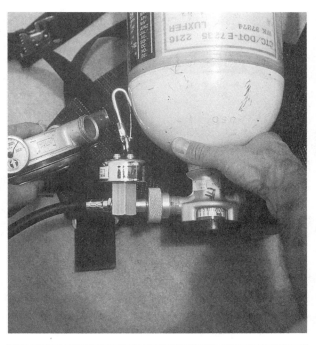

(b) Turn the cylinder "on" and ensure that the unit is full and that the cylinder gauge and sight gauge register the same pressure. A discrepancy may indicate a regulator or other air line problem.

(c) With all straps loose, don the unit. There are several methods for accomplishing this task-choose which one works for you.

(d) Pull the chest straps tight. The base of the unit should rest on your shoulders and the small of your back.

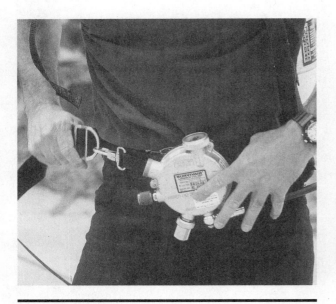

(e) Attach and tighten the waist strap. Tuck all loose ends away so they do not get tangled on anything.

(f) Perform a negative and positive pressure test and attach the head straps. Ensure the face piece is properly fitted and attach the low pressure hose to the regulator. Activate the air supply

Figure 6.6 *Continued*

Doffing Procedure The procedures for doffing, or taking the units off, may vary depending on the type of unit. Regardless of the type used, there are some general guidelines that are usually required.

1. Remove the backpack assembly. If an SCBA is used, the backpack assembly is usually the first component to be removed. This is usually due to the fact that the unit restricts the movement of the wearer and contributes to fatigue. As we will learn later in our discussion of decontamination, the mask is one of the last things taken off.

   To remove the backpack, disconnect the mask from the regulator. If the SCBA has a flexible hose and other chemical protective clothing needs to be removed, it is usually advisable to hold the low pressure hose high and away from the body to ensure that residual contaminants are not drawn into the mask.

2. Remove the face piece. Once the backpack is disconnected from the mask, the removal of the mask needs to be accomplished. In most cases, it is best to remove the unit without trying to loosen the straps, and simply grasp the face piece in the area of the chin. This is best accomplished by lifting the unit upward and pulling it clear of your head. After doing this a few times, you will actually see that there is considerable flexibility within the straps which allow it to pull clear.

3. Service the unit. Maintenance issues are addressed later in this chapter but as a general rule it is recommended that you follow the manufacturers guidelines.

Air Line Respirators

An air line respirator involves the use of a supply line that is continuously attached from the air source to the worker. The air line is dragged behind the worker and acts as an umbilical cord. This line is then connected to a remote source of air, usually a tank, outside the affected area.

A review of the example in Figure 6-4 shows that the only hardware carried by the wearer is a small regulator where the line attaches, the face mask, and a small bottle of compressed air. The small air bottle is known as the *escape pack* and is used to provide an emergency source of air if the primary supply is compromised, or if escape from the area is necessary. The main supply can be lost due to equipment failure, kinking of the air line, or mechanical failure. Another use of the escape pack occurs when the worker needs to leave an area by a different route than the one used to enter. In this case, the user is unfortunately required to disconnect from the air line and activate the secondary air supply from the escape bottle. The escape bottle is used only for emergencies. Therefore, most escape bottles only provide a 5-minute air supply.

Because of their makeup, air line units offer one major advantage over any other type of unit. They provide the highest level of respiratory protection for the

> **Safety**
> The escape bottle is used only for emergencies.

> **■ NOTE**
> The main supply can be lost due to equipment failure, kinking of the air line, or mechanical failure.

longest period of time. No other unit can provide the high level of respiratory protection while offering almost unlimited work time. Air line systems are also beneficial in that they add very little weight to the worker.

> **NOTE**
> Air line systems are also beneficial in that they add very little weight to the worker.

Although it may seem that this type of unit is the final answer, a couple of major disadvantages limit their use and effectiveness. These include the limited distance that can be covered by someone in an air line system, and the fact that they are expensive and hard to maintain in many situations.

RESPIRATORY PROTECTION: MAINTENANCE, STORAGE, AND RECORD KEEPING OF APR'S AND SCBA

Proper maintenance of respiratory protection equipment starts with a commitment to provide adequate protection for all employees. This protection includes developing and implementing an organized respiratory protection program that complies with the requirements of OSHA 29 CFR, part 1910.134, and other relevant nationally recognized standards such as ANSIZ88.2. This ANSI (American National Standards Institute) standard is another guideline for the development of a comprehensive respiratory protection program.

The following information is taken from a review of the OSHA requirements as found in the regulation. For the sake of brevity, the requirements are summarized to make them more understandable and applicable to our study.

Respirators shall be regularly cleaned and disinfected. Respirators used by more than one worker shall be thoroughly cleaned and disinfected after each use. Cleaning and disinfecting procedures are available in the manufacturer's instruction and maintenance manuals. In general, a mild cleaning solution with a disinfectant is usually appropriate. Once cleaned, all respirators should be thoroughly rinsed with clean water and allowed to air dry. Avoid wiping off the face piece with paper towels or other abrasive fabrics. Doing so will scratch the lens and reduce visibility.

> **Safety**
> Respirators used by more than one worker shall be thoroughly cleaned and disinfected after each use.

Respirators shall be inspected during cleaning. Worn or deteriorated parts should be replaced if the person performing the inspection is qualified to do so. Respirators used for emergency use should be thoroughly inspected at least once a month and after each use. Thorough preuse inspections mean a reduced likelihood of failure when the unit is in use. It is important to follow the manufacturer's instructions for inspection and maintenance of each unit.

> **Safety**
> Thorough preuse inspections mean a reduced likelihood of failure when the unit is in use.

Basic inspection of the mask portion of a respirator includes a thorough examination for any evidence of damaged, broken, or worn parts. All valves should be clean and in proper working order. All straps and/or fasteners should operate to fully open and closed positions. All rubber or plastic parts shall be inspected for signs of deterioration.

Inspection of SCBAs and other air supplying systems is considerably more complex. In addition to the inspection of the mask, the cylinders need to be examined and refilled with air as needed. All fittings should be checked for

■ NOTE
After the maintenance has been completed, all straps should be left in the open and loose position to allow for rapid donning of the unit in case of an emergency.

!Safety
Storage should not be in an area where there is direct sunlight because damage to rubber or plastic components may result.

!Safety
Fiberglass wrapped cylinders have a maximum service life of 15 years.

tightness, while the hoses and all straps should be carefully inspected to ensure proper operation. Additionally, all valves should be operated to check for adequate and unrestricted air flow. The low pressure warning system should also be tested to ensure that it is within the manufacturer's guidelines. After the maintenance has been completed, all straps should be left in the open and loose position to allow for rapid donning of the unit in case of an emergency.

After cleaning and drying, all types of respirators shall be stored in a convenient, clean, and sanitary location. When possible, the mask should be placed into a sealed plastic bag to prevent damage and contamination. Storage should not be in an area where there is direct sunlight because damage to rubber and or plastic components may result.

Cylinders that are used to store compressed air for breathing are required to be tested and maintained as prescribed in the *Shipping Container Specification Regulations* of the Department of Transportation (49 CFR part 178). This regulation requires that each steel cylinder be hydrostatically tested every 5 years and fiberglass (composite) wrapped (full or hoop) cylinders be tested every 3 years.

Fiberglass wrapped cylinders have a maximum service life of 15 years. As part of the inspection program, the *service life* of each cylinder should be reviewed to ensure that the cylinder is within its useful life expectancy. Cylinders that go beyond the service life should be removed from active service.

Record keeping is an important element of a respiratory protection program. Records should be kept of all inspection, maintenance, and repairs to each respirator used. Proper record keeping will help to ensure that all required maintenance is obtained as it is needed.

Summary

- The five types of respiratory hazards you may encounter include particulates, gases and vapors, a combination of particulates and gases, oxygen deficient atmospheres, and thermal extremes.
- Oxygen deficiency occurs when ambient oxygen content drops below 19.5%.
- In many cases, engineering controls are a better way to provide for the control of airborne contamination.
- Elements of a respiratory protection program include: Procedures for respirator selection, training for those who wear respirators, inspection, maintenance, and sanitation procedures, storage procedures, and medical surveillance practices. Additionally, the program should provide for atmospheric monitoring, record keeping, and fit testing for all employees who routinely wear respiratory protection.
- Medical exams for those who wear respirators should be provided annually.
- The OSHA standard for respiratory protection programs can be found in 29 CFR, part 1910.134.
- Respiratory protection devices can be divided into two main categories: air purifying and air supplying systems.
- An air purifying respirator (APR) is a respirator that has special filters designed to remove particulates and/or vapors from the air. The units have canisters or cartridges that either physically or chemically purify the air before it is inspired by the user. These units do not add supplemental oxygen when used.
- All of the cartridges are listed and approved by NIOSH and MSHA (Mine Safety and Health Administration).
- In order to safely use an APR, the following conditions must be met:
  - The product for which the cartridge is used must be known.
  - The chemical for which the cartridge is to be used must have adequate warning properties.
  - The airborne contamination levels must not be above IDLH values.
  - The atmosphere must have oxygen levels at or above 19.5%.
  - There can be no thermal extremes in the environment to be entered.
- OSHA has established values that are referred to as *protection factors,* which denote the acceptable level of the airborne contaminants for which various types of respiratory protective equipment may be used. Such protection factors used by OSHA are based on the PEL of the material.
- APRs are usually found in the full-face and half-face styles.

- The questions to ask prior to selecting a respirator include:
  1. Is there any type of respiratory hazard present in the workplace?
  2. Do you know what specific materials are present in the area to be entered?
  3. Is there a suitable APR cartridge for the contamination present?
  4. What are the airborne contamination levels?
  5. Are there any thermal extremes?
  6. Is there adequate oxygen in the work atmosphere?
- Air supplying respirators include open and closed circuit, self-contained breathing apparatus (SCBA), and air line systems.
- Air supplying systems provide the maximum level of protection against the five types of airborne contamination.
- Air supplying respirators have the potential to provide *positive pressure* within the mask. Such positive pressure further increases the protection afforded to the user because there is less chance that hazardous contaminants can be drawn into the mask even when a leak is present.
- An open circuit system is one in which the expired air of the user exits the mask through an exhalation valve. Closed circuit systems capture the exhaled air, chemically treat it to remove the CO_2, add oxygen, and return it to the system.
- All air line systems must have an escape cylinder included in the system. These cylinders typically prove 5 minutes of breathing air to the user.
- OSHA mandates that the quality of the breathing air shall meet at least the requirements of the specifications for Grade D breathing air as described in Compressed Gas Association Commodity Specification G-7.1, 1966. In order to verify the quality of air as Grade D, the air source should be tested regularly. NFPA Standard 1404 states that the air quality of all breathing air systems shall be tested at least every 3 months by a qualified laboratory.
- All respirators should be thoroughly cleaned and inspected after each use. Once cleaned, all respirators should be rinsed and allowed to air dry.
- Cylinders that are used to store compressed air for breathing are required to be tested and maintained as prescribed in the *Shipping Container Specification Regulations* of the Department of Transportation (49 CFR part 178). This regulation requires that each steel cylinder will be hydrostatically tested every 5 years while fiberglass (composite) wrapped (full or hoop) cylinders be tested every 3 years. Fiberglass wrapped cylinders have a maximum service life of 15 years.

Review Questions

1. Under normal operations, how much breathing time will a 2,216 psi air pack provide?
   A. 15 minutes
   B. 22 minutes
   C. 27 minutes
   D. 30 minutes
2. Name four conditions that must be present before deciding to use an air purifying respirator.
3. What are the two main types of respiratory protective equipment?
   A. APRs and particulate masks
   B. Filters and canisters
   C. Air purifying and supplied air
   D. SCBAs and air line systems
4. According to OSHA, an oxygen deficient atmosphere is one that is
   A. At or above 20.9%
   B. Below 19%
   C. Below 19.5%
   D. Above 23%
5. List the five types of respiratory hazards.
6. Which of the following OSHA standards contains the requirements for respiratory protection programs?
   A. 29 CFR, part 1910.134
   B. 29 CFR, part 1910.120
   C. 29 CFR, part 1930.120
   D. 29 CFR, part 1910.100
7. True of False? Engineering controls should be looked at to remove airborne contaminants whenever possible.
8. List two advantages of an APR.
9. Identify the basic components of an SCBA and describe how it differs from an air line system.
10. True or False? Medical exams for those who must wear respiratory protection should be performed annually.
11. Steel SCBA cylinders must be hydrostatically tested
    A. Every 2 years
    B. Every 3 years
    C. Every 4 years
    D. Every 5 years
12. True or False? Fiberglass wrapped cylinders have a maximum service life of 15 years.
13. OSHA mandates that breathing air shall meet at the requirements and specifications for _____ breathing air as described in the Compressed Gas Association Commodity Specification _____.
    A. Grade A, AB 970
    B. Grade A, G-7.1, 1966
    C. Grade D, G-7.1, 1966.
    D. Grade D, AB 970
14. Which of the following NFPA standards requires that all breathing air supply systems shall be tested at least every 3 months by a qualified laboratory?
    A. NFPA 1404
    B. NFPA 1500
    C. NFPA 704
    D. NFPA 1400
15. True or False? Any respirator cartridge will protect the wearer against all airborne contamination.
16. True or False? After cleaning and rinsing, an SCBA mask should be dried off with paper towels.
17. True or False? Warning properties are those characteristics of a material that would alert a person wearing an APR that the canister has become saturated.
18. True or False? APRs offer a high degree of thermal protection.
19. List two disadvantages of full-face respirators.
20. List three advantages of SCBAs.

Chapter 7

Personal Protective Equipment

Learning Objectives

Upon completion of this chapter, you should be able to:

- Understand the importance and need for personal protective equipment.
- Identify the conditions for use and components of level A protection.
- Identify the conditions for use and components of level B protection.
- Identify the conditions for use and components of level C protection.
- Identify the conditions for use and components of level D protection.
- Understand the differences between permeation, penetration, and degradation.
- Define breakthrough and apply the term to any suit selection.
- Understand the health considerations of wearing chemical protective clothing.
- Understand the importance of the use of compatibility charts.

INTRODUCTION

Although there are many different types of chemical protective equipment, not all are equal in providing protection to the wearer. Some chemical protective suits are more effective at protecting the wearer against acids, whereas others are good at repelling a hydrocarbon exposure. This "one suit does not cover all hazards" reality poses several interesting questions to the novice responder or site worker. Workers and responders alike have commonly asked, "Which suit will offer the most protection from the chemical?" and "How do I know it will give me the protection I need for the duration of the job I am about to do?" These questions are not always easy to answer and may further complicate a decision on gloves, boots, or respiratory protection. PPE stands for personal protective equipment, including but not limited to respiratory protection chemical resistant suits, boots, gloves, splash aprons, face shields, and goggles.

This chapter is designed to help you understand when and why you wear PPE and the limitations of each different combination. Before digesting this information, however, you must accept this one very important concept. PPE is not worn to enable or encourage you to intentionally contact or get otherwise mixed up with the chemicals you are working with. The PPE you are wearing has not made you invincible and impervious to a chemical exposure! If you think you are beyond danger, you will eventually get hurt. The suits and gloves and boots and everything else you are wearing are for one reason and one reason only. If the chemicals happen to get on you by some unfortunate circumstance, you have taken measures to reduce the chance of being exposed and potentially injured. Stomping around in puddles or vapor clouds or intentionally letting your suit get covered with product does not make you the coolest responder or waste site worker on the job; it only shows that you have some desire to risk your life for some silly chemical release. The cleanest person exiting the work site is the winner in the hazmat business because that person has reduced the work of decontamination and helped ensure that he or she will see family and friends at the end of the day. This chapter looks at several different configurations of PPE and the situations that may affect the choice of one level of protection over another.

This text approaches PPE from a practical approach applicable to workers in the field. Additional technical reading on design criteria, performance criteria, test methods, and so forth may be found in the following National Fire Protection Association (NFPA) documents:

NFPA 1991 (1994 Edition)	*Vapor Protective Suits For Hazardous Chemical Emergencies*
NFPA 1992 (1994 Edition)	*Liquid Splash Suits for Hazardous Chemical Emergencies*
NFPA 1993 (1994 Edition)	*Support Function Protective Garments for Hazardous Chemical Operations*

These standards are very rigid and contain information on strict testing and

■ **NOTE**
Although there are many different types of chemical protective equipment, not all are equal in providing protection to the wearer.

■ **NOTE**
PPE stands for personal protective equipment, including but not limited to respiratory protection chemical resistant suits, boots, gloves, splash aprons, face shields, and goggles.

! **Safety**
PPE is not worn to enable or encourage you to intentionally contact or get otherwise mixed up with the chemicals you are trying to clean up.

construction requirements. These requirements must be met by any suit manufacturer hoping to earn NFPA approval and must be available to the purchaser of the garment. Some of the technical information you may likely see in the data package include suit construction fabric, zipper or closure assembly information, seam construction and technique, visor and face shield specifications, cleaning instructions, testing data, repair instructions, and warranty information. Read technical data packages for the specifics on suit performance. If the suit that a field worker selects bears the stamp of approval from the NFPA, it will have been thoroughly tested and inspected before anyone wears it in combat. In short, the suit will be tried and proved to offer protection against the chemicals it is designed to shield you from.

> **!Safety**
> ● Read technical data packages for the specifics on suit performance.

POSITIVE IDENTIFICATION OF PRODUCTS AND ASSOCIATED HAZARDS

All the best personal protective equipment in the world may be worthless unless you can answer "yes" to the following two questions.

1. Have I positively identified the hazardous product and all the hazards associated with it?
2. Have I chosen a type and level of protection appropriate for the hazard?

A "no" answer to either of the above questions may indicate that you are putting yourself at risk and that you do not have a systematic approach to solving your PPE challenge. Unfortunately it may not always be possible to answer "yes" to question 1, which makes the other question subject to much speculation and heartburn. Knowing and understanding your chemical hazards is the key to sound PPE selection. The reality is that there are times when you cannot absolutely identify the chemical you are working with, especially in response work when approaching unknown drums or other containers. In the next section we discuss this potential problem and look at both of these questions individually.

> **!Safety**
> ● Knowing and understanding your chemical hazards is the key to sound PPE selection.

Have I Positively Identified the Hazardous Product and All the Hazards Associated with It?

Much of the information contained in this text deals with basic chemistry and hazard classification. This information is quite important in the big scheme of things as we now come face-to-face with a very brutal reality in hazmat: If you do not know anything about the material you may potentially contact, how can you protect yourself from an incidental exposure? In most cases there is enough information on the scene to determine what the substance is, you just have to pay attention to the clues!

How do you choose the equipment that may save your life if you do not even know the first thing about the chemical? This is a huge issue when you face an unknown drum and have only made a guesstimate as to what the hazards may be. It is scary business because not all suits, boots, or gloves, are created equal. Some construction materials for these items are not compatible with certain types of

degradation
decomposition of a substance at a molecular level

> ⚠ **Safety**
> The suit you put on to protect yourself from a chemical exposure should be designed to give you the best protection possible against the chemicals you expect to encounter.

specific gravity
the weight of a material compared to the weight of an equal volume of water

> ⚠ **Safety**
> Stack the deck in your favor by getting as much information as possible on your chemical enemy!

> ■ **NOTE**
> Be on the lookout for visible labels or other markings that could indicate what the released chemical is.

> ⚠ **Safety**
> More information translates to reduced risk.

chemical compounds. As an example, a certain kind of suit may be good for a particular kind of acid, but may be prone to total **degradation** if exposed to a strong hydrocarbon-based solvent. The reverse may also be true, with a suit protecting well against hydrocarbons but poorly against acids or bases. The point is that suit compatibility is an important component in PPE selection. The suit you put on to protect yourself from a chemical exposure should be designed to give you the best protection possible against the chemicals you expect to encounter.

Do you really understand the nature of your problem? Ideally, a responder or worker should know the chemical or mixture of chemicals that may be encountered. If this is not possible, an attempt should be made to identify the chemical classification that the material belongs in. There are various field testing kits on the market today and local custom dictates which one you use. Some test methods utilize paper strips that show positive for chlorine, or fluorine, or possibly oxidizing properties, but do not indicate an actual chemical name. Other tests require extracting a sample of the chemical and completing a battery of reactions that will point toward a chemical class or characteristic of the sample. An illustration of this concept may be a sample testing positive for flammability, or the presence of sodium in the compound, or confirming **specific gravity**. If you are lucky, you might find multiple characteristics, but in most cases the information yielded in this fashion is sketchy. Work on fact, not fiction, whenever possible. Stack the deck in your favor by getting as much information as possible on your chemical enemy! The bottom line may be that you are working on less information than you would like and find yourself hoping to at least nail down a hazard class.

Are there any witnesses or facility representatives who can be of assistance? Are there any bystanders or anyone else who may be able to shed light on a released substance? If responding to a fixed facility, utilize the site workers whenever possible to positively identify drums or tanks. In many cases someone can usually identify at least what is supposed to be stored in most containers. Take this information as "nice to know" until you verify it for yourself.

Is there a need to make entry into an unknown scene or atmosphere to take a sample of a material for field identification or lab analysis? This is the granddaddy of all the things you do not really want to do as a hazmat responder. Going into an area that gives no real clue as to what the product may be leaves even the most seasoned veteran with concern. PPE selection is merely an educated guess, which becomes very disconcerting when it is you in the suit. The reality is that it may not be entirely possible to get all of the facts, and in some situations, actions may need to be taken with incomplete information. It is therefore vital that personnel working at waste cleanup sites or those involved in emergency response gather as much information about the chemicals involved as possible. More information translates to reduced risk. Once a positive identification is made, consult at least three reference sources to verify as many of the physical properties of the chemical as possible. Look at such characteristics as pH, flammability, dermal and respiratory toxicity, state of matter, specific gravity, and vapor density. This information is very important as you may have to simply reason out your PPE selection. Unfortunately most reference sources do not have clear recommendations on PPE selection. Two common reference sources that do have PPE recommendations include the appropriate material safety data sheet (MSDS) and the National Institute for Occupational Safety and Health Administration (NIOSH) *Pocket Guide to Chemical Hazards*. We will later find that our chemical compatibility charts are somewhat limited in the numbers of chemical substances they are tested against. In many cases it may be necessary to apply a "family" approach to suit selection. Compatibility charts provide technical data on how well a particular material will hold up against a test battery of chemicals. An example of reasoning out your PPE selection may go like this. You are the first responder to a spill of tertiary butyl alcohol on the floor of the manufacturing area. After the response team has been assembled, a perimeter has been set up and secured to keep anyone from being exposed. The decision is made to enter into the spill site and clean up the puddle. You are assigned the task of researching the MSDS and coming up with an appropriate level of protection. There is no doubt that the material is tertiary butyl alcohol but when you look up that substance on the compatibility chart, it does not appear! This may cause your pulse to quicken just a bit, but do not fear. The next best thing is to determine if the substance falls into a wider class or family that *is* represented on the compatibility chart. A quick scan finds good solid testing data on isopropyl alcohol. You realize the two substances are related and that the suit will offer roughly the same resistance to the tertiary butyl alcohol as it would to the isopropyl alcohol. Essentially you were forced to make a subjective decision on incomplete data. Are the physical properties the same for the two substances? Definitely not, but if the two substances are closely related placing the chemical in a broad classification may be the best thing you have going. Be careful when using this "family" approach—it is not definitive and only represents an educated guess!

> **Safety** Compatibility charts provide technical data on how well a particular material will hold up against a test battery of chemicals.

> **Safety** Be careful when using this "family" approach—it is not definitive and only represents an educated guess!

Information regarding the type of equipment that needs to be utilized can be obtained from a variety of sources including the material safety data sheets, reference libraries, and the equipment manufacturer. Again, care must be taken to ensure that the protective gear used is compatible with the hazards that are anticipated.

Have I Chosen a Type and Level of Protection Appropriate for the Hazard?

This question may seem easy to answer after the product has been identified but it is not always a clear-cut event. Once again you should use big-picture thinking. This means that you need to consider all of the circumstances surrounding the spill as well as the material itself. A spill of a volatile hydrocarbon presents a different picture when it occurs outside than it would inside a closed room with poor ventilation. Not only may your tactics change, but what you choose to wear to the event may be different. The state of matter released may also have some bearing on the level of protection you desire. Remember—truly understanding the chemical hazard is the key to effective PPE selection! Would a small pile of magnesium turnings inside a building require a different level of protection than a large puddle of vaporous hydrochloric acid in an enclosed lab? The answer is a resounding "most likely" because the circumstances and hazards of each release are so different. The bottom line is that you need to look at everything going on when it comes time to choose a level of protection. Do not get caught up only in the reference books to decide which gloves work best with a particular spill or what suit offers the best protection against a substance. Use common sense thinking and look around at the setting—it may give you some good clues.

TYPE AND LEVEL OF PROTECTION

Essentially, there are three basic categories of personal protective equipment. Within these groups there are different levels or variations, but we explore only some of the more commonly encountered. Most of these types are available for use by the employee who works with the various hazards encountered in the workplace. Others are for use by the emergency response personnel who attempt to stabilize an emergency incident. Unfortunately, while the types of clothing listed below do provide a wide range of protection, there is not one single type of personal protective system that will shield you from all hazards. Most types of chemical protective clothing are made up of various plastics that will quickly melt in the event of exposure to high heat. Most structural fire fighting clothing, while providing considerable protection from high temperatures, is not designed to repel most chemicals, especially corrosives, because most turnouts are made of fabric and leather, which tend to absorb the various chemicals and off gas them slowly, thereby exposing the wearer well past the initial exposure.

Structural Firefighting Clothing

This equipment is designed to protect against extremes of temperature, hot water, hot particles, and the ordinary hazards encountered in fire fighting. It is not designed to provide significant protection from chemical hazards. In general, it is composed of a fire retardant fabric and at times trimmed in leather to prevent abrasion. Some of these fabrics, particularly the leather, may actually increase the likelihood of chemical exposure because they tend to absorb many types of chemicals. Figure 7-7 illustrates a typical fire fighting ensemble. You may encounter this level of protection worn by virtually all public fire agencies as well as industrial fire brigades. Once again, the limitation of the PPE lies in the fact that it does little to completely seal off the wearer from harmful chemical vapors.

High-Temperature Protective Clothing

This equipment is designed for short-term exposures to high temperatures. It is also divided into two subgroups called **proximity suits** and fire entry suits. Like the structural firefighter clothing, it is not designed to provide chemical resistance

---

▌Safety
● Remember—truly understanding the chemical hazard is the key to effective PPE selection!

■ NOTE
Most types of chemical protective clothing are made up of various plastics that will quickly melt in the event of exposure to high heat.

proximity suits
a type of firefighting gear used when operating in areas of extreme heat

> **■ NOTE**
> Some of these fabrics, particularly the leather, may actually increase the likelihood of chemical exposure because they tend to absorb many types of chemicals.

> **■ NOTE**
> Although there are other standards, such as those listed by the EPA, we refer to the NFPA standards in this text.

encapsulating
description of a one-piece chemical resistant garment that completely covers the entire body of the wearer

and may be seriously degraded by corrosives and strong solvents. This piece of equipment is very specialized and you may very well never see a fire entry suit in your career. These suits are not used for anything other than extreme fire conditions such as those seen in foundries and steel mills. Personnel using these suits are very well trained in the use and limitations of the garment.

Chemical Protective Clothing

This category of equipment is divided into three subgroups according to NFPA standards. Although there are other standards, such as those listed by the Environmental Protection Agency (EPA), we refer to the NFPA standards in this text. In general terms, personal protective equipment can be broken down into three categories:

- Vapor protective suits
- Liquid splash suits
- Support function garments

The vapor and liquid suits may also be subdivided into **encapsulating** and nonencapsulating, but we investigate the differences in the section on Level A Protection. Advantages and disadvantages are discussed as well as some guidance on the topic of suit selection criteria.

To help us navigate the various chemical protective clothing (CPC) available we reference the NFPA 471 standard, *Recommended Practice for Responding to*

Figure 7-7 *Typical fire fighting ensemble.*

> **Safety**
> Remember—NFPA 471 refers to levels of protection. NFPA 1991, 1992, and 1993 refer to standards of suit construction.

Hazardous Materials Incidents. This document covers everything from work zones to decontamination to Department of Transportation (DOT) placarding, but our interest here lies in the use and selection of PPE. Some paraphrasing of the actual verbiage has been done to give the reader an easier understanding of the intent of the document.

Once again, this PPE decision making is based on the hazard of the material involved, along with the setting of the spill. Remember—NFPA 471 refers to levels of protection. NFPA 1991, 1992, and 1993 refer to standards of suit construction. Personnel will choose one of these levels of protection in either the course of their job assignment or in the event of a chemical emergency.

We begin our discussion with the least restrictive level of protection that is commonly found and used in the **support zone.**

Level D

Level D is composed of a normal work uniform and is considered the lowest level of protection available. This level is somewhat misleading, however, in that it can also include a wide variety of chemical resistant gloves, boots, face shield, hard hats, and other chemical clothing. What it cannot include is the routine use of any type of respirator or respiratory protection equipment. Occasionally, emergency air supplies may be available but should not be used during normal work functions. The following list illustrates some common items found in Level D protection. Some items may become optional depending on need.

- Coveralls
- Chemical resistant safety shoes or boots with a steel toe and shank
- Chemical resistant booties over work shoes or boots if necessary
- Gloves—These may vary with the hazard encountered
- OSHA-approved hard hat
- Face shield or safety goggles
- Emergency air supply if applicable

This level of protection is utilized when there is no danger of exposure to the worker from any airborne materials, or when the chemicals used do not pose a hazard by inhalation or skin contact. If any type of respiratory hazard is present, level D is not appropriate. The worker shown in Figure 7-8 is dressed in one version of a level D ensemble. Remember that level D can be different than the illustration depending on the need.

> **support zone**
> area of no contamination where logistic and support functions are performed, also known as the cold zone

> **Safety**
> If any type of respiratory hazard is present, level D is not appropriate.

Level C

The types of chemical resistant clothing used in Level C can also be used in Level B and may very well be identical. What separates Level C from other levels is the use of a NIOSH-approved air purifying respirator (APR). We now get into some

Figure 7-8 *Worker in level D work uniform ascending a storage vessel.*

■ **NOTE**

The types of chemical resistant clothing used in Level C can also be used in Level B and may very well be identical.

gray areas because it is the degree of respiratory hazard that is paramount in our thinking. Here we are placing a high priority on an oftentimes elusive question, "How much of a respiratory threat is there?" Keep in mind that Level C is selected only when the hazards of the material are known and are of such a magnitude so as to pose a moderate level of danger for the employee or emergency responder. It is important to note that when this level is utilized, both the type of chemical and the concentration of the material should be known. Because of the limitations of the respiratory protection associated with this level, careful monitoring of the scene is necessary to ensure that the airborne contamination stays below the immediately dangerous to life and health (IDLH) level and that the oxygen level is above 19.5%. Generally speaking, chemical resistant clothing meeting the requirements of NFPA 1993, *Standard for Support Function Protective Garments for Hazardous Chemical Operations,* is applicable for this level of protection.

A typical Level C ensemble includes but is not limited to the following equipment:

- NIOSH-approved full-face or half-face air purifying respirator.
- Chemical resistant clothing compatible with the material being used or present. This could range from a simple splash apron to a one-piece coverall or jump suit.
- Chemical resistant outer gloves with chemical resistant inner gloves.

- Chemical resistant safety shoes or boots with steel toe and shank.
- Disposable **overbooties** if needed.
- OSHA-approved hard hat.
- Face shield or safety goggles if half-face respirator is used.
- Recommended equipment includes a two-way radio, fire resistive coveralls over or underneath the chemical protective clothing (if a threat of fire is suspected), and escape air supply if applicable.

Level C protection is denoted by the use of an air purifying respirator, not the clothing worn with it.

Figure 7-9 shows a worker in a typical level C dress.

Level B

Level B protection is used whenever the hazards of the material and situation encountered are such that a significant level of respiratory and dermal risk is present. This determination is very subjective, but this level is customarily used when a chemical has a high potential for injury due to a serious inhalation hazard, but only a moderate potential for skin absorption. It may appear strange to think about a material having only a "moderate" threat of causing damage to your skin. Most people would agree that if the chemical has any possibility of damaging skin they are going to be totally buttoned up! This level of protection affords the highest level of respiratory protection along with an acceptable degree of skin protection. Some level B suits are actually zipped over the entire body like the more protective level A suits but without some of the expensive features like attached boots and gloves. An entry into a situation where an unknown chemi-

overbooties
disposable covers slipped over heavier rubber boots

> **Safety**
> Level C protection is denoted by the use of an air purifying respirator, not the clothing worn with it.

Figure 7-9 *Worker functioning in level C.*

cal hazard is encountered requires no less than level B protection. Level B is probably the most commonly selected level due to its versatility and wide range of acceptable suits, boots, and gloves. This level of protection often utilizes the ultimate closure device—duct tape, for sealing gloves, boots, and face pieces to the chemical protective clothing. This technique becomes very obvious when you look over Figure 7-13a–m later in the chapter. Level B attire should adhere to the NFPA 1992 standard for *Liquid Splash Suits for Hazardous Chemical Emergencies*. An appropriately dressed level B hazmat worker will be clad in at least the following equipment:

> **Safety**
> Entry into a situation where an unknown chemical hazard is encountered requires no less than level B protection.

- NIOSH-approved positive pressure air supplying respirator, usually a self-contained breathing apparatus (SCBA), although an air line system (see Figure 7-10) may also be used.
- Chemical resistant clothing compatible with the material being used or present. This clothing may consist of long-sleeved chemical resistive overalls; a hooded chemical splash suit; nonencapsulating chemical suits; or encapsulating chemical suits. Figures 7-11 and 7-12 show two examples of typical level B protection utilizing SCBA.
- Chemical resistant outer gloves with chemical resistant inner gloves.
- Chemical resistant safety boots with steel toe and shank. Disposable overbooties are always recommended when contamination levels are expected to be high.
- Recommended equipment includes a two-way radio, OSHA-approved hard hat, and fire resistive coveralls underneath the chemical protective clothing.

> **Safety**
> The level of protection is based on the use of an air supplying respirator, not necessarily the type of clothing worn. A full range of clothing, from splash aprons to encapsulating suits, when worn with an air supplying system constitute level B protection.

Again, the level of protection is based on the use of an air supplying respirator, not the type of clothing worn. A full range of clothing, from splash aprons to encapsulating suits, when worn with an air supplying system constitute level B protection.

See Figure 7-13a–m for an illustration of a Level B dress out.

Level A

Level A protection is the highest level of protection utilized in dealing with a chemical release. It is required when the chemical that is being used or that has been released presents a significant threat by inhalation and/or absorption. In this case significant means anything that will kill you if it gets in you or on you! this level is sometimes referred to as "gas tight" or "fully encapsulating" because it eliminates any type of exposure of the worker to the chemical even in airborne or gaseous form. Standards for construction of these garments are found in the NFPA standard 1991 *Vapor Protective Suits for Hazardous Chemical Emergencies*. Keep in mind that level A means that you are zipped into the suit. This requires that you have an air supply inside the suit with you! All Level A suits have attached outer gloves, attached booties (which are usually

Figure 7-10 *Worker in level A attaching an umbilical air system to the suit.*

Figure 7-11 *Worker in Level B performing air monitoring.*

Figure 7-12 *Heavy equipment operator working in level B protection.*

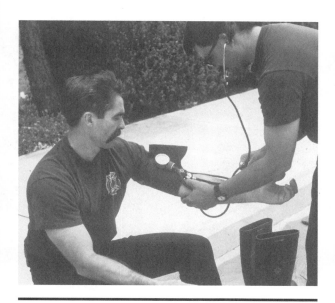

Figure 7-13 *Level B dress out. (a) Medical monitoring prior to donning PPE is required.*

(b) The lower part of the suit is applied followed by the boots. Under gloves can also be applied during this phase.

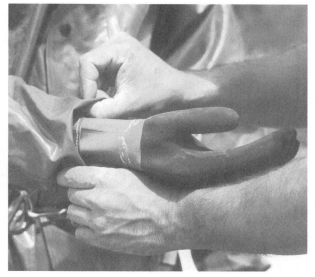

(c) After the upper half of the suit is applied, outer gloves are put on.

(d) Several inches of the suit are tucked into the outer glove leaving enough slack for easy elbow bending.

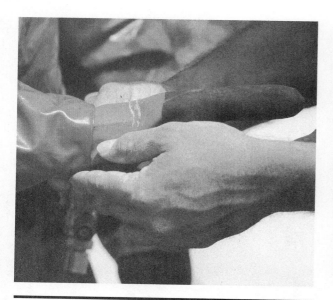

(e) A flap of suit is folded over the top of the outer glove or "bloused" creating a lip over the entire circumference of the glove.

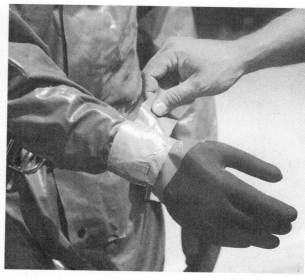

(f) Tape or other suitable closure methods are used to snugly seal the point where the suit is bloused over the glove.

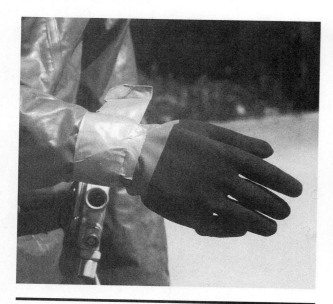

(g) Ensure total coverage of the seam, leaving a pull tab at the same point on both arms.

(h) Don the SCBA or other air-supplying respirator if applicable and prepare to secure the lower part of the suit to the boots.

Figure 7-13 *Continued*

Chapter 7 Personal Protective Equipment

(i) Tuck a generous amount of the suit into the boot and blouse it just as the arms were done. Leave enough slack in the blousing to allow for good flexibility of the knee.

(j) Secure the suit to the boots with tape of other suitable seam closure techniques.

(k) The facepiece is applied and secured to the suit with tape or other suitable seam closure techniques to reduce exposure to the contaminates.

(l) It is desirable to keep tape off the visor so as not to restrict the vision of the wearer.

(m) With nonencapsulating garments it is important to adequately seal the neck area of the suit. Make sure the coverage is complete but not restrictive to the wearer.

Figure 7-13 Continued

> **!Safety** Level A means that you are zipped into the garment, which requires that you have an air supply inside the suit with you!

> **!Safety** If duct tape is used to attach gloves or boots to the suit, then it is not a level A dress out!

worn under heavy duty outer boots), pressure relief valves, heavy duty zippers, and visors. If these are not in place on the suit, it is not a level A garment. Oftentimes confusion arises between the concepts of encapsulating and fully encapsulating. Some level B suits are encapsulating, which means that the wearer is in fact totally zipped into the suit. One significant difference however, is that an encapsulating level B suit does not have gloves attached to the arms of the suit. This is the quickest telltale sign of what type of suit you may be looking at. Table 7-1 may be of some help when attempting to tell the difference between an encapsulating level B suit and a fully encapsulating, gas tight level A suit. If duct tape is used to attach gloves or boots to the suit, then it is not a level A dress out!

The best way to understand what level A looks like is to see Figure 7-14 Level A includes but is not limited to the following:

- NIOSH-approved positive pressure air supplied respirator, usually a self-contained breathing apparatus or supplied air system with escape cylinder (refer to Figure 7-10).
- Vapor protective suits. These will be totally encapsulating, gas tight garments that ensure that the wearer is completely enclosed by the suit and provided with a suitable air supply.
- Chemical resistant glove system. Some manufacturers employ a three-glove system, which includes a cotton underglove covered by a layer of chemical resistant fabric. Over that a flame-retardant glove may be worn if the threat of fire is present. The entire system must pass the construction standards as listed in NFPA Standard *1991 Vapor Protective Suits for Hazardous Chemical Emergencies*.

Table 7-1 *Encapsulating versus totally encapsulating gas tight suits.*

Level B	Level A
No gloves attached to the suit.	Gloves attached to the suit.
Booties may or may not be attached.	Booties always attached to the suit.
Relief valve near the hood may be nothing more than a hole covered by a flap of the suit.	Relief valve is one way only and of ample construction.
Zipper may only have an adhesive-backed flap closure to ensure suit integrity.	Zipper is heavy duty with Velcro for extra closure protection.
Suit visor usually only sewn onto the suit.	Suit visor is sewn in with heavy duty tape covering the seam.
	This suit will have a tag inside stating that it is a fully encapsulating or gas tight suit.

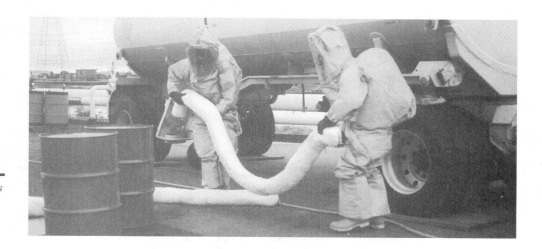

Figure 7-14 *Students practicing spill control in Level A suits.*

> **Safety**
> Some responders wear bicycle helmets in place of hard hats as they have a lower profile and seem to fit better. This piece of equipment is not OSHA approved, but is nonetheless effective.

- Chemical resistant safety boots with steel toe and shank. At times additional overbooties are worn when high levels of contamination are anticipated.
- Recommended equipment includes a two-way radio, OSHA-approved hard hat, fire retardant coveralls underneath the chemical protective clothing, cotton or fire retardant long underwear, cooling vests.

Some responders wear bicycle helmets in place of hard hats as they have a lower profile and seem to fit better. This piece of equipment is not OSHA approved but is nonetheless effective.

Donning a level A ensemble has some unique steps and local custom may dictate your exact procedures for actually getting into the suit. The step-by-step donning process as shown in Figure 7-15a–g is meant as a rough guideline to illustrate the basic procedure for a level A dress out.

Flash Protection

It is important to acknowledge the need for some form of flash fire protection when using PPE. Most often this flash protection is employed with the level A ensemble but can be incorporated anytime an entry is made into a potentially flammable atmosphere. Flash suits or flash protection is intended to only protect the wearer for a few seconds—enough time to rapidly escape the area or survive the initial flash of fire. So-called flash fires are actually vapor phase explosions capable of generating high heat and pressure waves—not a good place to be. The safer thinking is to ensure that the atmosphere being worked in is below the lower explosive limits (LEL) of the material you suspect is present. If the vapors are positively controlled, the flash suit just becomes a fancy piece of equipment that makes you look like a spaceman. Flash protection should be in place if there is a

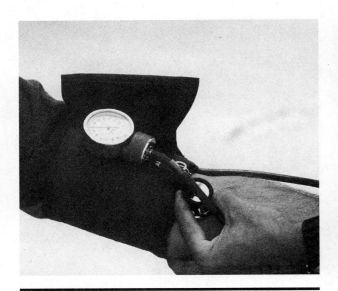

(a) Medical monitoring is done prior to getting into the suit.

(b) The lower half of the suit is applied with the feet going into the attached booties.

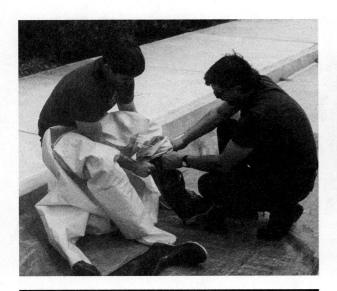

(c) Chemical resistant outer boots are put on.

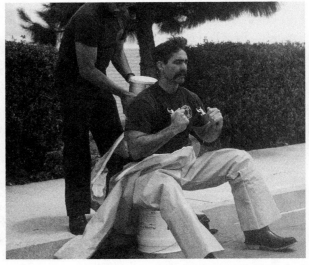

(d) The SCBA is put on or preparations are made to use an umbilical air system with escape cylinders.

Figure 7-15

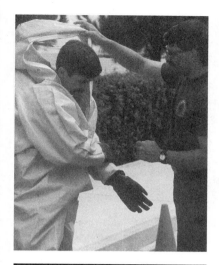 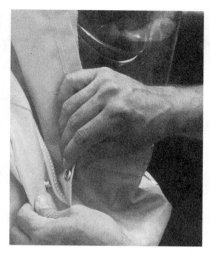 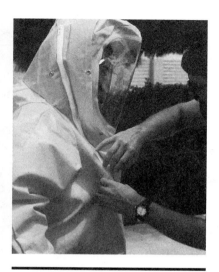

(e) The suit is fitted over the wearer and the arms are placed in the suit. Make sure the user is ready to go on air before completing this phase.

(f) The suit is fully zipped.

(g) All additional closures are securely fastened.

Figure 7-15 *Continued*

> **Safety**
> Flash suits or flash protection is intended to only protect the wearer for a few seconds—enough time to rapidly escape the area or survive the initial flash of fire.

> **Safety**
> Flash protection should be in place if there is a threat of fire during the entry.

threat of fire during the entry. These suits are typically made of a flame resistant material capable of withstanding heat for a short duration. Some older systems incorporate a fire resistive oversuit that is snapped onto a standard level A suit. Newer systems incorporate the flash suit as part of the level A ensemble and require no additional steps to use. This flash protection is not designed to be the main protective barrier against chemical exposure and is in place to protect against transient fire. Figure 7-16 shows a flash suit in place over a level A suit.

PPE—Conditions for Use

The following conditions for use are not all encompassing and are designed only to give you a guide in your thinking. They are broad guidelines and could encompass any number of variables and extenuating circumstances. Use common sense when selecting your levels of protection and if there is any doubt, be conservative.

Select Level A when

- Entering confined spaces or poorly ventilated areas where significant airborne hazards and low oxygen concentrations are expected.
- The chemical substance offers the threat of immediate skin destruction no matter what the setting.

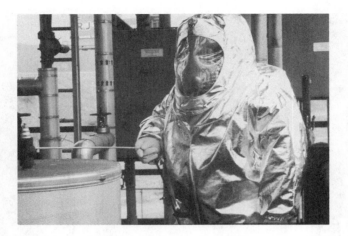

Figure 7-16 *An example of a flash suit worn over a Level A protective garment.* Courtesy of Mine Safety Appliances.

- A high potential for splash, immersion, or exposure to a dermally toxic or irritating substance is present.
- An IDLH atmosphere of a dermal threat is present.
- High concentrations of vapor, gases, or particulates with any threat of skin irritation or dermal toxicity are present.
- When you are uncomfortable dealing with a total unknown. You may recall from earlier reading that level B is the minimum level of protection for dealing with unknowns. Some responders favor a level A entry with unknowns and depending on the setting this line of thinking may be sound.

Select Level B when

- There is a high inhalation hazard from a material that is *not* capable of being absorbed through intact skin.
- Oxygen concentrations are below 19.5 % in the setting of a chemical release.
- Assessing spills of unknown origin.
- Atmospheres have been confirmed by direct read instruments confirming a non-IDLH dermal atmosphere.

Select Level C when

- Any atmospheric contaminants, splashes, or direct contact will not adversely affect or be absorbed through intact skin.
- All contaminants are known and have been confirmed by direct read instruments to be below any IDLH value.
- Oxygen levels are above 19.5%.

- All criteria for the use of air purifying respirators are met. Review Chapter 6 for details.

Select Level D when

- The atmosphere contains no airborne hazards.
- Work functions preclude chemical exposure except by splash hazards that are essentially non-hazardous by dermal exposure.
- Providing for basic safety at incidents or work sites. This base level of protection may provide a base layer of protection for a quick upgrade into higher levels of protection.

Some Points To Ponder Regarding Chemical Protective Clothing

The performance of PPE as a barrier to chemicals is determined by the material and quality of its construction. Chemical protective clothing is based on a plastic and/or elastomeric material overlaid on a paper or fabric substrate. Many of the names used to describe these materials are registered trademarks of the individual manufacturer and the exact chemical compositions are trade secrets. Many different materials are used for suit construction. Investigate the PPE you will be using and understand the benefits and limitations. You are encouraged to investigate trade magazines and local equipment vendors for specifics on the PPE you will be using. Typically each chemical interacts with a given suit fabric in a unique manner. The situation becomes even more complex when multicomponent solutions are involved. Essentially, you are responsible for knowing the specifics about the PPE you will be wearing. The following points may assist you in choosing a safe level of chemical protective equipment.

- No one clothing material will be a barrier to all chemicals.
- For a given clothing material type, chemical resistance can vary significantly from product to product. For example, not all brands of nitrile gloves provide equal protection.
- Some chemicals are so aggressive that few gloves or clothing will provide more than an hour's protection following contact. In these cases, it is recommended that clothing be changed as soon as it is safely possible after any contact with the chemical or chemical mixture.

In general, there is no such thing as impermeable chemical protective clothing. Knowing this fact, we must understand an important concept that influences our PPE selection. That concept deals with **permeation,** or an ability of a material to breach our chemical protective clothing. This property may be estimated by simple immersion tests wherein the PPE or a portion thereof is exposed to the test chemical in laboratory conditions. The material is then examined for obvious signs of degradation, swelling, or weight changes. This has been the traditional method for generating the chemical resistance tables included in many

---

Safety
Many different materials are used for suit construction. Investigate the PPE you will be using and understand the benefits and limitations.

permeation
the slow movement of a chemical from the outside to the inside of the particular item of PPE

PPE brochures. It is important to note, however, that permeation may occur with little or no visible or physical effect on clothing materials.

Permeation of a liquid or vapor through a given material is a three-step process involving:

- The chemical exposure at the outside surface of the PPE
- The diffusion of the chemical through the PPE material, and
- The appearance or detection of the chemical at the inside surface (i.e., toward the wearer) of the PPE.

Figure 7-17 depicts the process of permeation.

There are two principal points of importance when selecting PPE. The first deals with the rate at which chemicals permeate the clothing materials. Secondly, the time elapsed between the contact with the chemical and the appearance of the chemical inside of the PPE is important to note. This is known as **breakthrough time** and is a key factor when selecting your PPE.

In most cases, breakthrough time is one of the most important criteria for PPE selection. Measured breakthrough times are readily determined by permeation testing and are dependent on the sensitivity of the analytical method used in the test and the test procedure. These factors should be considered when comparing breakthrough time data.

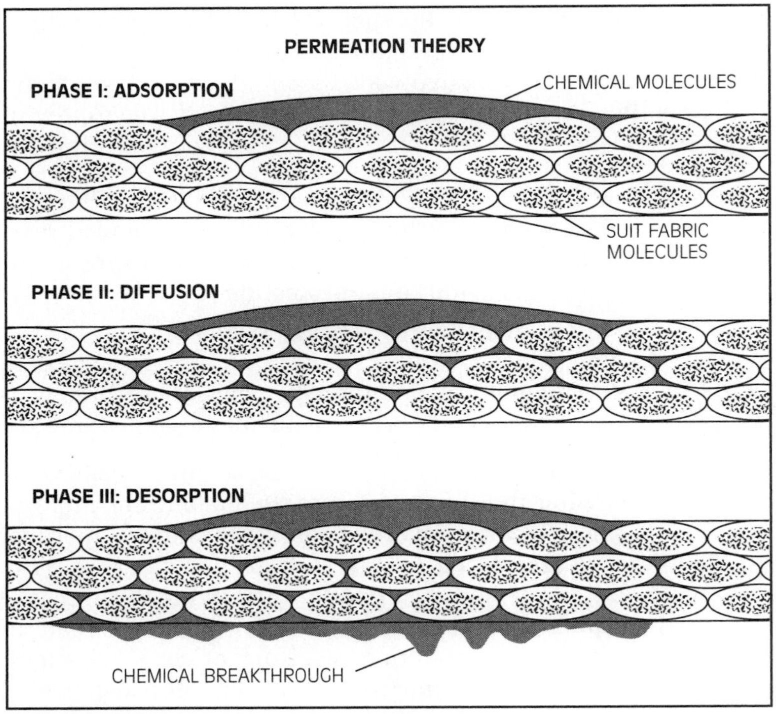

breakthrough time
elapsed time from initial contact of the outside surface of the PPE with chemical to the first detection of chemical on the inside surface

Figure 7-17
Permeation, the movement of liquid through a fabric.

PERMEATION THEORY

PHASE I: ADSORPTION — CHEMICAL MOLECULES / SUIT FABRIC MOLECULES

PHASE II: DIFFUSION

PHASE III: DESORPTION — CHEMICAL BREAKTHROUGH

Most PPE permeation data and other chemical resistance information are generated at 20°–25°C (80°F). Permeation rates increase and breakthrough times decrease with increasing temperatures. The extent of the reduction in barrier performance with increasing temperature is dependent on the chemical/material pair. Figure 7-18 illustrates the concept of permeation.

Permeation rate is often expressed in terms of the amount of a chemical that passes through a given area of clothing per unit time. (Common units are micrograms per square centimeter per minute.) Thus, the total amount of chemical permeating an article of clothing increases as the area exposed to the chemical is increased and also as the duration of exposure is lengthened. For a given chemical/material pair, the permeation rate decreases as the material thickness is increased. The thicker the fabric, the slower the permeation rate.

> **Safety**
> The thicker the fabric, the slower the permeation rate.

Permeation rate is also a direct function of the solubility of the chemical in the PPE material. Solubility, the amount of chemical that can be absorbed by a given amount of PPE material, is often expressed as grams liquid/ per gram of suit material. This absorption may be accompanied by swelling or other physical changes in the suit fabric. Immersion testing used to determine solubility is an effective means for evaluating the permeation of PPE.

Multichemical solutions represent a difficult problem relative to the selection of appropriate PPE. Rarely is there any prior experience with the solution in question, and often the components of the solution are not known. Furthermore, mixtures of chemicals can be significantly more aggressive toward the suit fabric than any one of the components alone.

> **NOTE**
> Immersion testing used to determine solubility is an effective means for evaluating the permeation of PPE.

Once a chemical has begun to diffuse into a suit material, it will continue to diffuse even after the chemical on the outside surface is removed, because a concentration gradient has been established with the material, and there is a natural tendency for the chemical to move toward areas of lower concentration. This phenomenon has significant implications relative to the reuse of PPE. For example, a possible field scenario is as follows:

- A chemical contacts and absorbs into a piece of protective equipment.
- Breakthrough does not occur during the workday because the PPE has low permeability relative to the chemical.
- Prior to removal, the PPE is washed to remove surface contamination.
- But the next morning some fraction of the absorbed chemical has reached the inside surface of the gear due to continued diffusion.

Figure 7-18 *An example of permeation.*

- If the piece of protective equipment is reused in the same contaminated atmosphere, the subsequent exposure will move more rapidly through the already established permeation pathway.

Of course, similar scenarios could occur over both shorter and longer time frames, for example, morning to afternoon or over a weekend. The user must take this possibility into account when reuse is considered.

Permeation, however, is not the only concern when discussing the failure of PPE. Another avenue chemicals take to defeat your gear is by direct penetration, which occurs as a substance moves directly through the closure points of a suit. These points could be through zippers, buttonholes, seams, flaps, or any other design feature specific to the garment. Penetration also occurs through tears or abrasions in the outer covering of the suit or gloves. Figure 7-19 illustrates the concept of penetration. Always be aware when working near jagged objects—a torn suit leaves you very vulnerable. In order to avoid a PPE failure due to penetration, pay attention to the following points:

- Do not kneel or crawl in the suit. The action of grinding knees, elbows, or hands on the pavement unduly abrades PPE and may lead to failure.
- Use heavy outer gloves over chemical protective gloves when moving drums or working with sharp objects such as glass, jagged metal, valve stems, or piping.
- Ensure that all zippers are fully closed and covered with a protective flap before entering contaminated areas. If duct tape is used as a seam sealer, inspect the taping to verify that no openings or gaps exist.
- Fully inspect any piece of PPE for holes, tears, thin areas, poor seaming, or any other defect that may compromise the garment. Garments fresh from the manufacturer's carton can have flaws. It is always better to inspect twice than have the garment fail once!
- PPE does not last forever! Respect the listed shelf life of the item and always observe the manufacturer's recommended storage practices. Old and outdated suits can easily crack or peel, thereby causing a route for penetration.

A third way that PPE may fail is by total degradation. This process is very rapid and generally causes the immediate failure of the plasticized outer coating of the garment. Essentially, the fabric dissolves, or otherwise fails, leaving behind the stringy fibers of the paper substrate. This really causes concern for the person in the suit as they have completely lost the integrity of the garment. Degradation is best illustrated by pouring gasoline into a styrofoam cup. Almost instantly, the gasoline begins to eat the styrofoam and will totally degrade whatever parts of the cup it touches. This degradation occurs because the two substances have similarities in their chemical makeup. It falls under the "like dissolves like" concept. Similar chemical compounds have a tendency to be soluble in one another. This point should be considered when deciding on PPE. Suits, gloves, boots, and

> **!Safety**
> Always be aware when working near jagged objects—a torn suit leaves you very vulnerable.

> **!Safety**
> Garments fresh from the manufacturer's carton can have flaws. It is always better to inspect twice than have the garment fail once!

Chapter 7 Personal Protective Equipment

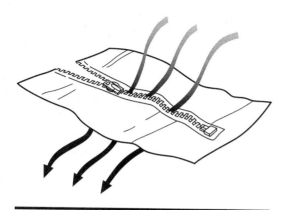

Figure 7-19 *Penetration of a chemical through a suit zipper.*

Figure 7-20 *Degradation, the quickest and most complete type of suit failure.*

■ **NOTE**
Similar chemical compounds have a tendency to be soluble in one another.

❗Safety
● Stay as clean as possible when working with chemicals. Boots and gloves are easily contaminated and require a conscious effort to keep clean.

❗Safety
● The reuse of PPE that has been contaminated is not advisable.

other items that are constructed of the same type of material that they will be exposed to may be subject to the degradation phenomenon. See Figure 7-20 for an illustration of degradation.

Protective clothing decontamination and reuse are controversial and unresolved issues at this time. Stay as clean as possible when working with chemicals. Boots and gloves are easily contaminated and require a conscious effort to keep clean. Often surface contamination can be removed by scrubbing with soap and water. In other cases, especially with highly viscous liquids, surface decontamination may be practically impossible, and the PPE should be discarded. Most chemical protective equipment is considered disposable and is generally removed from service after a single use. It is a fact that once the chemical is absorbed into the fabric, some of the chemical will continue to diffuse though the material. For highly resistant clothing the amount of chemical reaching the inside may be insignificant. However, for moderately performing materials, significant amounts of chemical may reach the inside. This may not occur during the work shift but can take place while a glove or another item of equipment is stored overnight. The next morning when the worker dons the glove, he may be putting his skin into direct contact with a hazardous chemical.

In addition to chemical resistance, the duration of the exposure and the surface area exposed affect the amount of chemical that may reach the inside surface. The reuse of PPE that has been contaminated is not advisable. Other factors affecting the performance of PPE are as follows:

- Stitched seams of clothing may be highly penetrable by chemicals if not overlaid with tape or sealed with a coating.
- Lot-to-lot variations do occur and may affect the effectiveness of the PPE. They may go undetected due to quality control in the formulation or in the manufacturing process.

- Pinholes may exist in fabrics due to the manufacturing process.
- Thickness may vary from point to point on the clothing item. Depending on the manufacturing process, the finger crotch area of the glove is particularly susceptible to thin coverage.
- Garment closures differ significantly from manufacturer to manufacturer and within one manufacturer's product line. Attention should be paid to button and zipper areas and the number of fabric overlaps in these areas.
- Do not wear jewelry, watches, badges, name tags, earrings, or any other trinket not directly related to your existence in the garment. Not only can these items tear or abrade the suit, they may become contaminated and end up in some landfill instead of your immediate possession.
- Facial hair other than closely trimmed mustaches may impede a proper seal around a face mask. This is hotly contested point around the United States and many people claim to get a perfect seal with bushy beards and gigantic mustaches. These claims may be true, but currently it is not a recognized practice and should not be allowed.
- The degree of protection provided by an item of protective clothing is also a function of its intended use. Factors such as abrasion, puncture and tear resistance, and reaction to perspiration and crumpling should be considered. Temperature, and to some extent, humidity influence the performance of some types of PPE. It is important to recognize that protective clothing can be cumbersome and restrictive and thereby hasten the onset of worker fatigue. A result is that the period of safe and effective worker activity may be reduced.
- Clothing performance is determined by its type, the specific formulation of that plastic or elastomer, and the clothing manufacturing process. For example, materials classified generically as nitrile rubber can differ significantly in composition and chemical resistance. Testing is the only means for identifying the superior products for a particular application.
- Duct tape has not been tested as to its chemical resistance in all instances. Because of this, duct tape used to seal PPE should be at least two layers thick. Although it is the standard issue for seam sealing in the field, it is not classified as a chemical resistant item of clothing.

Chemical Compatibility Charts

Chemical compatibility charts are guides that illustrate how a particular suit fabric will hold up against a certain type of chemical exposure. All manufacturers of suits, boots, gloves, and other items publish these guides and make them available to all purchasers of their garments. They are very useful in determining suit selection because they essentially contain testing data obtained from exposing their suit material to a battery of test chemicals. This test group of substances has a standard list that must be tested against with the manufacturer electing to add

> **⚠ Safety**
> Chemical compatibility charts should be consulted any time PPE is to be used.

> **⚠ Safety**
> A mental game goes on inside the suit. Staying calm and focused helps control breathing rates, reduce stress, and reduce fatigue.

vital signs
a set of manipulative evaluations reflective of a patient's general medical status including measurement of pulse rate, respiratory rate, and blood pressure.

Tyvek
suit fabric made of a lightweight paperlike substance

hyperthermia
an increase in body temperature caused by heat transfer from the external environment

heat stroke
serious medical condition caused by the inability of the body to regulate its temperature, usually brought on by high environmental temperatures

chemicals if they so desire. Obviously it is in the manufacturer's best interest to display suit fabrics with a wide range of chemical resistance. Chemical compatibility charts should be consulted any time PPE is to be used. Most guides base their testing on the characteristics of permeation, penetration, and degradation as it relates to the exposure of the chemical to the selected fabric. Most guides are also color coded for quick reference as to chemical resistance and can be easily read and understood.

A review of the *North Chemical Resistance Guide* as shown in the color insert should provide you with a clear idea of how to use these guides. While each chart may have unique properties, they all will give the basic information that you will so desperately seek. The best way to understand these charts is to use them and we will further acquaint you in an exercise at the end of this chapter.

HEALTH CONSIDERATIONS AND CHEMICAL PROTECTIVE CLOTHING

Another factor in deciding which level or type of PPE to be worn is the person who will actually be getting into the gear. Getting zipped into level A gear can be quite dramatic, and has been known to cause great concern to many folks getting into a suit. Claustrophobia and a sudden fit of breathing difficulty can be observed on a regular basis. Few people appreciate the mental toughness that is required to work in extremely uncomfortable, hot, restrictive equipment while being exposed to dangerous chemicals. Sound like fun? The grim reality is that anyone should recognize that if they need to have a plastic suit on to protect them from injury or death, they are engaged in a serious endeavor. In addition to this stark realization of potential danger, anyone in almost any level of protection will get hot, sweaty, and cranky. It is important to recognize that a mental game goes on inside the suit. Staying calm and focused helps control breathing rates, reduce stress, and reduce fatigue. The best part of the job occurs when you zip out of the suit and actually breathe fresh air.

In addition to the mental stress that may be encountered, many physical effects may be experienced. Becoming slightly overheated is the most common effect felt, but associated problems such as dehydration, altered mental status, rapid weight loss, and abnormal **vital signs** also occur. In normal conditions working in Level D clothing presents few health problems, but other levels of PPE in adverse weather conditions could affect the user. On a hot, sunny day, even the wearing of a lightweight **Tyvek** suit can raise body temperature to potentially alarming heights. As each higher level of protection is donned, the hazards posed by heat increases dramatically. Any individual in Level B, or especially Level A, should be closely monitored for changes in physical conditions possibly attributed to **hyperthermia**. This state of elevated body temperature can cause a number of medical conditions such as heat cramps, heat exhaustion, or the most alarming condition, **heat stroke**.

To combat these heat-related illnesses some response and cleanup agencies employ cooling devices underneath or inside PPE. These may include cascade fed air systems attached directly to the suit, chilled air systems, vests with ice packs in them, or actual circulating water systems. These systems have come in and out of vogue over the years but there is little conclusive data to prove they actually reduce core temperature. Whether or not you employ one of these cooling systems in the field, it is important to recognize the following heat-related problems. Figure 7-21 shows an example of a cooling pack used under PPE.

Heat Cramps

Heat cramps are quite common in warmer climates and work environments, but primarily affect people working outdoors. Heat cramps result in intermittent, painful contractions of skeletal muscles due to extracellular fluid shifting secondary to a loss of water and sodium. In lay terms, you lose water and the goodies that make your muscles and organs happy, possibly giving you a crampy feeling in the arms, legs, fingers, toes, or belly. Sweating brings sodium ions and other **electrolytes** to the surface of the skin where evaporation occurs and aids the cooling mechanism of the body. Sweating, therefore, not only results in water loss, but also an electrolyte imbalance that causes the characteristic muscle cramping. Working in hot environments can result in the loss of up to 2 liters of water per hour, and results in substantial sodium loss (up to 50 millequivalents).

A person suffering from heat cramps may exhibit:

- A normal body temperature.
- A rapid heartbeat. Any pulse rate over 100 beats per minute is considered tachycardic.
- A blood pressure normal for that person.
- A normal mental status. The person should be oriented to person, place, time, and purpose. This means they should be able to state their name, where they are, what day of the week it is, and what they have recently been doing.

electrolytes substances such as sodium, potassium, or calcium that dissociate into charged particles in water

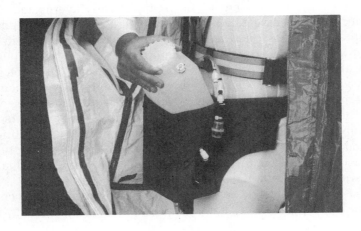

Figure 7-21 *An example of an ice vest used for cooling the wearer.* Courtesy of Mine Safety Appliances

- Cramping sensations in the hands, feet, arms, legs, or abdomen.

Treatment includes:

- Seeking medical attention.
- Removing the person from the hot environment.
- Rehydration with fluids. Water is acceptable but not the best source of rehydration. Because the cramping is due to electrolyte loss, electrolytes need to be replaced. A sports drink may be a better substitute but will still offer slow relief. If the condition of the person does not improve, medical treatment should be found. Part of the therapy will include an intravenous line (IV) supplying fluids such as normal saline (0.9% concentration) or lactated ringers. Lactated ringers is an **isotonic** solution containing potassium chloride, sodium chloride, calcium chloride, and sodium lactate in water. IV therapy is an advanced life support skill to be accomplished by paramedics and hospital personnel.

isotonic
a solution containing a like concentration electrolytes or nonelectrolytes, exerting the same osmotic pressure as the solution with which it is compared

■ **NOTE**
IV therapy is an advanced life support skill to be accomplished by paramedics and hospital personnel.

Heat Exhaustion

Heat exhaustion is also a very common heat-related condition and is a step worse than heat cramps. The symptoms are also created by excessive dehydration and salt loss due to sweating. The mechanism by which it occurs is similar to heat cramps but the signs and symptoms have some telltale significance. In addition to the dehydration and sweating, blood tends to pool in the vasculature while the body attempts to shed heat. This venous pooling can lead to insufficient blood return to the heart, which can cause dizziness, fainting, or collapse. The history of prolonged exposure to hot weather or inside work environments is a sure giveaway, and personnel operating in PPE for long periods are susceptible to heat exhaustion. Additionally, the heat exhaustion patient will most likely have had a low intake of fluids within the prior couple of days. It is then vitally important to hydrate as much as possible before beginning work in PPE. Fluids such as coffee and alcohol have a detrimental effect and should be avoided at all costs prior to the work period.

A person suffering from heat exhaustion may exhibit all the signs and symptoms of heat cramps in addition to:

orthostatic hypotension
a decrease in blood pressure when quickly changing position from supine to sitting or supine to standing

supine
description of a person lying flat on the back; face up

- Headache
- A dizzy feeling or an episode of passing out
- Diarrhea
- Low urine output due to dehydration. Adequate urination is frequent in occurrence and relatively clear in color.
- **Orthostatic hypotension** or positive orthostatic vital signs. Orthostatic vital signs can be quickly taken in the **supine** to sitting and/or sitting to standing position. A blood pressure and pulse rate are taken after the patient is supine or sitting for at least two minutes. The results are

hypotension
blood pressure with a systolic value below 100 mm Hg as measured with a blood pressure cuff and stethoscope

> **⚠ Safety**
> If taking orthostatic vitals with the patient in the standing position, keep a hand on them as they may pass out.

hypothalamus
a gland located in the brain that regulates many bodily functions such as metabolic rate and other functions of the endocrine system

metabolic acidosis
a medical condition characterized by an increase in the production of lactic acid within the body

lactic acid
an acid formed in the muscles by the breakdown of glycogen

recorded and measured against another blood pressure and pulse taken after the patient has been repositioned to the sitting or standing position for a two minute period. If sitting, the legs should be hanging below the torso as if properly sitting in a chair. If the systolic blood pressure (top number) drops by more than 20–30 points or the pulse rate increases by more than 20–30 beats per minute, the person is considered to have a positive orthostatic change in vital signs. In the setting of suspected fluid loss this could indicate a more serious medical problem that may require aggressive fluid replacement via IV therapy. In cases of substantial postural **hypotension,** the person will not be able to tolerate standing for any length of time. If taking orthostatic vitals with the patient in the standing position, keep a hand on them as they may pass out and hit the deck. Additionally, the patient may experience dizziness, a feeling of the room spinning, and possible nausea with vomiting.

Treatment includes:

- Seeking medical attention
- Removing the person from the hot environment.
- Application of cool compresses to the arm pits, back of the knees, and/or back of the neck.
- Rehydration with fluids. Water is acceptable but not the best liquid. If water is administered it should not be ice cold as this can have an adverse effect on the patient. Because the cramping is due to electrolyte loss, electrolytes need to be replaced. A sports drink may be a better substitute but will still offer slow relief. If the condition of the person does not improve, medical treatment should be found. An intravenous line (IV) may need to be established with fluids such as normal saline (0.9% concentration) or lactated ringers. Administration of these fluids feeds the electrolytes back to the body and corrects the temporary imbalance.

Heat Stroke

This medical condition constitutes a true life-threatening emergency, which occurs when the body's temperature regulation is lost. The **hypothalamus** regulates body temperature and when its ability to function properly is lost, a person can experience uncontrolled hyperthermia. This condition can be fatal and is thus critical to watch for. While true heat stroke is relatively uncommon, it is a condition that hazmat responders and workers should be aware of. Historically, a person will experience heat exhaustion prior to going into a true heat stroke, but in almost all cases the core temperature of the body will be at or above 105°F. In the response or waste site setting, the heat stroke will be due to exertion, possibly in a hot environment. This combination can make the medical condition worse as **metabolic acidosis** occurs secondary to the **lactic acid** produced during muscular activity. Hyperkalemia can also occur, which may lead to lethal heart rhythms

hyperkalemia
medical condition characterized by an abnormally high level of potassium in the body

> **⚠ Safety**
> If you suspect heat stroke, get immediate medical attention—time is of the essence in treating heat stroke!

bradycardia
very slow heart beat, usually slower than 60 beats per minute

> **⚠ Safety**
> Beware when cooling a patient—too rapid of a temperature drop can be just as dangerous as the overheated condition.

and possibly death. **Hyperkalemia** results from impending kidney failure and resulting muscle injury. If you suspect heat stroke, get immediate medical attention—time is of the essence in treating heat stroke!

A person suffering from heat stroke may exhibit:

- A core temperature above 105°F. A rectal thermometer is the most accurate way to check for true core temperature.
- Possible disorientation, mental confusion, coma.
- Seizure activity from the increased temperature.
- A very low blood pressure. Blood pressures below 90 systolic (top number) may be cause for alarm.
- A period of tachycardia followed by a period of **bradycardia**.
- Fast, panting respirations.
- Hot, dry skin except for heat stroke in the setting of exercise and then the person may have hot, sweaty skin.

Treatment includes:

- Call your local 911 service for an ambulance!
- Physicians and paramedics have some treatments, but not many. Lay persons and emergency medical technicians on the scene essentially can do two things: Give oxygen at high concentrations and remove all the persons clothing while employing rapid cooling measures. Beware when cooling a patient—too rapid of a temperature drop can be just as dangerous as the overheated condition. The oxygen has no down sides in this setting, but if the cooling measures are too extreme, adverse effects can occur. Other than that, your hands are probably tied from a medical standpoint and the best bet is to get the person to a hospital as soon as possible.

This leads to the next logical question of "When should someone not be allowed to continue working?" Obviously, a person should not return to work if they experience any of the medical conditions listed above. Regardless of the desires of the worker in question, he or she should not be allowed to keep working if you observe or suspect any of the following conditions:

- They show signs and symptoms of dizziness, disorientation, nausea, or fainting.
- They lose more than 3–5% of the body weight they started with.
- The core body temperature is above 100.4°F.
- The pulse rate is at 85–90% of their maximum heart rate. The maximum heart rate is arrived at by subtracting 220 from the age of the person. Additionally, persons not steadily recovering should be watched closely. Pulse rates should steadily drop while at rest and if a person has been resting for 20–30 minutes and still has a pulse rate of 110–120 they may need to receive formal medical attention.

Summary

- PPE stands for personal protective equipment, including but is not limited to chemical resistant suits, boots, gloves, splash aprons, face shields, and goggles.
- It is important to remember that one suit does not cover all hazards.
- PPE is not worn to enable or encourage you to intentionally contact the chemicals you are working with. It is worn to protect you in the event of an *unplanned* exposure.
- The following standards are relevant to personal protective equipment.

    NFPA 1991 (1994 Edition)   *Vapor Protective Suits for Hazardous Chemical Emergencies*

    NFPA 1992 (1994 Edition)   *Liquid Splash Suits for Hazardous Chemical Emergencies*

    NFPA 1993 (1994 Edition)   *Support Function Protective Garments for Hazardous Chemical Operations*

- In order to more effectively choose an appropriate level of protection, ask yourself these questions:

    Have I positively identified the product and all the hazards associated with it?

    Have I chosen a type and level of protection appropriate for the hazard?

- Knowing and understanding your chemical hazards is the key to sound PPE selection.
- Chemical protective clothing is divided into three subcategories according to the NFPA. They are vapor protective suits, liquid splash suits, and support function garments.
- Level D is composed of a normal work uniform and is considered the lowest level of protection available. The following list illustrates some common items found in Level D protection. Some items may be optional depending on need.

    Coveralls

    Chemical resistant safety shoes or boots with a steel toe and shank.

    Chemical resistant booties over work shoes or boots if necessary.

    Gloves—These may vary with the hazard encountered.

    OSHA-approved hard hat.

    Face shield or safety goggles.

    Emergency air supply if applicable.

- Level D is utilized when there is no danger of exposure to the worker from any airborne materials or when the chemicals used do not pose a hazard by inhalation or skin contact.

- What separates Level C from other levels is the use of a NIOSH-approved air purifying respirator.
- A Level C ensemble includes, but is not limited to the following equipment:
  NIOSH-approved full-face or half-face air purifying respirator.
  Chemical resistant clothing compatible with the material being used or present. This could range from a simple splash apron to a one-piece coverall or jump suit.
  Chemical resistant outer gloves with chemical resistant inner gloves.
  Chemical resistant safety shoes or boots with steel toe and shank.
  Disposable overbooties if needed.
  OSHA-approved hard hat.
  Face shield or safety goggles if half-face respirator is used.
  Recommended equipment includes a two-way radio, fire resistive coveralls over or underneath the chemical protective clothing (if a threat of fire is suspected) and escape air supply if applicable.
- Level B protection is used whenever a chemical has a high potential for injury due to a serious inhalation hazard, but only a moderate potential for skin absorption.
- Level B attire should adhere to the NFPA 1992 standard for *Liquid Splash Suits for Hazardous Chemical Emergencies.*
  Level B will include at least the following equipment:
  NIOSH-approved positive pressure air supplied respirator, usually a self-contained breathing apparatus, although an air line system may also be used.
  Chemical resistant clothing compatible with the material being used or present. This clothing may consist of long-sleeved chemical resistive overalls; a hooded chemical splash suit; nonencapsulating chemical suits; or encapsulating chemical suits.
  Chemical resistant outer gloves with chemical resistant inner gloves.
  Chemical resistant safety boots with steel toe and shank. Disposable overbooties are always recommended when contamination levels are expected to be high.
  Recommended equipment includes a two-way radio, OSHA-approved hard hat, and fire resistive coveralls underneath the chemical protective clothing.
- Level A protection is the highest level of protection utilized in dealing with a chemical release. It is required when the chemical that is being used or that has been released presents a significant threat by inhalation and/or absorption.
- Level A includes but is not limited to the following:
  NIOSH-approved positive pressure air supplied respirator, usually a self-contained breathing apparatus or supplied air system with escape cylinder.
  Vapor protective suits. These are totally encapsulating, gas tight garments. Ensure that the wearer is completely enclosed by the suit and provided with a suitable air supply.

Chemical resistant glove system. Some manufacturers employ a three-glove system, which includes a cotton underglove covered by a layer of chemical resistant fabric. Over that a flame retardant glove may be worn in the event the threat of fire is present. The entire system must pass the construction standards as listed in NFPA standard 1991 *Vapor Protective Suits for Hazardous Chemical Emergencies.*

Chemical resistant safety boots with steel toe and shank. At times additional overbooties are worn when high levels of contamination are anticipated.

Recommended equipment includes a two-way radio, OSHA-approved hard hat, fire retardant coveralls underneath the chemical protective clothing, cotton or fire retardant long underwear, and possibly cooling vests.

Flash protection should be in place if there is a threat of fire during the entry.

- Level A protection should be used under the following conditions:

    Entering confined spaces or poorly ventilated areas where significant airborne hazards and low oxygen concentrations are expected.

    The chemical substance offers the threat of immediate skin destruction no matter what the setting.

    A high potential for splash, immersion, or exposure to a dermally toxic or irritating substance is present.

    An IDLH atmosphere of a dermal threat is present.

    High concentrations of vapor, gases, or particulates with any threat of skin irritation or dermal toxicity are present.

    When you are uncomfortable dealing with a total unknown. Level B is the minimum level of protection for dealing with unknowns. Some responders favor a level A entry with unknowns and depending on the setting this line of thinking may be sound.

- Level B protection should be used under the following conditions:

    There is a high inhalation hazard from a material that is *not* capable of being absorbed through intact skin.

    Oxygen concentrations are below 19.5 % in the setting of a chemical release.

    Assessing spills of unknown origin.

    Atmospheres have been confirmed by direct read instruments confirming a non-IDLH dermal atmosphere.

- Level D includes but is not limited to the following conditions:

    Any atmospheric contaminants, splashes, or direct contact will not adversely affect or be absorbed through intact skin.

    All contaminants are known and have been confirmed by direct read instruments to be below any IDLH value.

    Oxygen levels are above 19.5%.

    All criteria for the use of air purifying respirators are met. Review Chapter 6 for details.

- Level D protection should be used under the following conditions:
  The atmosphere contains no known hazards.
  Work functions preclude chemical exposure except by splash hazards.
  Providing for basic safety at incidents or work sites. This base level of protection may provide a base layer of protection for a quick upgrade into higher levels of protection.
- No one clothing material will be a barrier to all chemicals.
- In most cases, breakthrough time is one of the most important criteria for PPE selection.
- Permeation is the slow movement of a chemical from the outside to the inside of the particular item of PPE.
- The time elapsed between the contact with the chemical and the appearance of the chemical inside of the PPE is known as breakthrough time.
- Do not kneel or crawl in the suit. The action of grinding knees, elbows, or hands on the pavement unduly abrades PPE and may lead to failure.
- Fully inspect any piece of PPE for holes, tears, thin areas, poor seaming, or any other defect that may compromise the garment.
- Heat cramps are common in warmer climates and work environments, but primarily affect people working outdoors. Heat cramps result in intermittent, painful contractions of skeletal muscles due to extracelluar fluid shifting secondary to a loss of water and sodium. Working in hot environments can result in the loss of up to 2 liters of water per hour and may cause substantial sodium loss.
- Heat exhaustion occurs due to excessive dehydration and salt loss due to sweating. A person suffering from heat exhaustion may feel dizzy, have fainting spells, or collapse.
- Heat stroke constitutes a true life-threatening emergency which occurs when the body's temperature regulation is lost. If you suspect heat stroke, get immediate medical attention—time is of the essence in treating this condition!
- A person should be removed from work if they exhibit any of the following symptoms.
  Signs and symptoms of dizziness, disorientation, nausea, or fainting.
  Loss of more than 3–5% of the body weight they started with.
  Core body temperature is above 100.4°F.
  Pulse rate is at 85–90% of their maximum heart rate. The maximum heart rate is arrived at by subtracting 220 from the person's age. Additionally, persons not steadily recovering should be watched closely. Pulse rates should steadily drop while at rest and if a person has been resting for 20–30 minutes and still has a pulse rate of 110–120 they may need to receive formal medical attention.

Review Questions

1. Define PPE.
2. One of the ways in which PPE fails is by permeation. What are the other two most common means of PPE failure?
   A. Penetration and tearing
   B. Penetration and degradation
   C. Degradation and melting
   D. Penetration and palliation
3. True or False? More information on the chemical hazard translates to reduced risk when selecting PPE.
4. Compatibility charts give you
   A. Information on cleaning and maintenance of PPE.
   B. A detailed description of level A protection.
   C. Information on how well a piece of PPE will hold up against certain chemicals.
   D. An idea of which MSDS sheet to read for chemical information.
5. Describe a typical level A ensemble.
6. True or False? It is acceptable to wear an APR for level B protection.
7. Which NFPA standard defines the different levels of protection?
   A. NFPA 471
   B. NFPA 472
   C. NFPA 473
   D. NFPA 1992
8. List three differences between an encapsulating suit and a fully encapsulating gas tight suit.
9. True or False? Level B can be worn to investigate a spill involving a total unknown chemical.
10. List three conditions that may require you to choose level A protection.
11. Describe a typical level C ensemble.
12. Level B attire should adhere to the requirements of which NFPA standard?
    A. NFPA 1990.120 Standard for *Level B Protection*
    B. NFPA 472 Standards for *Professional Responder Attire*
    C. NFPA 1991 Standards for *Chemical Protective Equipment*
    D. NFPA 1992 Standard for *Liquid Splash Suits for Hazardous Chemical Emergencies*
13. List three conditions of use for level B protection.
14. Define breakthrough time.
15. True or False? Permeation is the slow movement of a chemical from the outside to the inside of a particular item of PPE.
16. True or False? Level C provides as much respiratory protection as Level B
17. True or False? It is advisable to wear jewelry underneath your PPE.
18. List two observable signs of heat cramps.
19. Describe the typical signs and symptoms of heat stroke.
20. True or False? Heat stroke is a true medical emergency.

Compatibility Chart Exercise

Objectives:

- To promote an atmosphere of teamwork within the class.
- To encourage classroom participation by each student.
- To provide an opportunity for the student to become familiar with using chemical compatibility charts.
- To identify the limitations of chemical compatibility charts.
- To reinforce the concept of breakthrough time.
- To bring together several of the concepts learned thus far.

Time: 30–60 minutes
Materials Needed:

- Chemical compatibility chart for gloves as found in the color insert.
- Pens, pencils, or highlighters for each group or student.
- Additional paper for notes and answers as needed.
- An adequate reference source (Hawley's chemical dictionary, MSDS's, MERCK index, etc.) for the following chemicals: cyclohexane, aniline, carbon tetrachloride, acrylic acid.
- DOT Guidebook.

Instructions:

Individually, or in groups, do the following:
- Read the attached scenarios and research the listed chemical compound.
- Understand the physical properties of the chemical.
- Refer to the compatibility chart in the color insert and determine which material will provide the greatest chemical resistance for the chemical in the scenario. Pay special attention to breakthrough time.
- Record your results. You may be asked to answer several questions relating to the scenario. Some may require information learned in previous chapters.

1. Carbon tetrachloride spill

The response team has been called to the scene of a carbon tetrachloride spill in the laboratory. Upon arrival, you find approximately 5 gallons of liquid on the level tile floor. The student who was pouring the chemical has a minor splash exposure to the right forearm. The liquid has a sweetish odor and the room has poor ventilation.

1. Size-up this situation. List the applicable PPE and justify your thoughts.
2. Make sure to address at least the following points:
   - What are the properties of carbon tetrachloride?
   - What level of protection would you choose to deal with the problem?
   - What is the recommended glove?
   - Are any glove materials not recommended?
   - Is respiratory protection advisable for this situation?
   - How would you handle the exposure of the student? What are the signs and symptoms of overexposure?
   - What is the UN number for carbon tetrachloride?

2. Odor Investigation

The response team has been called to investigate an odor coming from the storage room. Upon arrival you look through the glass door and see a four-pack of containers sitting on the floor. You see a red label on the side of the box that reads "Flammable" and the words "cyclohexane" on the side of one of the spilled bottles. You have been assigned to research the chemical and recommend a glove for the entry team.

Size-up this situation and use your reference sources to learn about cyclohexane. Make sure to address at least the following points:
- What are the properties of cyclohexane?
- What is the recommended glove?
- Are any glove materials not recommended?
- Is respiratory protection advisable for this situation? If so, what type of respiratory protection would you recommend?
- What respiratory protection would you recommend if the airborne concentration of cyclohexane vapors were above 2,000 ppm?
- What is the UN number for cyclohexane? What guide number would you go to for emergency actions?

3. Unknown Spill

The response team has been called to investigate an odor emitting from the cargo area of a large delivery truck. The truck is parked at the rear of the building by the loading ramp. The driver tells you that he started feeling dizzy as he was unloading some packages. Other people have also stated that they can smell something in the air. The driver is a little shaken but tells you the boxes he was unloading said "acrylic acid" on the side. The incident commander assigns you to the technical reference team and your job is to find a glove that will offer good protection against acrylic acid.

Make sure to address at least the following points:
- What are the properties of acrylic acid? Is it a strong corrosive?
- What is the recommended glove?
- Are any glove materials not recommended?
- Is respiratory protection advisable for this situation? If so, what type of respiratory protection would you recommend?
- What are the signs and symptoms of overexposure to acrylic acid?
- Would an APR be acceptable if the airborne vapor concentration were below the IDLH for acrylic acid?
- What placard would you expect to find on the side of the truck?

4. Aniline Exposure

The response team has been called to the scene of an exposure resulting from a suspected aniline exposure. Upon arriving on scene, you are told that one of the workers in the receiving area handled some contaminated materials from a supplier. The victim has been totally decontaminated and sent to the hospital. You are assigned the task of gathering up the heavily contaminated clothing. There are strong vapors coming off the clothing and the area is poorly ventilated. Your job is to determine what level of protection is necessary to pick up the contaminated clothes.

Make sure to address at least the following points:
- What are the properties of aniline? Is it corrosive?
- What is the recommended glove?
- Are any glove materials not recommended?
- Is respiratory protection advisable for this situation? If so, what type of respiratory protection would you recommend?
- What level of protection would you recommend?

Chapter 8

Principles of Decontamination

Learning Objectives

Upon completion of this chapter, you should be able to:

- Define decontamination.
- Describe the importance of decontamination.
- Describe the location of a decontamination corridor and the importance of control zones.
- Understand when and how emergency decontamination procedures are to be used.
- Describe various methods used for decontamination.
- List seven factors that determine the extent of decontamination.
- Identify major factors in selecting a decontamination site.
- Describe the process of setting up a decontamination corridor.
- List the seven step method of Level B decontamination.
- Describe the responsibilities of the decontamination team leader regarding decontamination.

INTRODUCTION

Decontamination is one of the most critical steps in handling a hazardous materials emergency. Whether taking a shower at work before going home or setting up an elaborate array of pools, hoses, and scrubbers, decon is an important step in controlling an incident, preventing the further spread of contaminants, and protecting the health and safety of involved personnel. Indeed, protecting the health and safety of involved personnel is the most important reason for conducting decon.

Decon procedures may change from job to job. Varying factors determine how much and what type of decon is required. Taking a systematic approach allows decon to be complete and thorough, no matter how involved the process. Whether washing your hands with a safe degreaser to remove oil or setting up a corridor of step-by-step washing stations to clean off a corrosive, decon should be carefully thought out and well managed to ensure it is done correctly and with the right materials. Do not become complacent about decontamination. Always be thorough as many contaminants may not be visible to the naked eye. Even if the presence of a contaminant is only suspected, the procedure is: When in doubt, decon.

In most cases the detail given to decon is determined by the safety and health hazards posed by the contaminants. A flammable liquid such as ethyl alcohol may present an immediate life hazard if it ignites but can be decontaminated relatively easily by flushing the liquid from protective clothing. In contrast, a viscous liquid such as diesel fuel requires a more aggressive scrub and cleaning program. Therefore, it is important to have concrete facts regarding the chemicals involved prior to establishing a decontamination area, called the decon corridor.

Establishing Control Zones

This chapter focuses on setting up and using this **decon corridor**. A decon corridor usually consists of a series of pools to rinse and scrub off contaminated personnel exiting a **hot zone,** and then removing their equipment and chemical protective clothing. Observe in Figure 8-1 the position of the decon corridor (and entry corridor) in relation to the **warm zone.** The decon corridor shall always be in the warm zone (also known as the contamination reduction zone). Thus, before a decon corridor is set up, control zones have to be established.

After you have established a decon corridor (CRZ), steps can be taken to further prevent the spread of contaminants. For example, using a drop cloth with rolled edges as a platform for the decon corridor will act as a secondary containment system As much as possible, the decon corridor should be placed uphill and upwind of the hot zone to allow the runoff of excess contaminants or cleanup materials (usually water) to flow back into the hot zone if spilled.

Once the zones have been established and the area of the spill or discharge isolated, then the decon corridor is set up. Because no contamination has passed

---

decontamination
the complete and systematic removal of contaminants from people and property

■ **NOTE**

Taking a systematic approach allows decon to be complete and thorough, no matter how involved the process.

■ **NOTE**

Even if the presence of a contaminant is only suspected, the procedure is: When in doubt, decon.

decon corridor
designated area in which decontamination takes place

hot zone
area of highest contamination and hazard, to be entered only by properly trained and protected individuals; also known as the exclusion zone

warm zone
area where the decon corridor is placed, also known as the contamination reduction zone

Chapter 8 Principles of Decontamination

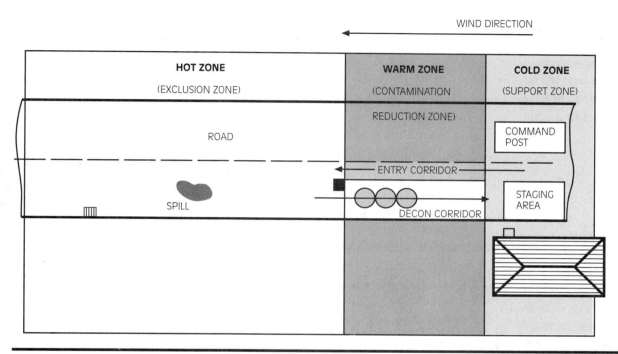

Figure 8-1 *Decontamination corridor setup between the exclusionary and contamination reduction zones.*

■ **NOTE**
The decon corridor shall always be in the warm zone (also known as the contamination reduction zone).

■ **NOTE**
A general rule is that the decon team is usually one level of chemical protective clothing below the entry team.

from the hot zone into the warm, the warm zone can usually be considered uncontaminated until used for the first time. This allows personnel to set up the decon corridor without undue worry about chemical contamination and thus in Level D clothing. While this is being done, other personnel can start dressing up in the required level of chemical protective clothing (CPC). A general rule is that the decon team is usually one level of chemical protective clothing below the entry team. For example, if the entry team is in Level B, the decon team is in Level C. Be aware that factors such as the toxicity of the chemical(s) involved may change this rule. Because decon is the systematic removal of contaminants, the beginning of the decon corridor may be more contaminated than the end. Thus, the decon team member closest to the hot zone and the area of highest contamination may be in the same level (in this example, Level B) as the entry team.

Setting up and staffing a decon corridor is a procedure that is improved only by practice and experience. In many cases, the entry team is ready to go into the hot zone well before the decon corridor is established and the decon team is ready. It is important to note that under "normal" circumstances, entry should not be made into the hot zone until the decon corridor is ready. As a general rule the entry team should go to work expecting the decon corridor to be ready when they exit—at any time. If for some reason, such as equipment failure or injury, the entry

> **■ NOTE**
> Under "normal" circumstances, entry should not be made into the exclusion zone until the decon corridor is ready.

team has to come out immediately, the decon corridor and team must be ready to receive them.

The final factor to keep in mind when setting up control zones and establishing a decon corridor is controlling the routes of entry and egress from the hot zone. A general rule is to have one way in (the entry corridor) and one way out (the decon corridor). This allows for control of access and accountability of personnel entering the hot zone. There is no quicker way to lose control of zones if personnel are wandering in and out at random. In addition, by placing the entry corridor and decon corridor side by side, the decon team can perform a final inspection on the personal protective equipment (PPE) of the entry team as it progresses into the hot zone. In some response activities, the entry team may go into and out of the decon corridor in lieu of a separate entry corridor. This example is just one of the many different ways to organize a response.

Emergency Decontamination Procedures

> **■ NOTE**
> The only time that entry into the exclusion zone is made without a complete decon corridor being set up is in time of immediate rescue or unusual hazard.

In some situations, formally establishing zones and setting up a complete decon corridor before entry is made may not be applicable. The only time that entry into the hot zone is made without a complete decon corridor being set up is in time of immediate rescue or unusual hazard. If there are victims involved that require immediate medical assistance, do not delay in setting up a basic decon area. Remember however, that you do not do any good in injuring yourself to help others. Ensure that at least some form of PPE is worn to protect you from whatever harmed the person being rescued. Have a basic decon available such as an emergency eyewash station or a shower.

It is important to keep in mind that people have different levels of modesty. Even when you are doing your best to decontaminate someone by removing a chemical from them and their clothes, they may object to being stripped in public. Reassure involved people you are performing decon for the best of intentions. If necessary, set up screens such as plastic sheeting or blankets. If someone does have their clothing removed, have a clean Tyvek suit to step into or rescue blanket to cover up with.

All of these basic factors must be kept in mind when setting up and then using a decon corridor. This chapter discusses the basics of decon and lists some examples to assist in preparing a decon corridor. In addition, concepts of teamwork and delegating authority under the incident command system will help everyone to work together to control the incident and prevent it from spreading.

METHODS OF DECONTAMINATION

Every hour of every day hundreds of chemical compounds are used in commercial businesses, industrial production, and even for everyday household purposes. Therefore, it is important to have knowledge and equipment for multiple methods of decontamination. Fortunately there are reference guides that can assist us in determining the right method for performing decontamination. It is best to avoid guessing or making an "estimate" of what hazards are posed by these hundreds of chemicals; use reference sources. For example, a material safety data sheet (MSDS) may inform you as the responder of the right type of decon material to use. In most cases, water is the accepted medium for decon, but this is not always the case. Water can be less than effective when washing off thick, viscous oils and should be used carefully when applied to strong acids and bases. In the case of oils, water mixed with a commercial detergent can break up the oils and in the case of acids and bases, any water reactive materials should be treated with caution. Even such water reactive acids like concentrated sulfuric acid are decontaminated with water. One item to consider however, is that heat is generated when water is mixed with most acids and bases.

As you can see, there are many ways of "skinning the decon cat." The process of decontamination can be divided into three acceptable techniques: dilution, absorption, and chemical degradation. No designated order exists for these techniques because hazardous materials and availability of decontamination resources vary.

Most dilution processes use water to flush the contaminate from protective clothing and equipment. Because water is readily available and relatively economical, this method is generally used first. By using dilution you risk possible reactivity with some materials and create contaminated water that must be contained and disposed of properly. By adding water to some chemicals, such as dry powders, you may only create a bigger problem instead of reducing the hazard. The simple application of water alone is not an effective means of decontamination in these cases. You must use a brush or sponge to scrub and scrape the contaminates from protective clothing prior to using water to rinse. It is important to understand that the application of water to hazardous materials may only reduce concentration, not necessarily change the strength of the material. An example is trying to dilute sulfuric acid with enough water to change the pH. Eventually the pH will be close to neutral, but a huge amount of water will be required. Subsequently, the contaminated decon water must be contained and properly disposed of. In the final analysis you may have solved your decon problem but created a larger issue with the disposal of the runoff. It is best to think about decon in a big picture framework.

The use of **absorbents** or absorbent material may be effective when decontaminating equipment or property but less effective for decontaminating personnel. Absorbent materials include soil, anhydrous fillers, clean dry sand, and commercially available products such as expanded clay. One advantage of the absorption technique is its ability to minimize the surface area of a liquid spill. Absorbents are also inexpensive and readily available. Disadvantages are that absorption materials are commonly limited to flat surfaces such as paved areas. Hazardous materials confined to an absorbent remain chemically unchanged and retain their original health and safety hazards. Finally, the use of absorbents usually creates a larger amount of contaminated material that again has to be properly disposed of.

■ **NOTE**
Decontamination can be divided into three acceptable techniques: dilution, absorption, and chemical degradation.

■ **NOTE**
Most dilution processes use water to flush the contaminate from protective clothing and equipment.

■ **NOTE**
The application of water to hazardous materials may only reduce concentration, not necessarily change the strength of the material.

absorbents
any materials capable of soaking up a spilled liquid

> **■ NOTE**
> Chemical degradation alters the chemical structure of the hazardous material.

For example, absorbent spill pads are very efficient in picking up oils, but response personnel tend to overuse them and are then stuck with many dirty spill pads that need to be properly cleaned up and properly disposed of.

Chemical degradation alters the chemical structure of the hazardous material. Some of the more commonly used chemicals for degradation are sodium hypochlorite (household bleach), sodium hydroxide as a saturated solution, sodium carbonate slurry (washing soda), calcium oxide slurry (hydrated lime), liquid household detergent, and ethyl alcohol. This type of degradation should not be confused with the degradation of PPE. Conceptually it is similar, but in these cases degradation is employed to reduce a chemical hazard in a controlled fashion. Knowing the material that caused the exposure allows you to choose the appropriate chemical for degradation. A common practice is the neutralization of corrosives. It is important to have the right information and facts before using neutralization techniques. You should refer to the chemical manufacturer, MSDS, or other reference materials to gain this information. For example, if an acid is involved, you would choose a base for degradation (reference Chapter 2 for a common neutralization reaction). When using chemical degradation as a method of decontamination, the most important safety factor is to make certain the chemical used is compatible with PPE. Remember, the goal of chemical degradation is to render the hazardous material less harmful than it was prior to decontamination. Doing so may also benefit in the final disposal of the contaminants.

> **■ NOTE**
> It is important to have the right information and facts before using neutralization techniques.

> **!Safety**
> When using chemical degradation as a method of decontamination, the most important safety factor is to make certain the chemical used is compatible with PPE.

SITE SELECTION AND MANAGEMENT

It is important to ensure that at waste site operations, a site safety plan is completed prior to entering an hot zone, and decontamination procedures must be included with the site safety plan. Therefore, it is ultimately the incident commander's responsibility to ensure that decontamination is set up and managed properly. When selecting a site for decontamination, the incident commander, or an assigned **decon team leader** must evaluate current and expected physical conditions. Factors such as wind speed and direction, incident access, water supply, and environmental surroundings directly affect the location of decon at an incident and must be taken into account. The person in charge must also evaluate if resources are immediately available to perform decontamination or if additional resources are needed. For example, portable water supplies may be used in remote locations or on waste sites.

decon team leader
the management position responsible for the operations and support of the decontamination team

Ideally, decontamination sites are located upwind, uphill, and upstream. Sometimes this location is not always practical, such as between buildings or along a waterfront. Shifting winds and migrating gas clouds should be taken into consideration beforehand to avoid moving the decon corridor once it is functional. In many cases the direction of the wind can be predicted by either expecting a local change or by contacting a local weather information source such as an airport or the Coast Guard. Always be ready to move the location of the decon corridor if required and preplan other locations.

> **■ NOTE**
> Ideally, decontamination sites are located upwind, uphill, and upstream.

Decon corridors a long distance from the spill itself may require transportation to and from the incident. In addition, the greater the distance, the greater the stress put on the entry team in terms of carrying equipment. Cleanup procedures requiring the use of Level A protective equipment will further compound transportation problems. By wearing large bulky suits and requiring a constant supply of air, an entry team needs to be creative when looking for transportation. Common modes of transportation are open vehicles such as pickup trucks. Keep in mind that the operator of the vehicle must also have the same level of protective clothing as the entry team, and the vehicle will need to be decontaminated following the incident. There have been documented cases where vehicle decontamination is not practical and the vehicle required proper disposal, in other words, it was thrown away.

Therefore, a balance is required in establishing distances between the zones. On one hand you do not want to have the entry team separated by a large distance, thus creating the problems we have just discussed. On the other hand, you do not want to put the decon corridor and the decon team in too close a proximity to a potential chemical hazard. Try to strike a balance between the two extremes. Address the needs of both the entry team and the decon team. Never put the safety of one over the safety of another.

Managing the Decon Area

Determining the establishment of zones and how far to set up a decon corridor from the spill are just two decisions hazmat personnel have to make. In order to delegate the decision-making process into reasonable "chunks," most hazardous materials response teams have adopted the principles of the incident command system (ICS). The ICS is discussed at length in Chapter 9, but a basic understanding of it in terms of decontamination is important at this juncture. At the hazardous materials incident the incident command system is utilized for organization and management, accountability, and safety. If extensive decontamination is required, a decontamination officer or decon team leader is the person responsible for all decontamination operations and command. The decon team leader consults with chemical experts and determines the appropriate decontamination methods to use, how much decontamination is required, and how much will be completed at the incident.

To assist in the control of the incident, the person in charge of decon works with other personnel to establish and determine where the control zones are to be set up. When a suitable decontamination site has been selected, an isolated perimeter should be established to mark the area where contaminated personnel should go for decontamination. As stated before, this area serves as the decon corridor in the warm zone and provides the route that should be taken between the hot zone and the cold zone. Remember, the decon corridor is the designated area where decontamination procedures take place. The decon corridor should be easily identified perhaps with orange cones and barrier tape or rope. Additionally,

■ **NOTE**
If extensive decontamination is required, a decontamination officer or decon team leader is the person responsible for all decontamination operations and command.

■ **NOTE**
The decon corridor should be easily identified perhaps with orange cones and barrier tape or rope.

warning signs should be placed well in advance of the decontamination zone to alert all personnel of the present danger. This procedure also allows the decon officer to maintain a higher level of control for the area.

In addition to controlling the establishment and setup of the decon corridor, the decon team leader is also responsible for the decon team itself. No matter how much specialized equipment is used, no matter how involved and technical the decon process is, it takes trained people to staff it properly. People, not equipment, perform decon. The decon team members have many responsibilities. Not only do they have to decon the entry team, but any equipment that is contaminated, as well as themselves. The last people to go through decon are the decon team members themselves. The decon team leader has the responsibility to ensure the decon team is properly protected with the right type of PPE, has the right tools to do the job, and understands how to perform the decon process.

■ **NOTE**
The last people to go through decon are the decon team members themselves.

Planning the Extent of Decontamination

The overall goal of a decontamination procedure is to have a written standard to increase personnel safety and decrease contamination risk factors. When developing a decontamination procedure the incident commander must first determine how much decontamination is needed. Minor incidents require a relatively simple decon and minimal washing. Larger scale incidents require larger scale decon areas, more technical equipment, more room, and certainly more personnel. Observe in Figure 8-2 how a typical two-pool decon corridor is set up.

This basic corridor can be scaled up or down in amount of materials and steps depending on the need. It does, however, serve as a good foundation to work from. Additional factors to think about when setting up a decon corridor may include the following:

- Type of contaminant
- Amount of contaminant
- Level of protection
- Work function
- Location of contaminant
- Reason for leaving the site
- Type of protective clothing worn

■ **NOTE**
A general rule is that decon should use as much equipment and decon material (such as water) as required, but no more.

The following paragraphs discuss each of these seven factors in some detail to assist you in making the decision on what exact decon procedures are required and to what extent. A general rule is that decon should use as much equipment and decon material (such as water) as required, but no more. Try to keep all of your steps, processes, equipment, and personnel at the minimum to do the job properly. Remember, if it is contaminated, then it must go through decon.

The first factor is having prior knowledge of the type of hazardous material involved. This factor is crucial in determining the extent of decontamination

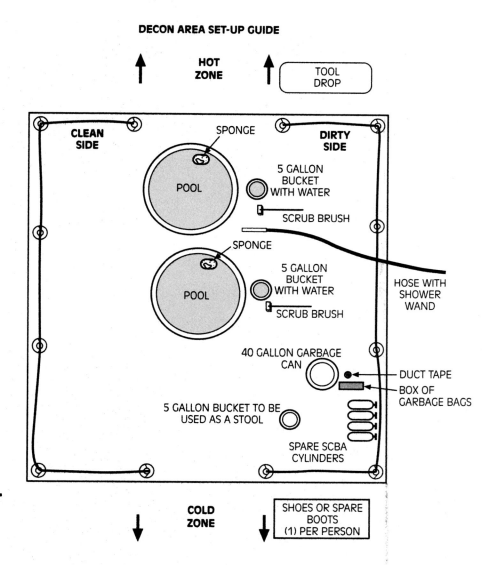

Figure 8-2 *Typical two-pool decon setup.*

required. Not all contaminants exhibit the same degree of hazard. Therefore, the more toxic a substance is, the more extensive and thorough decontamination must be. Whenever it is known or suspected that personnel can become contaminated with highly toxic or skin destructive substances, a full decontamination procedure should be implemented. If less toxic hazardous materials are involved, the procedure can be downgraded accordingly.

The next factor to consider is the amount of contamination. The amount cannot always be visually determined so treat all personnel and equipment accordingly. In other words, whatever comes through the corridor should be considered fully contaminated until *proved* otherwise. The decontamination team members need to consider that any hazardous material on protective clothing may penetrate, degrade, or permeate the clothing and harm the wearer. Therefore, team members must organize themselves and work rapidly, but thoroughly, through the decontamination process. Because the regulations under 29 CFR 1910.120 require the use of the **buddy system,** the decon team shall have to determine if one of the entry team members is more contaminated than the other. In this case, the entry team member who is most contaminated goes through decon first in order to reduce the possible impact of the contaminant upon their chemical protective clothing. Another way of determining who goes through decon first is by the amount of air remaining in the self-contained breathing apparatus (SCBA). The person with the least amount of air generally goes through the decon corridor first as he or she is more likely to run out during the decon process. Local custom may dictate which standard you use.

The third factor is the level of protection worn by the entry team, which further sets the standard for the decontamination layout and procedure. Each level of protection incorporates a different set of problems for the decontamination team. For example, a Level B entry team will usually have a self-contained breathing apparatus that requires extensive decontamination due to being outside of the suit. Entry team members in Level A suits are like "walking English muffins." The suits have the advantage of protecting the SCBA, but tend to be large and bulky with many "nooks and crannies" for contaminants to hide. It requires thoroughness to rinse and scrub contaminants from these suits.

The work functions of the entry team also determine the potential for contact with hazardous materials. If the entry team is simply on a fact-finding mission to determine the extent of the materials involved, they may be less contaminated than if they were assigned to clean up a spill. Knowing the job function of the entry team assists the decontamination team in setting up the decontamination corridor. In many cases, a non vaporous spill on a floor allows little contact with the material if the entry team uses some common sense. However, stopping a leak from a pressurized piping system (such as ammonia) can deluge the entry team member from head to foot.

The physical location of the contaminant on personnel is also of concern to the decon team. If an entry team member is contaminated on the upper portions of their PPE, they may run the risk of higher exposure than an individual contaminated on the arms or legs. Therefore, extra time must be taken when decontaminating the chest and head and removing the protective clothing from these areas. However, the chest and head are not the most commonly contaminated areas. The two areas most commonly exposed to high levels of contaminates are the hands and feet. Feet especially will be highly contaminated because they are the primary means of transportation for the entry team.

The reason for leaving the exclusion zone is another factor that determines the need and extent of decontamination. For example, if an entry team member leaves the spill to pick up or drop off tools or instruments but does not leave the

---

buddy system
a concept of teamwork among responders and site workers, requires a minimum team of two to operate in hazardous areas

■ **NOTE**
The decon team members need to consider that any hazardous material on protective clothing may penetrate, degrade, or permeate the clothing and harm the wearer.

■ **NOTE**
The two areas most commonly exposed to high levels of contaminates are the hands and feet.

hot zone, they may not require decontamination. A team member leaving to get a new air cylinder or change an air purifying respirator cartridge may require some degree of decontamination because they will enter at least part of the decon corridor. Sometimes entry teams encounter unforeseen problems or victims in the exclusion zone and need to leave immediately. In this event, all team members and victims need to go through emergency decontamination. The most common reason for leaving the hot zone before the task is completed is running low on air while wearing an SCBA. Due to different respiratory rates and the physical exertion of the job itself, some entry team members will run low on air before others. In this situation, both entry team members leave at the same time and the one with the lesser amount of air goes through decon first. Don't panic! SCBA systems are designed to give approximately 5-minutes warning before the air supply is exhausted, which is more than enough time to perform a thorough decon procedure. The decon team should reassure the entry team members that there is enough time to get them through before the entry team members run out of air.

Finally, most types of protective clothing are disposable and require a minimum of decontamination. For example, a Tyvek suit is inexpensive and designed for single use only. Why waste time and effort in decontaminating a suit that is to be thrown away? In addition, by decontaminating suits and equipment that are to be disposed of, more contaminated runoff may be created and that has to be properly disposed of as well.

All of these factors must be taken into account before establishing a decon corridor and performing decon. Determine beforehand who will make the decision as to who goes through decon first. When the entry team members enter the decon corridor, they are in the hands of the decon team. The decon team tells the entry team members exactly what they want them to do. For example, stepping from one pool into another requires lifting one foot, having the sole rinsed and scrubbed, and then stepping into the next pool. Entry team members may be hot and tired or they may be low on air. If they are in a Level A suit they have limited hearing and visibility. Talk loudly, but reassuringly to them. Have the entry personnel move around to gain access to all areas of the PPE instead of moving the decon personnel. This procedure limits movement of the decon team members, reducing possible contamination as much as possible.

> ■ **NOTE**
> When the entry team members enter the decon corridor, they are in the hands of the decon team.

DECONTAMINATION PROCEDURES

After the factors just described have been addressed and identified, the next step is to implement it all into setting up and running a decon corridor. An example of this can be seen in Figure 8-3. At this stage, the contaminant(s) have been identified and the right PPE has been chosen. Keep in mind that this process may require some time unless you are performing emergency decon. In many situations, the entry team will be ready to go well before the decon team is set up and ready. As stated, under normal operations, the entry team is not to enter until the decon team is ready.

Figure 8-3
Schematic top view of a two-pool decon pad.

■ **NOTE**
The cold zone is where all support services are located including the command post.

control point
the monitored area where the decon corridor enters the hot zone

To speed up the process in setting up a decon corridor, follow these three steps:
1. Select an appropriate decon site.
2. Assemble the equipment and personnel resources required.
3. Assign the decon team.

In the first step, the incident commander establishes the need for decontamination. After this has been accomplished, an appropriate site must be selected. It is preferred to select a site that borders the hot zone and is upwind, uphill, and upstream. The decon corridor serves as a walkway from the hot zone to the cold zone. The cold zone is where all support services are located including the command post. Next, the area should be marked off and isolated with specific entry and exit points using barricade tape and cones. The size of the decontamination area depends on the number of stations required by the incident, amount of space available, or the nature of the incident. A corridor of 75 feet by 15 feet should be adequate for a complete, multistep decontamination process. Whenever possible the decon corridor should be set up so an entry team member is able to flow from the **control point** through decontamination and to the cold zone without difficulty. Avoid dog legs, corners of buildings, or other factors that may inhibit the flow of the decon process.

Step 2 is assembling the appropriate resources required for decontamination. In some procedures, water, breathing air, and electrical power may be

accountability log
written record of information such as products involved in the incident, times of significant changes in the incident, names and entry/exit times of personnel in the exclusion zone, level of personal protective equipment used, and function of personnel in the hot zone

■ **NOTE**
A good rule of thumb is that for every two persons entering the hot zone at least one additional person should be assigned to the decontamination team.

needed to support decontamination efforts. Water is typically used for the initial rinsing of personnel, to supply portable showers, and decontamination of equipment. It can be found in fire engines, a nearby fire hydrant, or perhaps a residential/commercial water supply. If breathing air is required for the entry and decontamination teams, spare air cylinders or an umbilical air manifold should be available. Air cylinders must be decontaminated before they are taken out of the decon corridor to be refilled. Electrical power may be needed to operate pumps, vacuums, and portable lighting.

The third step is to assign the decontamination team. A decontamination officer is assigned by the person in charge to direct all activities in the decon corridor. This individual is responsible for monitoring the control point, coordinating information between the exclusion zone and the support zone, and establishing and maintaining an **accountability log**. When the incident is terminated, this log becomes documentation and is filed with the incident report.

In a typical two pool set up, three additional personnel are recommended for decontamination: two for initial wash down and to assist in removal of contaminated clothing inside the decontamination area and one in the support zone to distribute clothing and provide outside assistance. All personnel entering the area must have a specific assignment and must check in with the decontamination officer. On large emergencies, or if those being decontaminated are injured, more personnel may be necessary to facilitate decontamination. A good rule of thumb is that for every two persons entering the exclusion zone at least one additional person should be assigned to the decontamination team. An example of a properly staffed decon corridor can be seen in Figure 8-4.

Figure 8-4 *Students working as decon team members in mock scenario.*

!Safety
The decon team should avoid direct contact with contaminated equipment and personnel as much as possible.

The decon team should avoid direct contact with contaminated equipment and personnel as much as possible. Brooms, scrub brushes, and hoses with spray nozzles can be used to apply cleaning agents, move equipment, and guide contaminated personnel from one stage to the next. A good practice is to keep all the decon team on one side of the decon corridor in order to reduce the possibility of cross-contamination from overspray. Figure 8-5 shows an entry team member entering the decon corridor. Observe that a decon team member is using a broom handle to direct the movements of the entry team leader without making actual physical contact.

To further protect the entry team members, the appropriate level of personal protective equipment must be worn and maintained. Keep in mind that an acceptable level of protection is the same as the entry team or one level below the entry team. If the entry team is in Level A protection, then it is appropriate for the decon team to wear Level B. Keep in mind the seven factors that determine the extent of decon. If the chemical hazard present poses a significant risk, then the decon team may require a higher level of chemical protective clothing; perhaps the same as the entry team. On the other hand, the chemical may be readily decontaminated. Once

Figure 8-5 *Gross decon—getting the worst stuff off first.*

again, the first member of the decon team may be in the same level as the entry team and the other decon team members in one level below.

When the decontamination process has been completed the site should remain secure until isolated clothing and equipment in the **tool drop** can be removed for proper disposal. If contaminated materials are to remain at the scene for a considerable amount of time, provisions should be made to secure the area. Appropriate warning signs or labels should be placed near or attached to containers awaiting disposal. Security, support, and lighting may be required if the contaminated materials remain overnight. After all contaminated materials have been disposed of, all the equipment used for decontamination also needs to be cleaned, inspected, and tested. Equipment such as cotton jacket fire hose and vehicles that have been exposed to hazardous materials should be completely decontaminated, inspected, and tested before being placed back in service.

tool drop
area designated inside the hot zone, just outside the decon corridor, where grossly contaminated tools and other equipment are placed for later cleaning

Post Incident Analysis

The final step is to make arrangements for a post incident debriefing. Personnel involved should be provided with as much data as possible regarding the materials encountered during the incident. Follow-up examinations should be scheduled with medical professionals and records maintained for future reference.

■ **NOTE**
Follow-up examinations should be scheduled with medical professionals and records maintained for future reference.

Six-Step Level B Decontamination

As discussed in the factors determining how decon is to be performed, it is the responsibility of the entry team to determine who goes through decon first. As entry team members proceed through the step-by-step decon process, they shall be guided and instructed by the decon team members. Guidance and instructions can be verbal, by using hand signals, or by the use of portable radios. The following is an example of a typical six step process for decontaminating personnel in Level B protective clothing.

Step 1   Equipment drop
Step 2   Gross decontamination
Step 3   SCBA change (if needed)
Step 4   Breathing apparatus removal
Step 5   Safety suit and boot removal
Step 6   Field shower and redress

The first step is the equipment drop station. The tool drop is actually in the hot zone so contaminated tools can be left there for future use or decontamination. Entry team members should place all tools and equipment used in the hot zone at this designated tool drop. They should place any written information for the incident commander into a plastic bag or other isolation device for transportation through the decon corridor to the command post.

Figure 8-6
Scrubbing contamination off the PPE.

■ **NOTE**
Begin decon procedures at the head and proceed to the feet in a systematic and thorough manner.

The second step is referred to as gross decontamination. As shown in Figure 8-6, an entry team member is directed to step into the first wash and rinse pool. These pools should be large enough to contain all runoff that comes off the person being decontaminated. As you may have noticed, the pool seen in Figure 8-6 is too small for the job at hand. Be careful not to overspray and spread runoff or fill up the pool. Equipment such as scrub brushes or sponges shall be used to scrub the entire suit. Remember again that decon is a systematic procedure. It is a common mistake to first decon a person where the contamination is highest. This action can be self-defeating. Don't scrub off the chest when contamination remains on the head. The runoff from the head will again contaminate the chest.

Begin decon procedures at the head and proceed to the feet in a systematic and thorough manner. Tell the person being decontaminated to move around in a circle so the decon team member can clean all sides of the person without having to move around themselves. Once the entry team member has been scrubbed and washed they are directed into the second pool. Pools must be placed close enough together to step from one to the other without having to step on the tarp. Personnel with an SCBA on their back may feel off balance as they move through the decon corridor. Again, decon team members should tell them what they expect them to do and assist them by holding out brushes or having stands to hold on to. As team members move from one pool to the next, it is important to scrub the soles of the protective boots. This may put them off balance. Have the person raise one foot while holding on to a brush or stanchion, and scrub and rinse the foot off. The scrubbed and rinsed foot then steps into the next pool and the other foot is raised. You can see why the pools have to be close to each other! Now in the second pool a thorough rinsing and inspection should take place. While performing a visual inspection the decontamination team should make certain that all contaminants are removed before moving to the next step.

Step three is the tank change station. This process is only needed for personnel who are returning to the exclusion zone and have a low air supply. Keep in mind that even though the team member only needs a bottle change they must be decontaminated prior to the exchange.

Step four is the breathing apparatus removal station. The decontamination team removes the low pressure hose from the regulator if possible, and directs the entry team member to hold it high in the air and away from the body, using either hand. The decontamination team then removes the backpack harness and places it in a plastic bag for a later inspection. It is important to leave the face piece on to provide splash protection for the eyes and routes to the respiratory tract.

Step five is the safety boot and suit removal station. As seen in Figure 8-7, the decon team member directs the entry team person to step into a large plastic bag that has been rolled down. This allows the decontamination team to remove protective clothing systematically. Begin by removing all tape used to seal the gloves, boots, and hood. Next, beginning at the head and shoulders, remove the suit by turning it inside out, or rolling it off the person without touching them. A comparison is like peeling a banana by touching only the outer side of the peel and not the fruit itself. As the suit is rolled down toward the feet, tell the member to sit on a stool and continue removing the suit right into the plastic bag. The member steps out of his or her boots and swings his legs to the side. Then the team member is directed to step away from the bag and remove the breathing apparatus mask and inner gloves and place them in the bag with the suit.

Figure 8-7 *Decon team members removing an SCBA cylinder from an entry team member.*

The final step is completed in the cold zone and includes the field shower and redress station. All entry team members need to shower using clean warm water and lots of soap. It is important to ensure the privacy of all participants involved in the incident, so this station may be located away from the scene.

Putting all of these steps together and working as a team in a coherent manner takes time and practice. Do not expect to perform decon procedures perfectly the first or even second time. For this reason decon should be practiced as much as possible with drills and exercises before the real thing. Get used to the time factor in setting up decon and getting contaminated personnel through in a safe and thorough manner. No matter how well you practice, some stages will always go faster than others. Observe in Figure 8-8 how personnel have to wait for their turn to be undressed. While decon team members such as rinsers and scrubbers have done their tasks rapidly, the "bagger" has to work hard to remove all of the PPE from the person being decontaminated. Be patient and understanding. The bagger has the most demanding job of all. Finally, remember that the decon team has to go through decon as well. Since the bagger should be the least contaminated, he or she shall be the last to go through decon. When performing this process, do not move too far up the decon corridor. Go only as far as required. That is, "rinsers" start at the rinse station, baggers start at the bagging station. When everyone and everything is done correctly, the decon team has performed its task in preventing the contaminants from spreading and has protected the most important asset of any job—you.

Figure 8-8 *Suit removal.*

Summary

- Decontamination is the systematic process of removing contaminants from people and equipment.
- Decontamination is performed to protect personnel, the environment, and other resources.
- Dilution, absorption, and chemical degradation are three acceptable decon techniques used.
- Chemical degradation alters the chemical structure of a hazardous material.
- The decontamination corridor is located in the warm zone.
- A decon site should be located uphill, upwind, and upstream from the hot zone.
- A decontamination officer or team leader must be appointed and is responsible for operations at the decontamination site.
- Seven factors that determine the extent of decontamination are:
  1. Type of contaminant
  2. Amount of contaminant
  3. Level of protection
  4. Work function
  5. Location of contaminant
  6. Reason for leaving the site
  7. Type of protective clothing worn
- Emergency decon is the rapid removal of contaminants.
- Three steps in designing a decontamination procedure are:
  1. Select an appropriate decon site.
  2. Assemble the equipment and personnel resources required.
  3. Assign the decon team.
- Six steps in completing a full Level B decontamination are:
  1. Equipment drop
  2. Gross decon
  3. Tank change (if required)
  4. SCBA or APR removal
  5. PPE removal
  6. Field shower

Review Questions

1. What is decontamination (decon)?
2. What are three important reasons for conducting decon and which of these is the most important?
3. How are work zones important in relationship to decon procedures?
4. Can personnel in Level D clothing set up a decon corridor if Level A entry is to be made? If so, why?
5. Under normal operations should the entry team be sent into the hot zone before the decontamination corridor is set up? Why?
6. How and when should you use emergency decon procedures?
7. How do decon personnel determine what kind of decon materials to use? What decon material is most commonly used?
8. List the three acceptable decontamination techniques.
9. List the advantages and disadvantages of the absorption technique.
10. Who has the responsibility to ensure decon is set up and managed properly?
11. Describe the responsibilities of the decon team leader.
12. In your own words, discuss how distances are established between the zones.
13. List the seven factors for planning the extent of decontamination.
14. Identify some physical factors to keep in mind when setting up a decon corridor.
15. What are the three ways chemical protective clothing can break down?
16. List three steps that speed up the process of setting up a decon corridor.
17. What is the most common reason for the entry team leaving the hot zone before the task is completed? Why?
18. Describe how different work functions can vary the amount of contamination on entry team members.
19. List the six steps of a Level B decontamination procedure.
20. When removing contaminants from a chemical protective suit, where do you start and why?

Chapter 9

Incident Management and Scene Control

Learning Objectives

Upon completion of this chapter, you should be able to:

- Describe why the incident command system (ICS) is beneficial to emergency management.
- Describe the characteristics of the ICS including the aspects of common terminology, modular format, and span of control.
- Identify the job functions of the following command positions: Incident Commander (IC), safety officer, public information officer.
- Identify the "big five". Understand the role of the IC and the general staff positions of: finance, logistics, operations, planning
- List the responsibilities of the command positions assigned to the IC.
- Understand the terminology and methods associated with taking command of an incident.
- Define work zones and understand their importance.
- Identify supervisory functions that may exist beneath the general staff functions. Examples include security, decontamination team leader, entry team leader, and technical reference.
- Understand the need for a safety officer on all hazardous materials incidents.

INTRODUCTION

Any type of emergency incident, especially one involving hazardous materials, needs to be organized by a management structure. Without the use of a standardized and easily implemented system of management we run the risk of losing track of resources, missing a critical part of the problem, or endangering safety because there is no consolidated plan of action.

Public safety agencies have long recognized this problem and have worked to develop a system of emergency management incorporating the use of common terminology, span of control, resource tracking, and the concept of creating a modular system. Throughout the years, many different systems were developed and modified in an attempt to establish a standardized form of emergency management that could be used across the United States. Eventually, one universal framework of emergency management was established and accepted—the incident command system (ICS). The ICS is so applicable to the management of hazardous materials incidents that the Occupational Safety and Health Administration (OSHA) mandates its use in the Hazardous Waste Operations and Emergency Response (HAZWOPER) regulation.

This concept of a modular system is key to the success of incident management because not all incidents are the same size. In other words, some incidents are big and require many people to be coordinated and managed, whereas other incidents are relatively small and can be handled by a few responders. The greatest feature of the ICS is that it can be used for both types and every other kind of response in between. The ICS can be thought of as an elastic system that can grow or shrink depending on the need. This need is driven by the number of jobs needing to be done at any one time and by the concept of **span of control.** An effective span of control during an emergency is 5:1. This means that for every five responders, there is one direct supervisor.

The ICS effectively provides the framework for organizing job functions, tasks, and staff into a cohesive response effort capable of handling the event. It is the skeleton on which the **incident commander** or IC hangs the meat of the action plan. This flexible system helps transform incident confusion into action by answering the big questions of "Who is really in charge?" and "What am I supposed to be doing right now?" In emergency response, a visible and strong command presence is vital. Leaders must lead and followers must follow for the system to work. If the IC doesn't lead—someone else will!

THE INCIDENT COMMANDER

This section is designed to give you a working foundation from which to hone the skills of incident management. While running an incident is a difficult task, it is possible to understand and apply some common management principles to any type of emergency. One key principle is the concept of incident command and control. We devote the majority of this section to explaining the *concepts* of

■ **NOTE**
The ICS is so applicable to the management of hazardous materials incidents that OSHA mandates its use in the HAZWOPER regulation.

span of control
the maximum number of people that one can effectively supervise at a given time

incident commander
the person responsible for all operations at a hazardous materials emergency

■ **NOTE**
In emergency response, a visible and strong command presence is vital. Leaders must lead and followers must follow for the system to work. If the IC doesn't lead, someone else will!

incident command and how it can be successfully integrated into any emergency response system. Most people are familiar with the concept of command and control simply from working within a professional organization or corporation. A certain chain of command exists within any company and everyone from the top to the bottom has someone else to report to. Hopefully, the organization has goals and objectives and everyone involved understands the chosen direction and works to move the organization forward. Financial concerns and other strategic decisions are based on those goals and if all goes well, the goals are achieved. On the flip side, if not everyone shares the common goal or does not understand which way the organization is going, or does not care, the system begins to break down. Fragmentation of the group may occur and some members may begin to do their own thing based on their perception of what is right or wrong. Soon the group loses the common direction and some **freelancing** may occur. Ultimately, communications deteriorate, there is no effective control, and the organization may find itself expending volumes of effort with few tangible results.

freelancing
term used to denote the actions of a person or group of persons operating outside the organized plan of action

These same basic principles and pitfalls ring true for any emergency response. It becomes your duty as president of the ad hoc response company to stay running in the black, on task, and on schedule. This is no easy feat as many public sector incident commanders will quickly tell you. Veteran fire chiefs have had incidents spiral down the tubes on many occasions because they failed to get a good handle on their problem or simply lost control of the effort. This is not to say that simply taking charge of a response will make your troubles go away. This key concept is overlooked by many people who try to manage an emergency. The ICS does not make the problem go away—people *using* the ICS as part of their management tools will!

■ **NOTE**
The ICS does not make the problem go away—people *using* the ICS as part of their management tools will!

Additionally, ICs have the burden of wearing multiple hats during the course of an emergency. Aside from deciding the tactics and strategy during the response, an IC may be fielding questions from superiors, the media, facility representatives, city officials, and agents of the various governmental entities. These are very real demands made by others who may or may not appreciate what you are trying to do. It becomes your task as an IC to accomplish many things at once. The primary goal of an IC is to keep safety in mind at all times. You should also make it a high priority to not make the situation worse by your involvement or decision making. Your working motto should be "do no more harm" than has already occurred by the initial event: Accomplish this and you are already miles ahead of the game.

■ **NOTE**
The primary goal of an IC is to keep safety in mind at all times.

Your goal is to make the emergency go away as quickly and efficiently as possible in order to return the affected site to a business as usual state. How does all this happen? Essentially it boils down to you. You have decided to put on the IC hat and take charge of a chaotic, fast-moving situation that may have the potential to hurt someone or destroy an expensive facility. To perform this function takes courage, the willingness to accept responsibility, training, continuing education, and the qualities of a leader. To be *good* at this function requires a strong commitment to the previously mentioned requirements as well as a personal dedication to the team. Good leaders may be born, but great leaders work at refining their skills and learning as much as they can about the jobs they are required to do. You may be forced to make some very tough decisions and unless you are prepared to shoulder the burden, don't take the job.

Emergencies are so potentially diverse that it would be impossible to cover all bases or give a step-by-step guide for response work. Instead, it is better to equip the student with a few key concepts that can be applied to any situation. This format is relaxed, informal, and geared at providing a toolbox for managing an emergency. You will then be able to remove as many tools as necessary to solve whatever situation you are facing.

THE ROLE OF THE INCIDENT COMMANDER

At this point in your career you may or may not have had the opportunity to run an incident. If you have had the pleasure, you know it is a very stressful endeavor punctuated by moments of having no idea what you are going to do next. If you have never run the show, then you will soon experience this mentioned phenomena. Managing an incident is ever chaotic and never goes exactly as planned. Some people criticize your decisions while others praise you. Your actions may be quietly questioned and second guessed by many of your team members and you will ultimately earn some sort of reputation as an IC. This reputation may be good or bad but will certainly follow as well as precede you to every response you run. It is therefore critical to your team and yourself to start off on the right foot and understand some very basic concepts from the beginning. All incidents have a certain amount of chaos and disorganization. It is your job to manage this turbulence and direct the incident toward a safe resolution.

As an IC you will be making countless decisions on the emergency scene, ranging from determining **tactics** and **strategy** to incident termination and everything else in between. Many new ICs are overwhelmed with all the decisions and fall into the common trap of trying to decide everything at once. This course will lead to frustration of the responders as well as yourself and may ultimately cause disorganization throughout the scene. It is your job then, to prioritize the incoming flow of information and determine which items warrant your direct attention and which ones can be delegated. At this critical point an IC either gains a handle on the response or becomes the dog being wagged by the tail. Incident priorities vary depending on the nature of the emergency but in general an IC can assume responsibility in the following areas:

• *The IC Is Responsible for the Health and Safety of ALL Responders Working on the Scene.* As an IC you are the premiere safety person on the scene at all times. You are certainly responsible for your own safety first, but you also must ensure that anyone operating on your scene is accounted for and performing his or her function as safely as possible. As an IC you are there to make the situation better, not worse, by your involvement. You did not cause the incident and are therefore not responsible for what has occurred prior to your involvement. Once

■ **NOTE**
You may be forced to make some very tough decisions and unless you are prepared to shoulder the burden, don't take the job.

■ **NOTE**
All incidents have a certain amount of chaos and disorganization. It is your job to manage this turbulence and direct the incident toward a safe resolution.

tactics
methods or procedures used to achieve objectives

strategy
the overall theme or goal of the incident, usually either offensive or defensive

you arrive and take control you are 100% liable for all that transpires and all those operating under your direction. Even if you appoint a safety officer, it does not totally relieve you of the burden of overall scene safety. The IC is directly responsible for the safety of all responders. The overall aim of any commander is to make the problem go away as quickly as possible with the utmost concern for safety. It is your job to make sure everyone goes home at the end of the day!

> **■ NOTE**
> The IC is directly responsible for the safety of all responders. The overall aim of any commander is to make the problem go away as quickly as possible with the utmost concern for safety. It is your job to make sure everyone goes home at the end of the day!

- *The IC is Responsible for Deciding the Tactics and Strategy for the Entire Response.* "What are we going to do?" This simple question has tripped up all incident commanders at some point in their career. It is an easy question to ask, but sometimes difficult to answer. It is even more difficult when many pairs of eyes are on you and the stress level of those asking the question is off the meter. A good IC should be like a duck swimming on the smooth surface of a lake. Above the water, the duck appears serene and moving almost without effort. Below the surface of the water the duck is paddling like crazy! And so goes the life of an IC—calm and cool on the exterior while the brain and emotions are racing to solve the problem. So how does one remain so collected during a time of crisis? Good information and sound planning. Once you fully understand the problem and implement a plan, the rest simply becomes an issue of time management and working toward the desired objective. The million dollar question usually lies in deciding what the plan will be and how it will be carried out.

Initially, an IC must understand the difference between tactics and strategy. A good example of the use of tactics and strategy can be found by looking at a football team. The entire team should understand the basic strategies of the game—the offense is supposed to score points while the defense prevents them. The coach or some other key members of the team use individual plays as the tactics for accomplishing their particular goals. Strategy is big picture thinking for the incident. You may choose an offensive strategy of stopping a leak or a defensive posture of letting the release go until it stabilizes itself.

> **■ NOTE**
> Strategy is big picture thinking for the incident.

The role of the IC is to set the strategic tone for the incident and determine the tactics necessary to solve the problem. In many cases this strategy is developed using a team concept but the final call is made by the IC. One common pitfall here is the analysis-to-paralysis syndrome, which happens when the group "what if's" the situation to death and does not readily move to action. This is not to say that the team should rush into action without proper thinking, but many times developing and implementing a plan of action takes way too long. Be an aggressive planner as well as a decisive and action-oriented leader. Remember, action solves the problem.

> **■ NOTE**
> Be an aggressive planner as well as a decisive and action-oriented leader. Remember: Action solves the problem.

- *The IC is Responsible for Developing the "Battle Plan" for the Incident.* When commanding a hazmat incident, certain objectives must be met during the life cycle of the event. They are the milestones of a response achieved by good decision making and action by the IC. The five objective areas can be remembered by the mnemonic **H-E-A-R-T**.

H—Hazard identification. No matter what the problem is, it is the responsibility of the IC to get as much information as possible and to truly understand the nature of the emergency. It is important to know that a compressed gas cylinder is leaking but more important to understand *why* it is happening. Look at labels, placards, MSDSs, visible clouds, drums, puddles, or whatever clues are available. Always try to get a 360° view of the problem and obtain as much information from bystanders as possible. This first step is critical in coming up with a plan of action. Without truly identifying the problem you are basing your decisions on speculation and luck.

E—Evaluate your response. Consult preplans if they exist. If not, develop a response plan based on facts and probabilities. Once the problem is defined, define your response to it. Do you need additional resources? Do you have enough resources available to handle the incident? Is the event an emergency or a cleanup? This phase is accomplished by giving some clear, concise thinking to defining your goal for the event. Run the incident—do not let it run you!

A—Assemble work zones. No matter what type of incident you are running it is vital that you stake your claim so to speak and set up some work zones. In hazmat response we recognize the hot zone, warm zone, and cold zone as our work areas. Any type of incident, however, should have areas that are accessible only to the response team, thus limiting the exposure to others in the area and giving you some breathing room to think and work. Even a medical response benefits from securing an area in which to operate. This area allows for some privacy for the patient and keeps the team from being interfered with while they work. Do not take more real estate than you can manage, however, as it is detrimental to allocate the majority of your resources to keeping everyone away from the hazard. Take as much ground as you need to work safely. Be practical but generous when you decide how much territory is necessary.

R—Run the incident. It is critical for the IC to take charge of the incident, which means more than putting on the fancy command vest. Taking command is the paramount act that defines the course of the incident and sets the tone for the response. The IC is responsible for setting up the incident command system and presiding at its peak. Part of the job function is to keep a handle on the response and not get sidetracked with things that are not your responsibility. Knowing every detail of every system is not a requirement to be an IC. You can find someone who has the technical knowledge; it is your function to assemble the right team and manage what they do. An IC must strive to have a powerful and confident command presence at all times. Calmness and confidence are as contagious as fear and uncertainty in the face of adversity. By maintaining a calm demeanor and being thoughtful and decisive, you will inspire others to follow suit. The team tends to take on the demeanor of the person leading, and if the leader is strong, the team tends to be strong.

T—Termination of the incident. It is the IC's job to switch the mode of thinking once the incident is over and the cleanup phase has begun. Remember, there is a difference between emergency response and cleanup. They are different phases of a response and should be treated as such. This is not to say that you will simply walk away at the end of every emergency but just remember that your

■ **NOTE**
Without truly identifying the problem you are basing your decisions on speculation and luck.

■ **NOTE**
Run the incident: Do not let it run you!

■ **NOTE**
Take as much ground as you need to work safely. Be practical but generous when you decide how much territory is necessary.

■ **NOTE**
An IC must strive to have a powerful and confident command presence at all times.

primary job may be to terminate the emergency. Part of the termination procedures also include analyzing the incident for lessons learned, ensuring that all personnel are medically sound, performing thorough decontamination and, last but not least, documentation.

- *It Is Up to the IC to Delegate Supervision and Other Functions When Appropriate.* The incident command structure is built from the bottom up and is used to organize the overwhelming number of jobs to be done at a scene. Incident commanders must delegate some job functions to other responders in order to stay on task. If you are an IC, you should never be turning valves, putting on a suit, or resetting alarm panels. You are there to think and direct the operations. Stay focused on running the incident, not running around it! The next section describes the different job functions of the ICS, but for now, understand that delegation is very important for an IC. Delegation of tasks should occur early in the response and the IC should always try to maintain an effective span of control.

Essentially, the proper ratio of workers to supervisors should be established by the IC. For example, if a work party is sent to perform a specific function, the IC should appoint a leader of the group, give the group a name, and some means of communication. The IC could then manage five groups with up to five members each. The IC now deals with only five group leaders, which is much easier to keep track of. Remember, the incident command system is built from the ground up and should be expanded as the response grows. Figure 9-1 shows the

■ NOTE
The incident command structure is built from the bottom up and is used to organize the overwhelming number of jobs to be done at a scene.

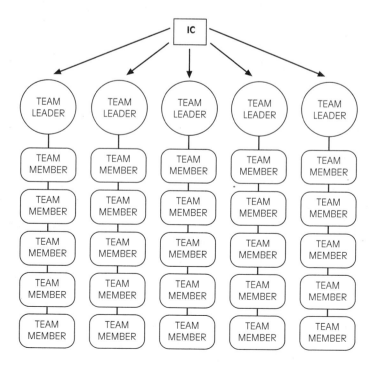

Figure 9-1 *A basic foundation for the incident command system.*

concept of how a small incident may get started and controlled using the ICS. This operation could be done a number of ways, but the important thing is that the foundation is being laid—it can always be expanded if necessary.

If more than five teams are needed or additional resources are called in to perform different functions, the IC has some management options. He or she must realize that the 5:1 ratio is met and it is time to delegate some supervision. The IC must now appoint a leader one level below who supervises the original five team leaders, who are supervising their five person teams. The IC has now bumped up a notch and is supervising only one person who is supervising five teams who are in turn supervising their own five-person team. Confusing? Think of it this way: The IC now has four open slots in his span of control. In addition to the search groups, the IC could now supervise a medical group leader or a hazmat group leader or a fire group leader who may be supervising five other team leaders who supervise five-person teams. Figure 9-2 illustrates the idea of delegation and increasing the span of control of the IC.

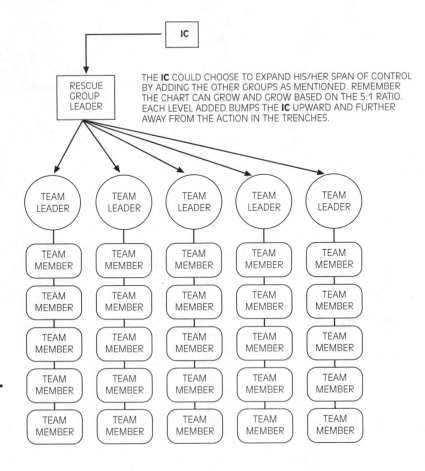

Figure 9-2 *The ICS-expanding the framework as the incident grows.*

CHARACTERISTICS OF THE ICS

One of the advantages of the incident command system is that it is very simple to understand and use. Think of the ICS as a big tool box with a bunch of drawers that can be pulled out at any time. Once the IC takes charge of the incident he or she is free to open drawers and use the management tools at will. In advance of opening the box, however, the IC should understand some of the basic characteristics of:

- Modular structure
- Unity of command
- Span of control
- Standardized terminology

■ **NOTE**
Once the IC takes charge of the incident, he or she is free to use the management tools at will.

Modular Structure

If the incident is of any size, or if it gains in complexity, the command structure will grow accordingly. One of the key characteristics of the system is that it is modular and can expand as the incident becomes more complex.

Breaking supervisory tasks and objectives into manageable chunks is a necessary part of the overall management program. It is not much different to manage an emergency than it is to manage any other complicated task. It often involves delegating key functions or areas of responsibility to others in order to accomplish the goals. If possible, all key functions or supervisory positions that are established should have radio communications with the IC. If this is not possible, other ways to communicate need to be developed. It is vitally important that the incident commander be kept informed of progress or problems to facilitate the decision-making process. The incident command system is a modular management system. Use whatever building blocks are necessary to create an effective framework for command and control.

Smaller incidents may require a scaled down command structure. It may be as simple as assigning one person to supervise a small workforce assembled to accomplish a specific task. When this is the case, ensure the group has radio communications, a call designator, and a system to track their location at all times. An example may be sending a small group to investigate an odor inside a building. The commander assigns the group a designator such as "search team" and gives them a specific task—Enter building 3 and make sure the entire structure has been evacuated. The search team would report directly to the IC and relay all pertinent information. Just remember to create a simple management structure for a simple task.

■ **NOTE**
The ICS is a modular management system. Use whatever building blocks are necessary to create an effective framework for command and control.

Unity of Command

A second important aspect of the ICS is that everyone works for someone at all times. The concept of one-person, one-boss is vital. Everyone involved needs to be responsible to someone, and someone needs to be responsible for everyone.

Otherwise, you run the risk of losing track of responders in the chaos that often accompanies an emergency situation.

The other part of unity of command deals with the idea that the incident has one person running the show—the IC. He or she develops the **operational goals** and plans for the incident. It would be a losing prospect to have multiple goals developed by different people who have action plans that compete with each other. You must control the incident—Never let it control you! This is accomplished through a strong command system that is unified with one plan and clearly understood goals and objectives.

operational goal
the end point of the incident as identified by the IC

■ **NOTE**
You must control the incident: Never let it control you! This is accomplished through a strong command system that is unified with one plan and clearly understood goals and objectives.

Span of Control

Span of control has been mentioned previously but is important enough to repeat. All of us have limitations as to the things we can do efficiently. In an emergency, it is critical that we observe those limits and do not stretch to the point of breaking.

This concept has been widely studied, and although there are a number of variables, including the type of situation, expertise of the subordinates, or skill of the supervisor, it is generally accepted that in emergency situations, the maximum amount of people an individual should lead is five. This is not a hard-and-fast rule that can never be violated, but, emergency situations are dynamic and can rapidly change. Do not be caught with too large a span of control. Break up the problem as needed, and take advantage of the skills of those in the organization.

Standardized Terminology

The fourth concept of the ICS is that it uses a standardized language. Without a common communication system, the incident would be destined for failure. This common terminology includes common descriptions of activities, common use of key terms to designate actions, and common names to denote job functions. Following are examples of some of the standard ICS positions. One of the advantages of the ICS is that these positions do not necessarily need to be filled in every incident. Because of its modular nature, it can grow as needed to meet the needs of the particular situation. The important point to consider is that if these positions are utilized, the job functions are standard. Some examples of positions that may be filled during an emergency are as follows:

- Safety officer
- Public information officer
- Decontamination team leader
- Entry team leader
- Medical group leader
- Scene security
- Hazardous materials group supervisor
- Technical reference
- Site access control

These functions are described in the following section but for now understand that there are some standard names for some common functions within the ICS.

ICS—The Big Picture

Now that the characteristics of the system are understood it is time to flesh out some of the job functions that may be designated during a response. These are the biggest functional areas of the incident command system and are only implemented in larger incidents. The exception may be the operations section which could be implemented when multiple areas need to be managed. The operations chief, for example, may run fire suppression, medical response, and hazardous materials activities by using division leaders. The big five in terms of functional sections include:

1. Incident commander
2. Operations section
3. Planning section
4. Logistics section
5. Finance section

general staff
the group of leaders heading the major divisions under the incident commander

The bottom four report directly to the IC, are referred to as **general staff,** and may supervise an infinite amount of activity under their particular discipline. Using span of control, each section chief could manage hundreds of people by maintaining a 5:1 ratio. How this occurs will become obvious once we show the actual structure. Following are the general responsibilities of the big five:

Incident Commander
- Is the ultimate authority and decision maker
- Sets policy for the incident and establishes the action plan
- Oversees incident safety even though a safety officer has been appointed
- Interfaces with public agencies

Operations Section
- Carries out tactical actions based on incident strategy
- Coordinates most of the people who are trying to make the problem go away
- Interfaces with the other public and/or private sector responders like battalion chiefs, police captains, and facility managers
- Reports directly to the IC

Planning Section
- Assists IC with deciding what the action plan will be
- Establishes the technical basis for action plans
- Analyzes and compiles information
- Reports directly to the IC

Logistics Section
- Provides the equipment and services called for in the action plan
- Keeps track of resources and locates new ones—these are the gophers
- Reports directly to the IC

Finance Section
- Manages the financial concerns of the incident
- Keeps track of payroll and pays the bills for the resources that logistics gets for the operations section
- Serves as the administrative branch of the big five
- Reports directly to the IC

In addition to these big five there are three important functions that are utilized by the IC: safety officer, public information officer, and incident liaison officer. These functions are part of the **command staff** and function alongside the IC.

Safety Officer
In a multiactivity incident, the IC may have to oversee numerous operations, thus creating a problem because the IC cannot move around the incident and ensure safe operations and practices. It is therefore necessary for the incident commander to appoint a safety officer to observe fire operations, rescue operations, or hazardous materials activities. This function is incredibly important on the emergency scene. For small incidents it may be filled by the IC. Appointing a safety officer does not totally absolve the IC of any responsibility regarding safety issues. Some of the duties of the safety officer include:

- Obtains briefing from the incident commander
- Obtains briefing from any other operational group on the scene
- Participates in the preparation of and implements an emergency response plan
- Advises other managers of any dangerous situations
- Has full authority to alter, suspend, or terminate any activity that may be judged unsafe if operating above the ILDH of the material released or is deemed to be imminently dangerous
- Ensures protection of all personnel from physical, environmental, and chemical hazards/exposures
- Ensures provision of required emergency medical services for assigned personnel and coordinates with any medical unit leader

Public Information Officer
One of the most overlooked positions of the entire incident command system is the public information officer (PIO). A good PIO can effectively work with the media to control the flow of information from the IC to the public. In general the PIO can be expected to perform the following:

---

command staff
staff who work for the IC but who are separate from the general staff, including safety officer, public information officer, and liaison officer

■ **NOTE**
Appointing a safety officer does not totally absolve the IC of any responsibility regarding safety issues.

- Obtains briefing from the incident commander
- Obtains briefing from other key personnel as necessary
- Prepares statements of materials to be released
- Advises the IC of any potential problem areas
- Designates an area to meet with the media
- Meets with the media as needed, giving factual and timely information
- Ensures that all related records and reports are maintained
- May assist in required notifications under federal and state laws

Liaison Officer

The liaison officer is responsible for providing a point of contact for assisting agencies that may respond to the incident. Essentially, the IC cannot meet with every represented outside agency and relay the events over and over to individuals as they arrive on the scene. The liaison officer represents the IC when dealing with the outside agencies and maintains a flow of information between the interested parties. A good liaison officer keeps everyone in the loop and frees up the IC to concentrate on his or her job—running the response. Some of the typical duties of the liaison officer include:

- Obtaining constant updates and briefings from the IC
- Maintaining a visible point of contact for the outside agencies

Another important function, but one that is not recognized by the ICS is that of the scribe. This position can be infinitely valuable if utilized and should be included whenever possible.

This function can be easily explained in one word—documentation. The scribe should follow the IC closely and document as much of the information that goes through that job function as possible. It is easy to get too hung up on writing everything down but what to omit simply becomes a judgment call by the scribe. Conversations that take place should be noted as well as who attend meetings and the times. This job becomes sort of a catchall aide for the IC and may at times be given some authority to speak for the IC. This is a good position for those wanting to become IC's as you get to see the inner workings of the command side of incident response.

Now that we have explored the larger job functions within the ICS, it is time to see where they each fit on the ICS flowchart. Figure 9-3 shows the location of the five general staff sections and the three command staff functions. Remember that the flowchart can get quite complex under each large section.

The IC should be thinking about some important items as he or she presides over this group. There must be a person in charge of every incident. No matter how big or small the event is, one person should be in charge and that person should be called the IC. There should be only *one* IC for the incident and all actions should be directed by that position. This is not to say that the IC shouldn't consult others or have some outside input while making decisions. It is always

■ NOTE
There must be a person in charge of every incident.

■ NOTE
The IC should be visible, relatively stationary, and organized. Therefore it is recommended that the IC establish a command post at or near the incident.

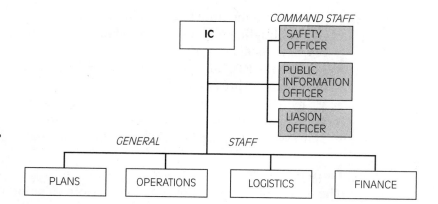

Figure 9-3 *Key supervisory functions of the ICS.*

prudent to run your ideas past someone else. This is often done in areas that cross jurisdictional boundaries. When the fire department and the facility run an incident together, decisions are often made jointly. The key point is that the IC should be a single person with a single unified plan of action. The IC should be visible, relatively stationary, and organized. Because of this, it is recommended that the IC establish a **command post** at or near the incident.

In many organizations, command vests are purchased and used to designate the responsibilities assumed by various people. This system is very effective especially when the standard ICS terms are used on the vests. The system alerts everyone on the scene that the ICS has been established and that someone is in charge. As responsibilities change, the persons wearing the vests simply exchange them for the one that designates their new role. The role of the IC is initially filled by the first person at the scene who may transfer the responsibility to higher ranking or more qualified personnel who arrive later.

The ICS flowchart can expand as the incident expands and you may find that everyone can become quite busy in a hurry. Figure 9-4 illustrates an example of what the ICS chart may look like in the event of a large-scale disaster. The functions listed below the general staff positions are shown in concept, because there is no limit to the type of function that occurs below finance, logistics, planning, and operations as long as they are pertinent to the group.

ICS—The Smaller Picture

Now that we have illustrated the main functions of the ICS we can break the system down into smaller components. These smaller components are useful for incidents that do not require the use of any of the general or command staff functions. In reality, you will most often deal with small incidents that require only a handful of responders. A management system still needs to be used, but now it looks more like the one shown in Figures 9-1 and 9-2. In these cases, an IC takes charge and supervises a few people. The general staff functions are not required

command post
the geographic location of the IC and the control center for the response; it should be clearly identified and its position known by all key staff members

■ **NOTE**
The role of the IC is initially filled by the first person at the scene who may transfer the responsibility to higher ranking or more qualified personnel who arrive later.

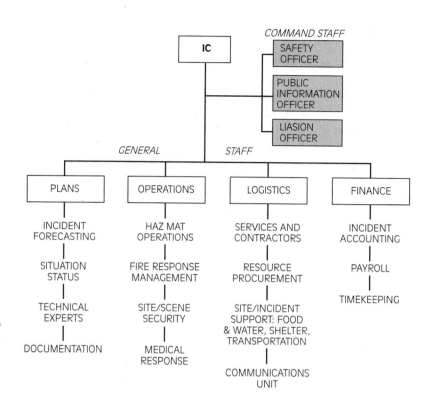

Figure 9-4 *Examples of the job functions working under the General Staff.*

■ NOTE
The ICS allows for considerable flexibility in developing the management structure.

and the management structure is basic. If necessary, some of the supervisory tasks are delegated but for the most part, the IC runs the entire show. As an example, let us look at a small spill of a chemical such as diesel fuel. The spill occurs at the receiving area of the facility when a delivery truck develops a slow leak in the fuel tank. In this case, let's say that you are the person in charge of the response. You arrive first at the scene with the other members of your emergency response group and immediately activate the ICS.

At the earliest onset of this example, you designate yourself as the IC and begin to direct the members to control the spill, shut off any ignition sources, and prepare for cleaning the mess up. Because this incident is very small, you elect not to appoint any of the command staff or general staff positions and instead, decide to fill the roll of each of them concurrent with your role as the IC. Is this allowed? Absolutely. That is the beauty of the ICS. The ICS allows for considerable flexibility in developing the management structure.

In our example, the IC can oversee the safety at the incident. Because only a few people are involved in the response and because the area involved is small, with a known chemical, the IC is filling the role of safety officer.

Also in our example, the IC elects to serve as the information officer. As the information officer, the IC deals with any media or information concerns, should

> **NOTE**
> Smaller incidents require smaller management. Delegate when necessary, but do not give away decision-making authority just for the sake of creating a big management system.

isolation
part of the concept of isolating a hazard and denying entry to unnecessary personnel in the area; a key step in identifying the initial area of the hot zone

stage/staging
the act of identifying and controlling a specific area for the computation of resources; oftentimes a staging officer is appointed to oversee the area and is in direct command with the IC or his/her designee

> **NOTE**
> Do not put yourself or others at risk and become part of the problem.

size-up
mental process used to evaluate the current or projected direction of the incident

they arise. If the incident escalates and the media responds, the IC may relinquish this responsibility and appoint a dedicated information officer, but for now, it is not necessary.

Do you see the flexibility of the system? It guides us to think about the functions that need to be activated, but it doesn't dictate the terms that we have to follow. Our ICS flowchart for the event would be minimal and the incident should be handled quite easily from a management standpoint. Smaller incidents require smaller management. Delegate when necessary, but do not give away decision-making authority just for the sake of creating a big management system.

TAKING CHARGE FROM THE BEGINNING

Taking charge of the incident from the beginning is important. Strong and visible command sets the right tone for the response and is critical to the overall success of the effort. We will investigate some initial steps you may take to get off on the right foot but remember, these are not absolute actions. Various circumstances may apply to any incident response. Use the information as a guideline to help you formulate an initial plan of attack.

Sample Command Procedures

Following is a suggested procedure for those who may be called upon to respond to incidents and institute the incident command system:

Step 1. Respond to the incident

A. Respond from an uphill, upwind direction and always maintain a safe initial distance from the problem. Note weather conditions, time of day, and other essential information.

B. Evaluate the situation and do some basic **isolation** of the problem.

C. **Stage** other responding resources if necessary—give yourself some room to think.

D. If outdoors, try to set up in an area where you are protected from the weather and have access to communications and technical reference materials.

E. Consider the location of your command post.

Do not put yourself or others at risk and become part of the problem.

Step 2. Size up the situation

A. Obtain a briefing from those involved in the incident prior to your arrival.

B. Do a rapid **size-up**.

C. Try to see as much of the incident as possible from a safe distance or get effective reconnaissance from team members.

D. Think about assigning tasks based on your initial assessment. Start thinking about how to match up the number of jobs with the number of responders.

Use the following questions to help assign priorities.
- What are the facts of the situation?
  Is something on fire?
  Is there a chemical spill? What is it and how much?
  Do I have a medical emergency? What type?
  Are people involved or trapped?
  Is evacuation necessary?
- What are the probabilities/possibilities of the situation?
  Is the fire spreading to other parts of the building?
  Where is the spill headed or likely to go?
  Will the problem require building evacuation?
  Is the weather likely to impact the situation?
- What is your current situation?
  Will immediate action solve any part of the problem? Examples: Starting CPR, shutting off a valve from a remote location
  Are you in a safe location?

Think before you act or give a command.
- What resources are available?
  How many responders can be expected to show up?
  Should the fire department, ambulance, or any other governmental agencies be notified?
  Do any other specialty response teams need to be activated?
  Do site security or the police department need to be called for scene control?

Step 3. Assume command and direct the incident

A. Using your radio or other communications and at some time early in the incident, make sure your team has the following information:
- Your name or other radio designator
- Statement that you are taking command of the incident
- Verification of the actual location of the emergency
- Confirmation in your own mind and for the other responders of the type of emergency you are facing
- Location of the stating area, if needed

Do not let team members waiting for an assignment push you for decisions. Take time to formulate a plan of attack.

> ■ **NOTE**
> Think before you act or give a command.

> ■ **NOTE**
> Do not let team members waiting for an assignment push you for decisions. Take time to think and formulate a plan of attack.

B. Following the initial transmission, further information can be given as needed
- Any other vital information for the other responders including life hazards
- Location of command post if pertinent
- Statement of what you are doing or planning to do including calling outside response agencies.

By this point in the incident, other team members will be arriving at the staging area and reporting in for assignments. As IC you may deploy team members immediately or wait for a workforce to assemble. Keep in mind that any offensive action should be carried out in pairs of team members. Think of the incident as you would a game of chess—contemplate all moves prior to taking action.

Figure 9-5 is a flowchart that gives examples of some decision making that may occur during a hazardous materials incident. Use it as a loose guide for developing an action plan or focusing your efforts at the scene.

ESTABLISHING SAFE WORKING AREAS—ZONES

An important task to perform early in the hazardous material response is to establish physical control of the incident. This task is very simple but is often delayed. The importance of establishing control is to:

- Protect the responder
- Protect the environment
- Protect the civilians
- Isolate the product and problem

This task is most often accomplished by directing the team to designate a safe distance around the hazard. The word *safe* should be emphasized. This procedure is to identify a perimeter that provides more than enough area between its boundary and the hazard. Once determined, the boundary line for the safe zone is not going to be inside of any contaminated areas. Continuing emergency operations outside the contaminated area may then proceed without personnel having the fear of being exposed.

The purpose of identifying an initial perimeter around a hazard is to establish control over the incident. A perimeter helps identify a hazard zone. It also denotes a way of controlling the influx of workers into the area while warning others to keep out. Most responders utilize red or yellow barrier tape for identifying these zones. The implementation of this concept is excellent for hazardous materials incidents, but can prove useful for any type of response. A safe zone may initially be larger than the situation dictates; it is better to start big and reduce it once additional information is available.

■ **NOTE**
Think of the incident as you would a game of chess: Contemplate all moves prior to taking action.

■ **NOTE**
A safe zone may initially be larger than the situation dictates; it is better to start big and reduce it once additional information is available.

Chapter 9 Incident Management and Scene Control

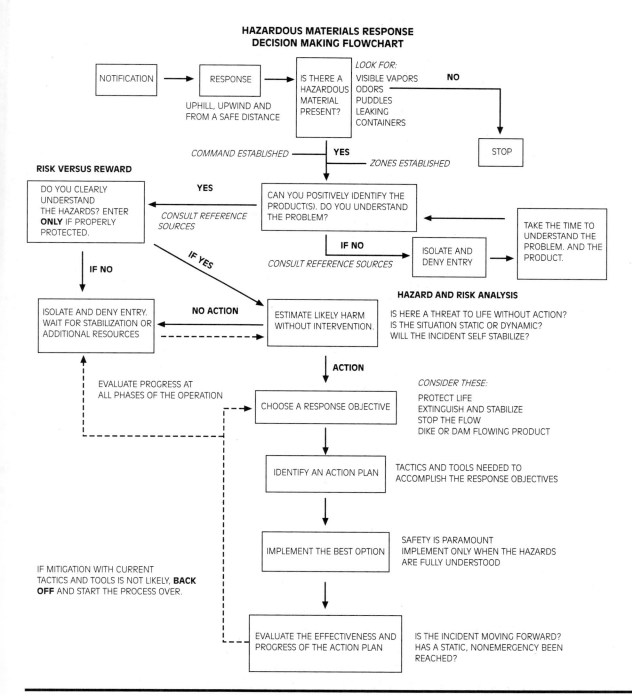

Figure 9-5 *A decision tree for managing a hazardous materials incident.*

In establishing zones, one of the first considerations is to establish the hot zone. This zone includes the area of highest hazard, both known and suspected; this area presents a clear and potential danger of product spread. It is likely that anyone entering this area will indeed become contaminated. Once established, the hot zone identifies the area of any product spread, container damage, or immediately exposed area to vapors, mists, dust, smoke, or runoff. The purpose of establishing a hot zone is to:

- Isolate the danger area from the work area
- Identify the area of greatest danger
- Identify an area prone to known contamination spread
- Restrict entry to only hazmat team personnel and specialists, and only then when properly clothed and equipped
- Minimize danger present to other emergency response personnel who are working in the area

Entry into the hot zone should be restricted to only those members of the emergency response team or other designated specialists as determined by the IC. Special tools and clothing appropriate for the type of hazard will likely be required.

The warm zone is established just outside the hot zone. Like the hot zone, the warm zone is an area where potential contamination might be present. The purpose of establishing a warm zone is to allow support activities to take place that could present some hazard to other personnel, property, or the environment. It should be large enough to allow for work to be accomplished in such a manner that contamination spread to the workers will be nonexistent if planned and executed correctly. Typically, all decontamination activities take place within the warm zone. The warm zone only becomes warm when the decon process is begun. Until that time, the only two zones that matter are the hot zone and the cold zone.

If a portion of this area does become involved by the product or the contamination is spreading due to wind or slope of ground, then this portion of the warm zone should be upgraded to a higher level of control.

The cold zone denotes an area that is totally free from contamination and hazard. It is an area of safe refuge from which support activities originate and should include anything outside the perimeter of the warm zone. The IC should conduct operations from the cold zone.

Access into a Hot Zone

Access into the hot zone is strictly controlled by the incident commander or his or her designee. No one enters without permission and entry should be controlled at a designated entry point. Since it is to be assumed that personnel wearing chemical protective gear will become contaminated, steps should be planned and

■ **NOTE**
Entry into the hot zone should be restricted to only those members of the emergency response team or other designated specialists as determined by the IC.

■ **NOTE**
The warm zone only becomes warm when the decon process is begun. Until that time, the only two zones that matter are the hot zone and the cold zone.

■ **NOTE**
The IC should conduct operations from the cold zone.

> **■ NOTE**
> Entry and exit into the hot zone can be through the decon or an area directly adjacent.

site access control officer
person who controls the movement of personnel and equipment into and out of the contaminated areas

implemented in advance to establish a complete decontamination corridor. Entry and exit into the hot zone can be through the decon or an area directly adjacent.

Identities of all those who enter shall be logged. This is usually done by a **site access control officer.** The length of time spent inside the hot zone shall also be logged. Each person entering shall be inspected to determine if they are properly clothed and the garments are in good condition. Everyone who enters will work on the buddy system with appropriate communications provided. Typically, it is the responsibility of a designated safety officer to know who is entering the zone, what level of protection is being used, and the time available for them to work in the hot zone. If the IC is acting as scene safety officer he or she should have a good method for personnel accountability.

OVERVIEW OF JOB TITLES AND DESCRIPTIONS

In order to correctly identify and assign personnel to supervisory functions the IC must be aware of the standard terminology and mission of those specific jobs. The job titles and descriptions listed below relate primarily to hazardous materials incidents but may also pertain to other responses as well. These job functions are modular and can be plugged into any command structure by the IC. Each job function has a description of duties as well as reporting responsibilities. You should be able to discern to whom each function reports and where it fits into the ICS. It should also be understood that these job functions and terms are recognized by almost all public response agencies in the country. In other words, the local fire department or regional hazmat team will be well aware of these job functions and terminology.

Site Security The site security officers or others who are responsible for the security function of the site on a daily basis need to respond to most types of emergencies occurring on the site. Depending on incident factors, security may be a partner in working with the IC as part of a unified command or may participate as an assisting group. Some functional responsibilities that may be handled by site security personnel are:

- Isolating the incident area.
- Managing crowd control.
- Managing traffic control.
- Managing site protective action.
- Coordinating with outside law enforcement agencies.
- Assisting in documentation as needed.

Hazardous Materials Group Supervisor In a hazardous materials emergency, it is often necessary to break off the hazardous materials part of the problem from the other parts. This is accomplished through the use of the hazardous materials group supervisor. This position reports directly to the incident commander. The hazardous materials group supervisor also participates in the formulation of tactics and strategy and coordinates the activities of the decontamination leader, technical reference, and equipment procurement. He or she

- Obtains briefing from the IC.
- Consults with the entry team leader and IC on tactics within the hot zone.
- Obtains information from the safety officer regarding accepted work practices and procedures.
- Conducts the team meeting with all involved personnel prior to entry into the hot zone.
- Recommends personal protective equipment and tool selection for entry team members. These recommendations are screened by the technical reference personnel and entry team leader prior to use.
- Maintains radio communications with the entry team.

Medical Group Leader In many types of emergencies, the functions associated with medical care are important enough to need its own leader. In this case, the IC may designate a medical group leader. Such would be the case in any multicasualty event such as a building collapse, bus accident, or large chemical release with exposures.

This position reports to the incident commander and is in charge of overall incident medical triage, treatment, and assistance with ambulance transport. The medical group leader also interfaces with the hazardous materials group supervisor in the event of a medical emergency involving chemical exposure. Other responsibilities include:

- Obtaining briefing from the incident commander in most cases. This position may also work with the hazardous material group supervisor when decontaminating a patient with chemical injuries.
- Coordinating triage and treatment operations at major emergencies.
- Coordinating with ambulance personnel during the movement of patients from the site to transport.
- Recommending medical equipment needs to the IC.

Entry Leader If a hazardous materials emergency is significant, requiring very complex activities, it is often wise to consider establishing responsibility for the entry portion of the incident. This is particularly useful when multiple entry teams are operating. He or she would report to the IC unless a hazardous materials group supervisor has been established. The entry leader is responsible for the overall entry operations of assigned personnel within the hot zone and also

- Obtains briefing from the hazardous materials group supervisor or IC.
- Supervises entry operations.
- Recommends actions to mitigate the situation within the hot zone.
- Carries out actions as directed by the hazardous materials group supervisor or IC to mitigate the hazardous materials release or threatened release.
- Maintains communications and coordinates operations with the decontamination leader.
- Maintains communications and coordinates operations with the site access control leader.
- Maintains communications and coordinates operations with technical specialist/hazardous materials reference.
- Maintains control of the movement of people and equipment within the hot zone, including contaminated victims.
- Directs rescue operations, as needed, in the hot zone.

Decontamination Leader This position may be used in those very complex hazardous materials to break off one of the major parts of the problem. The decontamination leader reports directly to the IC unless a hazardous materials group supervisor has been designated. The decontamination leader is responsible for the operations of the decontamination team and

- Obtains briefing from the IC or hazardous materials group supervisor.
- Establishes the decontamination corridor.
- Identifies contaminated people and equipment.
- Supervises the operations of the decontamination element in the process of decontaminating people and equipment.
- Maintains control of movement of people and equipment within the decontamination zone.
- Maintains communications and coordinates operations with the entry leader.
- Maintains communications and coordinates operations with the site access control leader.
- Coordinates the transfer of contaminated patients requiring medical attention (after decontamination) to the medical group.
- Coordinates handling, storage, and transfer of contaminates.

Site Access Control Officer The site access control leader reports to the hazardous materials group supervisor and is responsible for the control of the movement of all people and equipment through appropriate access routes at the hazard site and ensures that contaminants are controlled and records are maintained. This function is sometimes accomplished by facility security personnel. The site access control officer

- Obtains briefing from the hazardous materials group supervisor.
- Organizes and supervises assigned personnel to control access to the hazard site.
- Oversees the placement of the hot zone and the decon corridor.
- Ensures appropriate action is taken to prevent the spread of contamination.
- Ensures that injured or exposed individuals are decontaminated prior to departure from the hazard site.
- Tracks persons passing through the contamination control line to ensure that long-term observations are provided.
- Coordinates with the medical group for proper separation and tracking of potentially contaminated individuals needing medical attention.
- Maintains observations of any changes in climatic conditions or other circumstances external to the hazard site.
- Maintains communications and coordinates operations with the entry leader.
- Maintains communications and coordinates operations with the decontamination leader.

Technical Specialists–Hazardous Materials Reference A position that is often valuable in a hazardous materials emergency is one of technical specialist. This position is key to providing information needed in order to develop a plan of action. This person reports to the IC or the hazardous materials group supervisor if one is designated. This position provides technical information and assistance to the persons who will be making those key decisions regarding the incident through the use of various reference sources such as computer databases, technical journals, and facility representatives. The specialist also

- Obtains briefing from the IC or hazardous materials group supervisor.
- Provides technical support to the IC or hazardous materials group supervisor.
- Maintains communications and coordinates operations with the entry leader.
- Provides and interprets environmental monitoring information.
- Provides analysis of hazardous material sample.
- Determines personal protective equipment compatibility to hazardous material.

- Provides technical information management with public and private agencies, i.e., Poison Control Center, Tox Center, Chemical Transportation Emergency Center (CHEMTREC), State Department of Food and Agriculture, or National Response Team.
- Assists IC with projecting the potential environmental effects of the release.

INCIDENT TERMINATION

The entire purpose of incident management is to bring the problem under control. Thus far we have discussed the concepts of incident command and focused our attention on making the problem go away. The next logical step is to investigate the role of an IC after the incident has been stabilized. The main focus of this section is on event review and illustrating some common mistakes and pitfalls that an IC may encounter.

Following is a list of eight common mistakes a commander may make during an incident. This list has been informally compiled by countless commanders who have had incidents get away from them. Some have been minor and inconsequential to the success or failure of the incident. Others have had a huge impact on the response and have been quite detrimental. At any rate, a wise IC will take note of them and not duplicate someone else's errors.

Eight Common Mistakes Made by Incident Commanders

1. *Failure to plan.* Not planning for "the big one" or the worst case scenario is mistake number one! Incidents usually occur at the one place you do not think about and they happen at the worst possible time. Do not allow yourself to say "if" when preplanning an incident. This assumes you do not actually believe anything bad will happen. Always preface preplanning with the phrase "when an incident occurs at. . . ." This tends to make the threat a little bit more realistic and will help you to focus on real possibilities. The emergency that is the easiest to handle is the one you have already practiced for and run countless times. Whether it be in training or in real life, experience is a real benefit when managing an incident.

2. *Failure to take charge of the incident.* Setting up command and calling yourself the IC does not constitute taking charge of the incident. The IC must be a leader and someone that the team is comfortable following. Weak leadership encourages weak performance and confusion, which results in less than desirable results. Be strong and decisive and instill confidence in the people you are leading.

3. *Failure to understand the nature of the incident.* As the IC it is paramount that you truly understand the nature of the problem at hand. It is one thing to know that the toxic gas sensor is going off, but the astute IC focuses on finding out why the alarm is occurring. This concept seems to be very basic but it is one that

is often overlooked. Take the time to really know what you are faced with before committing people or resources to battle.

4. *Failure to base decision on fact rather than emotion.* Emergencies are great for getting everyone pumped up and ready to act. They are also great for making people behave in the most illogical and irrational fashion imaginable. Time after time commanders give orders or take action based on nothing more than emotion. Avoid getting caught up in the moment and be the one who is thinking clearly and is unaffected by the nature of the event. As an emergency responder it is your job to bring order out of chaos and it can best be accomplished by calm, cool thinking and good information. Once again, take the time to get as much information as possible about the situation and make your decisions based on fact, not emotion.

5. *Failure to anticipate changes.* A good IC formulates a plan of action and implements it with the anticipation that it will work as it was conceived. A great IC conceives the same plan but also determines alternatives as the original plan is being implemented. You should school yourself to think 10 to 30 minutes ahead of the incident and always be considering alternative courses of action. Even if the initial plan looks like it is working, take time to consider different approaches and strategies—just in case. This thinking will prove beneficial in the long run because eventually you will be faced with a plan that is not working. Do not resign yourself to sticking with a faulty plan—change your course of action and move on!

6. *Failure to set up effective communications.* It is the responsibility of the IC to ensure that some form of communication is in place and operating properly. Very few problems on the scene create more headaches than poor communications. When people cannot effectively talk to each other and goals are not understood by all, team members will not be working as safely as they should. Poor communications can slow the incident down. Make sure that the key players can talk with each other and the flow of information is unencumbered.

7. *Failure to accept blame.* This concept is simple. You are the IC—you are the sole set of shoulders upon which success and failure rests.

8. *Failure to learn from mistakes.* Most people never take advantage of this great luxury. A mistake does not have to be yours in order to learn from it. Look at the success and failure of previous incidents and figure out why they turned out the way they did. If someone made a bad decision, learn from that decision and avoid the same situation when it happens to you. Additionally, if you do make a mistake, admit it as your own, learn from it, and determine not to make it again. Experience is a good teacher if you are astute enough to pay attention!

Event Review

An IC should also be interested in conducting an event review. This action is essential as it brings closure to the incident. A properly conducted event review often provides insight into what went right and what went wrong at an incident.

■ **NOTE**
Take the time to really know what you are faced with before committing people or resources to battle.

■ **NOTE**
Poor communications can slow the incident down.

■ **NOTE**
Experience is a good teacher if you are astute enough to pay attention!

The action is not used to assign blame or unduly criticize the performance of any one member. The incident was a team effort so the review should be conducted as a team. However, it is crucial to be honest and open during the session. If a performance truly was below expectation, it should be noted as such. As long as all comments are positive and offered as a means to improve performance, honesty is the best policy. The event review also serves as a basis for fixing problems that may lead to further incidents. The following points may be used by ICs as a guideline for conducting an event review:

Collection of Information Obtain as much information from all parties involved prior to the event review. Responders may have had unique vantage points or perspective during the event. It is vital to compile as much factual information about the incident as possible. Some useful information may include:

- Dispatch times and the names of those involved.
- Current operational status of the systems or machinery prior to the event.
- History of those systems and failure points that caused the event.
- Communications information.
- Tools and equipment used.

Analysis of the Event and Cause Determination Essentially this phase is the reconstruction of the review. ICs should compile the facts and activities as they are recalled. Once again, system point failures should be documented in order to determine if the incident was caused by an unusual event or other failure. The actual or probable cause should be noted. There are several ways to determine cause. Some proven methods include event charting and walk-through reconstruction. Select a suitable method and use it as consistently as possible.

Corrective Action Should Be Planned and/or Taken Once again, the event review could function as a change agent under the right circumstances. When things go wrong and an IC can figure out why they went wrong, it is incumbent on him or her to point out the flaw and attempt to correct it. This idea applies to personnel performance as well as system performance and should be viewed as an opportunity for improvement. Some corrective acts may require specific action immediately or a longer term plan for follow-up correction. Poor responder performance may identify a training weakness that could take a period of time to rectify. Huge system changes may cost large sums of money and may require phase-in periods. The bottom line is that emergencies happen for a reason and it is partly the job of the IC to figure out why.

Summary

- The IC's job is to bring order and direction to the chaotic nature of an emergency.
- The overall scene safety is the responsibility of the IC.
- ICs need to set incident objectives and priorities on every incident.
- Incident action plans need to be developed for every incident. They can be formal or informal but need to be understood by all the key players at the incident.
- It is the IC's responsibility to drive the incident forward in an attempt to bring rapid control and closure to the event.
- Incident strategy is the overall theme or big picture thinking for the event. It is the larger scale goal setting for the incident and should be determined by the IC.
- Tactics are the individual actions or procedures used by the team to make the problem go away.
- H-E-A-R-T is a mnemonic that describes the objectives that should be met for all incidents

  H—Hazard Identification: Take the time to figure out what the nature of the problem really is.

  E—Evaluate Your Response: The IC must develop the action plan for the incident based on the nature of the problem, availability of resources, and skill level of the responders.

  A—Assemble Work Zones: All incidents benefit from having adequate room to work. Set your work zones in such a way that you protect the site personnel and give the team a secure and controlled environment in which to work.

  R—Run the Incident: The IC must lead the incident. He or she does not have the luxury of being a casual observer. ICs with a strong, calm decisive manner will instill confidence in their followers.

  T—Termination of the Incident: The IC is responsible for declaring an end to the emergency phase of the operation. In the event further clean up is needed, the IC may declare the emergency over and transfer command to another person responsible for guiding those efforts. Termination includes the postincident analysis, which should also be led by the IC who ran the response.

- ICs must delegate tasks and supervision when necessary.
- Span of control refers to how many persons a supervisor can effectively manage during an emergency.

- A 5:1 ratio of workers to supervisors is a recommended span of control for emergency response.
- The incident command system should be implemented at all incidents regardless of size.
- The ICS is a modular system characterized by common terminology, unity of command, and span of control.
- The general staff sections are: Finance, Operations, Logistics, and Planning.
- The finance section deals with accounting, payroll, and documentation.
- The operations section deals with getting the job done. The operations chief may supervise many different disciplines such as fire response, medical activities, and hazardous materials actions.
- The logistics section deals with procurement of services, equipment, and personnel.
- The planning section deals with directing the incident on a long-term basis and provides technical support.
- The IC presides over the general staff consisting of the scene safety officer, scribe, and public affairs officer.
- On smaller incidents, the IC can wear many hats and fill many of the ICS functions simultaneously.
- A scene safety officer should be in place for all incidents. The IC may fill that role for small incidents.
- Each incident must have some form of communication in place that is understood by all responders.
- Work zones should be established on every incident. They may include the hot zone, the warm zone, and the cold zone.

■ **NOTE**
The ICS is a modular management system. Use whatever building blocks are necessary to create an effective framework for command and control.

Review Questions

1. Which of the following is not one of the key elements of emergency management?
   A. Command and control
   B. An effective chain of command
   C. Strong leadership
   D. An IC that only works in the area where the release occurred

2. True of False? In emergency response, a strong and visible command presence is vital.

3. True or False? An IC may fill many or all of the command functions during small-scale incidents.

4. True or False? The IC is ultimately responsible for the safety and health of all responders on the scene.

5. True or False? The IC should allow anyone on the scene to determine an action plan and communicate it to all others on the scene.

6. Explain the difference between tactics and strategy.

7. Define each letter in the following management acronym.
   H-
   E-
   A-
   R-
   T-
8. True or False? It is advisable to have several ICs in charge of an incident to ensure that sound decisions are made.
9. True or False? Delegation is a key function in emergency management.
10. Which of the following is an accepted span of control for managing an emergency scene?
    A. 1 supervisor to 10 team members
    B. 2 supervisors to 5 team members
    C. 1 supervisor to 5 team members
    D. 2 supervisors to 7 team members
11. Describe the incident command system and explain why it is useful when managing an incident.
12. Which of the following is not a common aspect of the incident command system?
    A. Standardized terminology
    B. Modular structure
    C. Span of control
    D. Tactical consideration
13. Identify the "big five" general staff positions and describe their responsibilities.
14. What does unity of command mean?
15. Describe at least four duties fulfilled by the safety officer.
16. True or False? The safety officer, public affairs officer, and scribe all work directly for the IC.
17. Which of the following is not a general staff position?
    A. IC
    B. Logistics officer
    C. Operations officer
    D. Plans officer
18. List two ancillary functions that an operations chief may supervise.
19. True or False? Small incidents do not require an IC.
20. What does size up mean?

Chapter 10

Planning

Learning Objectives

Upon completion of this chapter, you should be able to:

- Understand the emergency planning requirements for hazardous substances.
- Know the various planning regulations and the interrelated and sometimes overlapping requirements at the local, state, and federal level.
- Know the specific elements of a HAZWOPER Site Safety Plan.
- Know how to complete an emergency operations site response plan.
- Understand about resources and references to assist in emergency response planning.

INTRODUCTION

Emergency planning is a very complex process of identifying hazards, analyzing risk, determining how to mitigate or reduce potential problems, developing response, and recovery plans, developing procedures, and conducting training then testing the plans with mock scenarios. Just keeping an emergency plan current for a facility with minimal hazards is difficult. Then add multiple layers of emergency planning because of the presence of hazardous materials or waste, which bring to bear all the regulatory requirements that exist at the federal, state, and local government level. Compounding this problem is the ever-changing web of regulations, which are sometimes conflicting and overlapping, and a lack of clear language on ways to combine or integrate these various planning requirements.

The challenge of how to adequately address this emergency response plan is to figure out how to develop a comprehensive document that will keep the regulatory community happy and still have a written document that can be implemented to provide an effective and safe response to an incident.

The Hazardous Waste Operations and Emergency Response (HAZWOPER) regulations have some very specific planning requirements and this chapter provides you with details on what must be included in these plans as well as providing guidance on how to develop an integrated plan to meet other regulatory health, safety, and environmental regulations. It also provides a sample emergency response safety plan that can be easily completed and focuses all the emergency response personnel on SAFETY FIRST. Even with all the overlapping emergency planning requirements, many elements of the HAZWOPER plan may have already been done for one of these other regulations. Use existing documentation whenever feasible.

■ **NOTE**
With all the overlapping emergency planning requirements, many of the elements of the HAZWOPER plan may have already been done for one of these other regulations. Use existing documentation whenever feasible.

PLANNING FOR HAZARDOUS SUBSTANCES RELEASE EMERGENCIES

If there are hazardous substances at a facility, according to Murphy's law there will be an accident or a spill.

Various local, state, and federal laws also recognize that spills are likely to occur and require sites that have any hazardous substances to develop emergency response plans. When developing these plans it may be helpful to utilize these following priorities:

- Life safety and public safety
- Environmental protection
- Protection of assets and business continuity

As guidance for developing emergency response plans, this chapter outlines the preplanning requirements and emergency response planning. Besides emergency planning being *required,* it makes common sense to protect the workers at the site, those that might be affected near the site, and to protect the environment and to get back to normal operations as quickly and safely as possible. Remember

> **■ NOTE**
> Remember that the safety of employees and the public are the first priority in emergency planning and response.

that safety of employees and the public are first priority in emergency planning and response.

Coordinating Planning Efforts—Public/Private Partnership

Many of the environmental protection laws require preplanning and coordinated response protocols for emergency response planning, which requires coordination between the private sector and public agency responders and emergency planners. Local government has historically controlled land use, and building and fire codes to protect the public from hazardous materials. This type of control for the public good now logically extends into emergency planning and coordinated response. With limited resources for emergency response at both the site and community level, a system of working together or **mutual aid** between local, state, and federal response agencies, in partnership with site emergency personnel will bring together the resources needed to quickly mitigate the problem.

mutual aid
pre-determined agreements between jurisdictions that allow for using shared resources in the event of a large-scale incident

Updating and keeping the various emergency plans current is a challenge. This challenge is further complicated by keeping all the key players involved. Keeping site personnel and management trained and informed about the plans and procedures is an ongoing process. Emergency planning just doesn't end with the development of a nice big plan on the shelf, but is an ongoing process that requires good communication to be effective.

There certainly is a recognition among the various federal, state, and local regulatory community to allow, and in some jurisdictions, encourage integration of the various hazardous materials site safety and emergency response plans. In general, a site plan is organized as a single document, with the various component sections and appendices covering all tasks, operations, and contractors/subcontractors or outsourced employees and can be used to promote efficiency and enhance completeness, clarity, and coordination.

For more information on emergency planning, there is a listing of resources and Internet Websites in Appendix D.

EMERGENCY PLANS

Besides the detailed requirements for a site specific health and safety plan and an emergency response plan, what other emergency response and subsequent training requirements are there?

For starters, the federal Occupational Safety and Health Administration (OSHA) requires all employers to plan and prepare for emergencies in the workplace and to train employees to perform whatever emergency response roles they are assigned. Emergencies include natural disasters such as severe weather, earthquakes, and fires. Incidents caused by humans include chemical spills, explosions, and workplace violence. It seems ironic, but emergencies fall into response for routine operations and nonroutine. These planning requirements as they

> **NOTE**
> OSHA's primary concern is employee safety; the other federal laws focus primarily on environmental protection.

relate to HAZWOPER are very specific, however this section provides an overview of the other regulatory drivers for emergency response planning. OSHA's primary concern is employee safety: the other federal laws focus primarily on environmental protection.

OSHA Emergency Action Plans

All employers, whether private or public are required to implement emergency action plans (EAPs) for emergencies in the workplace. The EAP must be in writing, and must cover those designated actions employers and employees must take to ensure employee safety from fire and other emergencies.

> **NOTE**
> The EAP must be communicated in writing unless there are ten or fewer employees. Then OSHA allows the company to communicate the plan verbally.

The EAP planning requirements include emergency response training for employees depending on the extent to which employees are involved in emergency response. Even if your company or agency does not permit employees to assist in handling emergencies and all your company does is to evacuate all employees and other site personnel from the danger area, an EAP is required to be prepared and implemented. The EAP must be communicated in writing, unless there are ten or fewer employees, then OSHA allows the company to communicate the plan verbally. If the plan has been verbally communicated there must be written documentation that this task has been communicated. In other words, in the case of dealing with OSHA, if it is not written down and documented, it did not happen.

The EAP must include at least the following elements:

- Emergency escape procedures and emergency escape route assignments
- Procedures to be followed by employees who remain to operate critical site operations before evacuation (safe shut down)
- Procedures for accounting for employees after emergency evacuation is completed
- Rescue and medical assignments for designated employees
- How fires and other emergencies are reported
- Names and job titles of persons/departments who are responsible for the plan and can explain employee job duties under the plan

OSHA Fire Prevention

The fire prevention standard requires an assessment of fire hazards in the workplace as well as the identification of prevention and suppression systems on site. While this standard does not specifically require an emergency response plan, it does provide valuable information on hazard identification and prevention, which are both key elements in the requirements for the HAZWOPER Site Health and Safety Plan. You will see a recurring trend of the interconnectivity of the planning requirements for environment, health, and safety. You can use this information in the development of the HAZWOPER Site Health and Safety Plan or the emergency response plan. Many plans incorporate reference information from other related plan documents and include these referenced plans as appendices.

OSHA standards require a written fire prevention plan containing the following elements:

- A list of all workplace fire hazards and procedures for handling and storing flammable materials; identification of potential ignition sources in the workplace; procedures for controlling the sources of ignition; and listing the types of fire protection equipment systems available for fire control
- Names or job titles of personnel responsible for maintaining fire prevention equipment or systems or controlling ignition sources
- Names or job titles of personnel responsible for the control of fuel source hazards
- Housekeeping procedures for controlling accumulation of flammable and combustible wastes
- Maintenance procedures for fire protection equipment and systems

HAZARD COMMUNICATION

The hazard communication standard, similar to the fire prevention plan does not specifically require an emergency response plan. It has valuable hazard analysis and site-specific chemical information that can be used in development of the HAZWOPER planning process.

The hazard communication standard is designed to ensure evaluation of the hazards of all chemicals present in the workplace and ensure that both employers and employees receive relevant information about these chemicals in the workplace.

All employers must develop, implement, and maintain at each workplace a written hazard communication program. The program must include procedures to ensure that requirements for labels and other warnings, MSDSs, and employee information and training are all met. It must include:

- A list of hazardous chemicals in the workplace, with each chemical referenced to its MSDS
- Methods to inform employees about the hazards of nonroutine tasks
- Hazards associated with chemicals located in unlabeled pipes in work areas

OSHA PROCESS SAFETY MANAGEMENT

■ **NOTE**
The OSHA standards are required by the Clean Air Act amendments.

The major objective of process safety management (PSM) of highly hazardous chemicals is to prevent unwanted releases of hazardous chemicals especially into locations that could expose employee and others to serious hazards. The OSHA standards are required by the Clean Air Act amendments.

Every employer must address actions that employees are to take when there is an unwanted release. These actions can range from handling and stopping small releases to significant releases, evacuation from the danger areas and allowing the local community emergency response organizations to handle the release or a combination of these actions.

Employers, at a minimum, must have an emergency action plan that facilitates the prompt evacuation of employees when there is an unwanted release of a highly hazardous chemical. If an employer wants specific employees to control or stop the minor emergency or incidental release, these actions must be preplanned and procedures developed and implemented. Handling releases in the process area must include advance planning, providing appropriate equipment for the hazards, and conducting training.

If the employer wishes to use plant personnel to render aid to those in the hazard areas and to control or mitigate the incident, employees must be trained to the appropriate level of the HAZWOPER standard. The PSM evacuation planning can be accomplished through the Emergency Action Plan. There are numerous federal requirements for environmental protection that require emergency planning and notification for hazardous substances releases to the air, water, or land.

■ **NOTE**
There are numerous federal requirements for environmental protection that require emergency planning and notification for hazardous substances releases to the air, water, or land.

Super Fund Amendments and Reauthorization (SARA) of 1986—Emergency Planning and Community Right to Know (EPCRA), Title III

Title III of the SARA establishes several different reporting and planning requirements for businesses that handle, store, or manufacture certain hazardous materials. These reports and plans provide federal, state, and local emergency planning and response agencies with information about the amounts of chemicals businesses use, routinely release, or accidentally spill. These also provide the public and local governments with information about chemical hazards in their communities, as well as providing information to enable emergency responders to respond safely to accidents.

President Clinton signed Executive Order 12856 in 1993, outlining a timetable for federal government facilities to begin to comply with all portions of the SARA Title III by 1995.

SARA Title III is actually four separate programs, with very specific reporting and planning requirements. They are listed in the following sections:

1. SARA Section 301-303—Planning for emergency response
2. SARA Sections 311-312—Reporting chemical inventory
3. SARA Section 313—Reporting on release of regulated toxic chemicals
4. SARA Section 304—Reporting leaks and spills

■ **NOTE**
Title III of the SARA established several different reporting and planning requirements for businesses that handle, store, or manufacture certain hazardous materials.

Familiarity with the spill reporting and emergency planning at the local level are important in the development of the HAZWOPER plans, especially in coordinating with public agencies.

Transportation of Hazardous Materials and Wastes

The U.S. Department of Transportation (DOT) has specific notification and documentation requirements for a company that is involved in shipping hazardous materials. A shipper must provide emergency response information useful in cleaning up a spill that must accompany each shipment.

The information is similar to that found in MSDSs and must be available on the transport vehicle but kept away from the hazardous materials. In addition, the shipper must provide a 24-hour emergency response telephone number, monitored by personnel able to provide callers with specific information and emergency procedures regarding the chemical being transported. Most shippers contract with the Chemical Manufacturers Association emergency service called CHEMTREC. This resource is used primarily for response but also provides some training materials to the general public. The 24-hour emergency number for CHEMTREC in the United States is 1-800-424-9300. There is a Canadian service much like CHEMTREC called CANUTEC. It is the Canadian Transport Emergency Centre and is operated by the Transport Dangerous Goods Directorate of Transport Canada. Their 24-hour emergency phone number is 613-996-6666. These are invaluable resources for emergency response information in transportation-related incidents.

The U.S. DOT has incorporated necessary emergency planning and notification requirements into their Emergency Response Guide (see Chapter 5).

Resource Conservation and Recovery Act of 1976 (RCEA)—Hazardous Waste

This act is commonly called the "cradle to grave" management of hazardous waste. In addition to regulating the accumulation, storage, tracking, and classifications of hazardous waste, it also contains a requirement of a Hazardous Waste Contingency Plan. Each facility must have its own contingency plan to deal with emergency situations if they are generating hazardous wastes as defined under RCRA. The plan must be designed to minimize hazards to human health or the environment from fires, explosions, or unplanned releases to the air, soil, or surface water.

A copy of the contingency plan must be kept on site and on file with the local emergency response agency, such as fire department and the local office of emergency management.

Release Reporting and Response

> ■ **NOTE**
> It is important to understand the importance of making all the proper agency notifications in a timely manner.

Emergency release reporting and emergency response requirements are included in five of the federal laws, and in some states more stringent reporting requirements are included. It is important to understand the importance of making all the proper agency notifications in a timely manner. There are very severe civil and criminal penalties associated with these reporting requirements. But the primary reason for reporting should be for protection of the public health and the environment, as well as mobilizing outside agency assistance in a prompt manner to prevent any further environmental damage,

The laws covering emergency reporting are:

- "Superfund" (Comprehensive Environmental Response, Compensations, and Liability Act of 1980, CERCLA)
- Clean Water Act (CWA)
- Resource Conservation and Recovery Act (RCRA)
- Hazardous Materials Transportation Act (HMTA)
- Emergency Planning and Community Right to Know Act of 1986, also known as SARA Title III

Each of these documents is too large to be included here but it is important to know that they exist. In the event they pertain to your operation it is imperative that you fully understand how they pertain to your site.

Releases can be categorized into the following:

- Releases of hazardous substances
- Releases of hazardous substances into surface waters
- Releases of oil or petroleum products
- Releases of wastewater
- Releases from underground storage tanks
- Releases of hazardous waste
- Transportation-related releases of hazardous materials

Clean Water Act

The federal Clean Water Act provides the basic national framework for regulating discharges of pollutants into the nation's navigable waters. In addition to providing a framework for permitting discharges, the Clean Water Act requires spill prevention and release reporting. All spills and unauthorized discharges must be reported to the National Response Center. They can be reached 24 hours a day in the United States and Canada at 1-800-424-8802.

Sites that might discharge chemicals into navigable waters must develop a Spill Prevention Control and Countermeasure Plan (SPCC). This plan must include an overview of construction and operating guidelines and other effective spill containment procedures. The SPCC must be reviewed and certified by a registered professional engineer.

Clean Air Act

Recent changes to the Clean Air Act (CAA) expanded SARA Title III reporting to create an accidental release prevention program to prevent accidental releases of substances that could have disastrous off-site consequences (CAA Section 112 (r)). The Risk Management Program requires detailed planning and analysis for facilities that have toxic, flammable, and explosive hazards above a threshold quantity. These substances have been determined by the U.S. Environmental Protection Agency (EPA).

One element of the Risk Management Plan is the development of an emergency response plan.

Elements of the Site Safety Plan—Hazardous Substance Cleanup Operations

A site safety plan, which establishes policies and procedures to protect the site workers, the public, and the environment from the potential hazards of the hazardous waste site must be developed before site activities proceed. These site safety plans range from documentation that might fill several large binders for a Superfund site to a few pages for a simple underground storage tank removal project. The documentation may vary in length and breadth, but all site safety plans must contain specific elements, and the safety information is the most important part of the plan.

> **■ NOTE**
> The documentation may vary in length and breadth, but all site safety plans must contain specific elements, and the safety information is the most important part of the plan.

The intent of the site safety plan is to minimize accidents and injuries that have the potential to occur during normal activities or that have the potential to occur because of the location of the site. This plan may be required in areas prone to severe weather, such as tornadoes, hurricanes, flooding, blizzards, or areas subject to geological activities such as earthquakes or volcanoes. It is readily apparent how the specific requirements of the site safety plan can incorporate the planning information that may have been already developed for the various other emergency planning regulations outlined previously.

The regulations 29 CFR 1910.120 (b) (4) (ii) (A)–(J) specify that a safety and health plan for hazardous substance cleanup operations, as a *minimum* shall address the following:

- A safety and health risk analysis for each site task and operation found in the workplace
- Employee training assignments to ensure compliance with all applicable regulations
- Personal protective equipment (PPE) to be used by employees for each of the site task and operations as determined by the personal protective equipment program
- Medical surveillance requirements as determined in the medical surveillance program
- Frequency and types of air monitoring and environmental sampling techniques and instrumentation to be used, including methods of maintenance and calibration of monitoring and sampling equipment to be used
- Site control measures consistent with the site control program

- Decontamination procedures consistent with the procedures determined for the conditions of the site
- An emergency response plan meeting the requirements for safe and effective response to emergencies, including the necessary PPE and other equipment
- Confined space entry procedures
- A spill containment program

While the regulations provide guidance on what elements must be provided in the Site Health and Safety Plan, the following outlines the practical steps to follow in development of the Site Health and Safety Plan:

- Identify key personnel and alternates responsible for site safety. The personnel can be designated by name or by function. (For example: site supervisor)
- Describe the risk associated with each operation conducted
- Determine that all site personnel are adequately trained to perform their assigned job responsibilities and to handle the specific hazardous situations they may encounter
- List and describe the personal protective clothing and equipment to be worn by site personnel during various site operations and/or locations
- Detail the specific medical monitoring requirements
- Outline the program for periodic air monitoring, personnel monitoring, and environmental sampling, if needed
- Describe the action to be taken to mitigate existing hazards (for example, containment of contaminated materials) to make the work environment less hazardous
- Define site control measures and include a site map
- Establish decontamination procedures for both site personnel and equipment
- Develop standard operating procedures (SOPs) for the activities that can be standardized
- Develop an emergency response plan for safe and coordinated response to accidents

All site personnel must receive a preentry briefing on these components of the Site Health and Safety Plan prior to beginning any work. This briefing can be integrated with other regulatory requirements, and cover programs such as emergency action, fire prevention, hazard communication, and respiratory protection.

A similar safety and health plan is also required for Section (p) Certain Operations Conducted Under the Resource Conversation and Recovery Act of 1976 (RCRA) for Treatment Storage and Disposal (TSD) facilities.

ELEMENTS OF THE EMERGENCY RESPONSE PLAN

The regulations require an emergency response plan for all three categories of workers covered by HAZWOPER: (1) for hazardous substances cleanup operations—Section (b) (4) H; (2) TSD workers—Section (p) (8) (ii); and (3) Emergency responders—Section (q) (2). The elements of the emergency response plan are similar for all three categories. As pointed out previously, the emergency response plan can be a section or appendix of the site-specific safety and health plan. OSHA defines emergency response as a response effort by employees from outside the immediate release area or by other designated responders to an occurrence that results or is likely to result in an uncontrolled release of a hazardous substance.

If employees at the company or agency are expected to respond to a hazardous substance emergency, OSHA expects a preplanning process. This process includes identifying the types, quantities, and location of hazardous substances, determining the risk and hazards of each substance, and developing response procedures and training for worst-case scenarios.

Response to an incidental release where substances can be absorbed, neutralized, or otherwise contained at the point of origin by employees in the immediate area or by maintenance personnel at the point of release are not considered to be emergency responders within the scope of the HAZWOPER standard. A response to a chemical where there is no potential safety or health hazards is not considered to be an emergency.

To assist in determining whether a spill is an incidental release or an emergency, federal OSHA has developed the following qualitative (subjective) and quantitative (numerical) determinants. These following parameters have been designed to help determine whether a hazardous substance release warrants an emergency response and is subject to the provisions of the HAZWOPER standards:

Qualitative determinants:

1. The release poses a life- or injury-threatening situation. This decision may be obvious or it may be a judgment call depending on the amount and type of hazardous substance released.
2. The release requires employee evacuation.
3. The situation requires immediate attention because of danger.
4. The release causes a high level of exposure to a toxic substance.
5. The situation is unclear or data are lacking.

Quantitative determinants:

1. The release poses or potentially poses conditions that are immediately dangerous to life or health (IDLH).
2. The hazardous substance release exceeds or could exceed 25% of the lower explosive limit.

■ **NOTE**

OSHA defines emergency response as a response effort by employees from outside the immediate release area or by other designated responders to an occurrence that results or is likely to result in an uncontrolled release of a hazardous substance.

3. The release exceeds the permissible exposure limit (PEL) by an unknown proportion.

In addition to preparing an emergency response plan, a response system must be developed. OSHA has incorporated the nationally recognized Incident Command System (ICS). ICS must be used in the response to a hazardous substance release that triggers the threshold to be considered an emergency response under the HAZWOPER standard. When in doubt in determining if an accidental release falls under the OSHA definitions for a spill, always err on the side of caution and use the HAZWOPER standard in response.

The following elements should be included in an emergency response plan:
- Preemergency planning and coordination with outside agencies
- Personnel roles, lines of authority, training, and communication
- Emergency recognition and prevention
- Safe distances and places of refuge
- Site security and control
- Evacuation routes and procedures
- Decontamination
- Emergency medical treatment and first aid
- Emergency alerting and response procedures
- Critique of response and follow-up

The emergency response plan is to be developed for all potential hazards at the site and is written in general terms. In the event of an accidental spill or uncontrolled release of a hazardous substance, a written, incident-specific emergency plan is developed for the particular response. At this point, all the preplanning and training should come together, and a coordinated incident-specific site safety plan is compiled by the trained response team.

SAMPLE INCIDENT-SPECIFIC SITE SAFETY PLAN

The site safety plan as shown in Figure 10-1 is for use in emergency response to a hazardous substance release. Although there is no standard or nationally recognized model, this sample is an easy fill in the blanks that can be used at an accidental release or spill. It uses the ICS as the basis for the organizational structure. It has been designed from a response perspective and can be rapidly completed at the scene and easily adapted to changing situations.

> ■ **NOTE**
> When in doubt in determining if an accidental release falls under the OSHA definitions for a spill, always err on the side of caution and use the HAZWOPER standard in response.

HAZARDOUS MATERIALS INCIDENT SITE SAFETY PLAN

Incident Name _____ Date _____ Operational Period _____

Site Information

Incident Location _____
Safe Access Route _____
Command Post Location _____
Control Zones _____
    Exclusion _____
    Contamination Control _____
    Support _____
Weather Conditions _____
Wind Direction _____ Speed _____ Temp/Time _____
Forecast _____

Organization

Incident Commander _____
Hazmat (HM) Group Supervisor _____
HM Technical Reference _____
Safe Refuge Area Manager _____
Safety Officer _____
Site Access Control _____
Entry Leader _____
Decon Leader _____
Entry Team
    1. _____
    2. _____
    3. _____
    4. _____
Backup
    1. _____
    2. _____
    3. _____
    4. _____
Decon
    1. _____
    2. _____
    3. _____
    4. _____

Figure 10-1 *Sample site safety plan.*

Hazard Evaluation

Chemical Name(s) _____
Hazards _____
General Hazards and Safety Precautions _____

Monitoring

LEL Instruments _____	( ) Continuous or _____
O2 Instruments _____	( ) Continuous or _____
Toxicity/ppm Instruments _____	( ) Continuous or _____
Radiological Instruments _____	( ) Alpha ____ Beta ____ Gamma ____

Protective Clothing

Entry _____

Decon _____

Decontamination

Decontamination Corridor Location _____
Decon Layout _____

Decon solution for Personnel _____
Decon solution for Equipment _____

Communications

Radio Frequencies Assigned _____
    Command _____
    Tactical _____
Cellular Telephone Numbers _____

Additional Communications _____

Emergency Procedures

Safe Refuge Area _____

Escape/Evacuation Alarm _____

Emergency First Aid Location _____

Figure 10-1
Continued

Plan Review

Safety Officer Signature _____ Time _____ Date _____
HM Group Supervisor Signature _____ Time _____ Date _____
Incident Commander Signature _____ Time _____ Date _____

Incident Name _____ Date _____ Operational Period _____

Plan Amendment

Check Amended Section:

- ☐ Site Information
- ☐ Hazard Evaluation
- ☐ Protective Clothing
- ☐ Emergency Procedures
- ☐ Organization
- ☐ Monitoring
- ☐ Decontamination
- ☐ Plan Review

Amended Information

Plan Review

Safety Officer Signature _____ Time _____ Date _____
HM Group Supervisor Signature _____ Time _____ Date _____
Incident Commander Signature _____ Time _____ Date _____

_____
_____
_____
_____
_____
_____
_____
_____

Figure 10-1
Continued

CHEMICAL INFORMATION SHEET

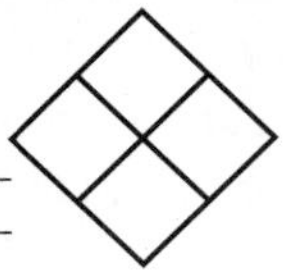

DOT Placard _____

Common Name _____
Chemical Name _____
Chemical Formula _____

Physical Description _____
Liquid _____ Solid _____ Gas _____
Chemical Structure _____
Vapor Density _____
Specific Gravity _____
Flash Point _____
Flammable Range _____
Solubility _____
Boiling Point _____
Melting Point _____
Vapor Pressure _____
Other _____
Incompatibilities _____

Primary Health Hazards:

Concentrations
(PEL, TLV, other)

Ingestion _____
Skin/eye absorption _____
Skin/eye contact _____
Carcinogen _____
Teratogen _____
Mutagen _____
Combustibility _____
Toxic by-products _____
Flammability _____
Explosivity _____
LEL _____
UEL _____
Reactivities _____

Corrosivity _____
pH _____
Neutralizing Agent _____

Recommended PPE: _____

Sources of Technical Information:

Figure 10-1
Continued

Summary

- Emergency planning is a process of identifying hazards, analyzing risk, determining how to mitigate or reduce potential problems, developing response and recovery plans, developing procedures, conducting training, and then testing the plans with mock scenarios.
- The safety of employees and the public are the first priority in emergency planning and response.
- In general, a site plan is organized as a single document with the various component sections and appendixes covering all tasks, operations, and contractors/subcontractors or outsourced employees and can be used to promote efficiency and enhance completeness, clarity, and coordination.
- OSHA requires that all employers plan and prepare for emergencies in the workplace and that all employees be trained to perform whatever emergency response roles they are assigned.
- All employers, whether private or public are required to implement emergency action plans for emergencies in the workplace.
- The emergency action plan must be communicated in writing, unless there are ten or fewer employees, then OSHA allows the company to communicate the plan verbally.
- The emergency action plan must include at least the following elements:
    1. Emergency escape procedures and emergency escape route assignments
    2. Procedures to be followed by employees who remain to operate critical site operations before evacuation (safe shut down)
    3. Procedures for accounting for employees after emergency evacuation is completed
    4. Rescue and medical assignments for designated employees
    5. Procedures for reporting fires and other emergencies
    6. Names and job titles of persons/departments who are responsible for the plan and can explain employee job duties under the plan
- All employers must develop, implement, and maintain at each workplace a written Hazard Communication Program that must include procedures to ensure that requirements for labels and other warnings, MSDSs, and employee information and training are all met.
- The Hazard Communication Standard does not specifically require an emergency response plan.
- The major objective of process safety management (PSM) of highly hazardous

chemicals is to prevent unwanted releases of hazardous chemicals, especially into locations that could expose employees and others to serious hazards.
- SARA Title III is actually four separate programs, with very specific reporting and planning requirements. They are listed in the following sections:
  1. SARA Section 301-303—Planning for emergency response
  2. SARA Sections 311-312—Reporting chemical inventory
  3. SARA Section 313—Reporting on release of regulated toxic chemicals
  4. SARA Section 304—Reporting leaks and spills
- Each facility must have its own contingency plan to deal with emergency situations if it is generating hazardous wastes as defined under RCRA. The plan must be designed to minimize hazards to human health or the environment from fires, explosions, or unplanned releases to the air, soil, or surface water.
- The laws covering emergency reporting are:
  1. Superfund (Comprehensive Environmental Response, Compensations, and Liability Act of 1980, CERCLA)
  2. Clean Water Act (CWA)
  3. Resource Conservation and Recovery Act (RCRA)
  4. Hazardous Materials Transportation Act (HMTA)
  5. Emergency Planning and Community Right to Know Act of 1986, also known as SARA Title III
- The federal Clean Water Act provides the basic national framework for regulating discharges of pollutants into the nation's navigable waters.
- All spills and unauthorized discharges must be reported to the National Response Center.
- The regulations 29 CFR 1910.120 (b) (4) (ii) (A)–(J) specify that a safety and health plan for hazardous substance cleanup operations, as a minimum shall address the following:
  - A safety and health risk analysis for each site task and operation found in the workplace
  - Employee training assignments to ensure compliance with all applicable regulations
  - Personal protective equipment to be used by employees for each of the site tasks and operations as determined by the personal protective equipment program
  - Medical surveillance requirements as determined in the medical surveillance program
  - Frequency and types of air monitoring and environmental sampling techniques and instrumentation to be used, including methods of maintenance and calibration of monitoring and sampling equipment to be used

- Site control measures consistent with the site control program
- Decontamination procedures consistent with the procedures determined for the conditions of the site
- An emergency response plan meeting the requirements for safe and effective response to emergencies, including the necessary PPE and other equipment
- Confined space entry procedures
- A spill containment program

- If employees at the company or agency are expected to respond to a hazardous substance emergency, OSHA expects a preplanning process. This process includes identifying the types, quantities, and location of hazardous substances, determining the risk and hazards of each substance, and developing response procedures and training for worst-case scenarios.
- Judgments as to whether a release warrants an emergency response are based on the following:

Qualitative determinants:

1. The release poses a life- or injury-threatening situation. This decision may be obvious or it may be a judgment call depending on the amount and type of hazardous substance released.
2. The release requires employee evacuation.
3. The situation requires immediate attention because of danger.
4. The release causes a high level of exposure to a toxic substance.
5. The situation is unclear or data are lacking.

Quantitative determinants:

1. The release poses or potentially poses conditions that are immediately dangerous to life or health (IDLH).
2. The hazardous substance release exceeds or could exceed 25% of the lower explosive limit.
3. The release exceeds the PEL by an unknown proportion.

- The following elements should be included in an emergency response plan:
  - Preemergency planning and coordination with outside agencies
  - Personnel roles, lines of authority, training, and communication
  - Emergency recognition and prevention
  - Safe distances and places of refuge
  - Site security and control
  - Evacuation routes and procedures
  - Decontamination
  - Emergency medical treatment and first aid

- Emergency alerting and response procedures
- Critique of response and follow-up.

Review Questions

1. List the three emergency planning and response priorities for hazardous substances releases?
2. There are four regulatory requirements for federal OSHA for employee health and safety. Name these four plan documents.
3. There are four separate programs within the Superfund Amendments and Reauthorization Act—Emergency Planning and Community Right to Know. List the sections and reporting requirement.
4. There are numerous hazardous materials release of spill reporting requirements in federal laws governing hazardous substances. Why are notifications required in a "timely manner"?
5. The requirements for development of spill response planning for releases into navigable water is found in which act?
6. The key elements of a Site Safety Plan for Hazardous Substances Cleanup Operations are outlined in 29 CFR 1910.120 (b) (4) (ii) A–J. List these ten elements.
7. Provide the OSHA definition of an emergency response.
8. There are two types of ways to determine whether a spill is an incidental release or an emergency. What are these determinants?
9. List the elements of an emergency response plan.
10. The nationally recognized incident management system is called _____.

Chapter 11

Air and Environmental Monitoring

Learning Objectives

Upon completion of this chapter, you should be able to:

- Understand the reasons for performing air monitoring.
- Understand the behavior of gases.
- Describe the basic gas laws.
- Define the role of an industrial hygienist.
- Understand the principles of when and how to operate air monitoring equipment including radiation detectors, oxygen meters, combustible gas indicators, photoionization detectors, flame ionization detectors and colorimetric tube systems.
- Describe the proper use of air monitoring equipment in confined spaces.
- Describe the conditions for use of air monitoring equipment in outside areas.
- Understand the effects of local weather.

INTRODUCTION

Gases, by the definition of the word itself, denote chaos. Whether by looking at the way they can move or how they physically react, gases represent a physical state of hazardous materials that requires specific attention. For centuries, early researchers understood the basic physical properties of liquids or solids, but vapors or "airs" remained misunderstood or generally unknown. Scientists realized that the air around them did have specific properties, but since most gases are invisible, little study was made. Until the eighteenth century, bad air, such as from swamps (methane, then known as marsh gas), was treated as a health hazard. Today, we still use the term malaria to denote a medical condition although the word was derived from the Italian language and literally means "bad air." Not until the experiments of Allesandro Volta (the creator of the first true battery), was marsh gas treated with scientific respect.

Further experimentation was done isolating specific gases in the atmosphere. With the successes of Priestly (oxygen, 1774) and Cavendish (hydrogen, 1766), science began to understand that not only was ambient air comprised of more than one gas, but gases came from other sources and represented specific hazards. Priestly for example, suspended objects in the "fixed air" (carbon dioxide) above beer vats. His subjects included mice, which died, and candles, which went out, thus giving the first evidence of one of the prime hazards of gases—asphyxiation. One workplace that has characteristically represented a hazardous atmosphere and indeed still does are mines, especially coal mines. Here the inherent dangers of **firedamp** and **afterdamp** are still prevalent today.

It is critical for the complete study of hazardous materials to include a chapter on gases and vapors and how to detect their presence through air monitoring. The use and production of gases in the post-Industrial Revolution society is manifold. Not only mines, but almost all major industrial processes such as petrochemicals and computer manufacture rely upon gases for production. Naturally produced gases such as methane and hydrogen sulfide also pose a problem as well as gases found in waste sites, drums, confined spaces, sewage treatment, chemical manufacturing, and agriculture. Gases and vapors, both man-made and natural represent a health hazard and require air monitoring. Figure 11-1 shows the use of air monitoring equipment, a combustible gas indicator, at an underground storage tank removal site. Here it can be seen that gases represent a specific danger to those who work with, store, or transport them in any manner. Due to the large variety of gases present in our society, it is beyond the scope of this chapter to explain how all may react. We will, however, understand how gases "work," where they may accumulate, and how their presence is determined in order to make the workplace safer and to protect human life.

In order to work with **gases,** it is important to understand their physical state and nature of movement. To understand the definition one must also recognize the three hazards that gases pose. First, a gas tends to occupy the space of its container. Thus, if the gas becomes free of its container, it will continue to expand

firedamp
methane gas

afterdamp
carbon monoxide

■ **NOTE**
Gases and vapors, both man-made and natural, represent a health hazard and require air monitoring.

gas
a substance whose physical state is such that it always occupies the whole of the space in which it is contained

Figure 11-1 *Testing for combustible gases in an underground storage tank.*

until it reaches the same pressure as the ambient atmosphere. This is why gases usually expand upon release at **standard temperature and pressure** (STP). The second hazard described in this definition is that manufactured gases are always in some sort of container, such as a cylinder. With a change in the physical characteristics of the gas, the cylinder can rupture or explode, becoming a missile hazard. Finally, each gas has a vapor density in relation to the normal atmosphere. Many gases such as hydrogen, methane, neon, and other gases will rise into the atmosphere and disperse. Other gases are heavier than air and will sink and collect in low spots, often taking time to disperse. An example of this is chlorine. It is a heavy, dense gas that dissipates slowly under normal conditions. For this reason it was used as the first major poison gas in April of 1915 in World War I. Soldiers exposed to the gas sought refuge out of the trenches while their colleagues choked to death a few feet below them in the bottom of shell holes and trenches.

GAS LAWS

■ **NOTE**
By altering the physical factors that act upon gases, such as excessive pressure or heat, they can become unstable.

Although gases were discussed previously in Chapter 3, it is important to emphasize the specific physical characteristics of gases as they relate to hazardous materials response. In most cases, gases are stored and transported under pressure in containers. By altering the physical factors that act upon gases, such as excessive pressure or heat, they can become unstable. Since these conditions may arise even under normal atmospheric conditions, gases must be treated with a high level of respect. There is a reason why in transportation gases (DOT class 2) are considered the second highest hazard after explosives (DOT class 1). They can rise, sink, flow, burn, explode, asphyxiate, poison, and corrode people and equipment.

Before going further into the physical hazards of a gas, it is important to understand the basic properties of gases. Understanding gases is dependent upon the relationship between the volume of a gas, its temperature and pressure. These principles were generally introduced in Chapter 3, but we may now validate those concepts with mathematical equations. Other factors are used to determine mass and volume of gases, but the scope of this section is to give the reader the understanding that gases are primarily affected by changes in temperature, pressure, and/or volume. The result of any significant change in these conditions may lead to container failure, one of the driving themes of this chapter.

The first of the gas laws to be demonstrated is Boyle's law, named after its creator, Robert Boyle. This law states that the volume of a gas varies inversely with pressure when temperature is a constant. Thus:

$$P \times V = \text{constant}$$
$$P_1 \times V_1 = P_2 \times V_2$$

standard temperature and pressure
0°C at 1 atmosphere of pressure

In this formula, P and V denote pressure and volume. P_1 and V_1 are the initial pressure and volume and P_2 and V_2 are the final values for pressure and volume. For example, if you want to know what volume 100 cubic feet (100 ft³) of a random gas occupies if its pressure is changed from 15 to 250 psia with temperature remaining fixed, Boyle's law can thus read:

$$V_2 = 100 \text{ ft}^3 \times \frac{15 \text{ psia}}{250 \text{ psia}} = 6 \text{ ft}^3$$

The second gas law to be determined was by the French scientist Jacques Charles. Charles's law demonstrates that if pressure is a constant, then volume increases with an increase in temperature. A simple example is the energy force behind a hot air balloon. Thus, if a gas sample is compared at a constant pressure but different temperatures and volumes, Charles's law can be demonstrated as follows:

$$\frac{V_1}{T_1} = \frac{V_2}{T_2} \quad \text{or} \quad \frac{V_1}{V_2} = \frac{T_1}{T_2}$$

In these formulas, V and T denote volume and temperature (absolute). V_1 and T_1 denote initial volume and temperature, and V_2 and T_2 denote final volume and temperature. Absolute temperature, or Kelvin (K) must be used to factor the temperature. [0 K = −273.15°C and 0°C = 273.15 K]. Thus, if a 250 cubic millimeter plastic bag of a random gas is heated from 25°C to 40°C, the final volume after warming can be determined as follows:

$$V_1 = 250 \text{ cm}^3$$
$$T_1 = 25°C \text{ or } 298.15 \text{ K}$$
$$T_2 = 40°C \text{ or } 313.15 \text{ K}$$

Substituting the above values for V_2, the following applies:

$$\frac{250 \text{ cm}^3}{298.15} = \frac{V_2}{313.15}$$

or,

$$V_2 = \frac{313.15}{298.15} \times 250 \text{ cm}^3 = 262 \text{ cm}^3$$

When neither temperature or pressure is constant both Boyle's law and Charles's law can be combined into the following formula known as the Combined Gas law:

$$\frac{V_1}{V_2} = \frac{T_1 \times P_2}{T_2\, P_1}$$

The final consideration of these laws is that the volume of a gas may be forced to remain constant while the other factors that drive Boyle's law and Charles's law change. In other words, the gas is in a container such as a metal cylinder. In this case the values of V_1 and V_2 remain the same. As the gas is heated, the pressure increases. This is referred to as Amonton's law

$$\frac{P_1}{T_1} = \frac{P_2}{T_2}$$

When the temperature of a cylinder is increased significantly (such as in a fire), the cylinder may rupture or explode.

DETECTION OF GASES

The amount of a gas that may adversely affect us may be extremely small. Arsine gas, for example, has an immediately dangerous to life and health (IDLH) of approximately 3 ppm. Gases may be invisible, colorless, or odorless. Never should your sense of smell be used to find a gas. Never should an environment be judged "clean" because it looks clean. The detection of gases present in an environment shall always be done with correctly calibrated instruments. Gases can pose a significant hazard to your health in various manners. Thus, the monitoring of an atmosphere can reduce your risk of exposure to a gas. To perform this function, a series of different monitors are used. Keep in mind there is no such thing as a universal gas monitor that performs all functions required to completely analyze a test atmosphere. This chapter describes each monitor and its purpose.

■ **NOTE**
The detection of gases in an environment shall always be done with correctly calibrated instruments.

There are various manufacturers and models of air monitoring equipment on the market. Some models may perform more than one function, test for different levels, or test for a variety of gases, but all work in the same manner to identify contaminants in the atmosphere.

When detecting gases, some groups such as public agencies may have to take steps to identify a completely unknown atmosphere because the source could be almost anything. In other situations, such as in a factory, a hazardous materials response team may know exactly what has been released or at least be able to narrow down the potential release due to knowing what is present at their facility and where it is. In other words, if you know what is present because it is the only possible chemical present, then air monitoring may be done only to confirm the presence of that material. Do not waste time in running tests for unknown gases. Before or even during air monitoring procedures, responders can be on the lookout for other methods of identification such as labeling and placarding.

The use of monitoring equipment not only protects the health and safety of the individual but allows waste workers and responders in the hazardous materials field to establish other factors such as selecting the proper level of personal protection equipment, establishing control zones and evacuation areas, defining a decontamination corridor, assessing the potential health effects of exposure, and determining the need for specific medical monitoring. Air monitoring is a continuous, ongoing process. Air monitoring shall be done not only upon first entry or investigation of an environment, but on a regular basis throughout any operation. Ongoing air monitoring allows for changes in the atmosphere due to heating or cooling during the day, movement due to natural or forced ventilation, the movement of toxic materials themselves, and any other situation that may alter atmospheric conditions.

In most cases, detection is made using hand-held, direct-reading instruments. The choice of a specific model is up to the individual, but any instrument used should be portable and rugged, easy to operate under field conditions, intrinsically safe (not a spark hazard), and produce reliable and useful results. Waste workers and responders using air monitoring equipment shall be trained on how to use them properly and correctly analyze their results. Never should an individual be expected to learn an instrument by using it as he or she "goes along." Everybody should have a strong level of confidence not only in the accuracy of their equipment, but in their ability to correctly use it and analyze the data.

This chapter describes the sequence to be used by members of a properly trained cleanup or response team who determine atmospheric hazards. In other words, use a step-by-step process to confirm the presence of a hazardous situation involving oxygen levels and/or the presence of gas hazards. Remember, you are there to rapidly determine the presence or absence of atmospheric hazards and to contain and control the release. Further testing may require the specialties of an industrial hygienist.

An industrial hygienist uses the same or similar tools described in this chapter, but does so in a more precise manner. This does not mean that results

> **■ NOTE**
> Air monitoring shall be done not only upon first entry or investigation of an environment, but on a regular basis throughout any operation.

> **❗Safety**
> Never should an individual be expected to learn an instrument by using it as he or she "goes along."

in the field are faulty. Those quickly obtained results tell whether a chemical is present and to what extent atmospheric contamination can be expected. For example, a properly trained waste worker or responder may test a storage tank for oxygen deficiency or the presence of a combustible gas. An example of the work of industrial hygienists is working with "sick buildings" or "sick rooms." In this situation, precise readings and levels must be determined to find out exactly what is present and exactly where the source is. In sick rooms or buildings it is not a spill or release of a chemical as is common with hazardous materials incidents, but dealing with offgassing from consumer articles such as carpet, chairs, and plastics. An industrial hygienist may also have to research the previous use of the land and work/life habits of involved personnel to further determine what may be causing the complaint.

THE SEQUENCE OF AIR MONITORING

As mentioned, there is no such thing as a universal gas detector. To properly determine the atmospheric conditions of a space, tank, cylinder, or area, various types of monitors are required. The following is the suggested thought process to be followed in testing an atmosphere:

- First, if radioactive materials are known or suspected in the test atmosphere, the use of radiation monitors is required. This action is more the exception than the rule due to the relative rarity of incidents involving radioactive materials.
- Thus, an oxygen indicator is generally used first, primarily in areas that may be oxygen deficient, such as a confined space.
- Next, a combustible gas indicator (CGI) is used to detect the presence of combustible gases or vapors.
- A flame ionization detector (FID) is used to detect organic gases and vapors. This device may include a gas chromatograph. In many situations, the use of this device is optional.
- An ultraviolet photoionization detector (PID) is used to detect organic and some inorganic gases and vapors. Like an FID, the use of this device may be considered optional.
- Finally, a system of colorimetric indicator tubes is used to identify specific gases.

THE USE OF AIR MONITORING EQUIPMENT

Oxygen Meters

The first monitor that should be used is the oxygen meter unless the presence of radioactive material is known or suspected, especially in a situation where there

■ **NOTE**
According to OSHA, any atmosphere less than 19.5% oxygen is considered oxygen deficient and any atmosphere more than 23.5% oxygen is considered oxygen enriched and thus an increased fire hazard.

zeroing
bringing a monitoring device to a baseline or zero readout on gauges or digital readouts

is the possibility of oxygen deficiency or oxygen enrichment such as a confined space. Oxygen meters work on the principle of allowing oxygen to diffuse into a detector cell and measuring a chemical reaction establishing a current between two electrolyte cells. The normal percentage of oxygen in the atmosphere is approximately 21% at sea level. According to the Occupational Safety and Health Administration (OSHA), any atmosphere less than 19.5% oxygen is considered oxygen deficient and any atmosphere more than 23.5% oxygen is considered oxygen enriched and thus an increased fire hazard. Oxygen meters may be used to determine the type of respiratory protection required, the risk of combustion, and if there may be variation on other instruments used such as a combustible gas indicator. Many models do not work accurately in less than 16% oxygen. As seen in Figure 11-2, oxygen meters are straightforward instruments that have a visual readout and an audible warning that alarms when the oxygen levels drops below 19.5% or rises above 23.5%. Like all instruments they require calibration and **zeroing** before use. This setting should be done in a normal oxygen level atmosphere before the instrument is exposed to the test atmosphere. The accuracy of an oxygen indicator may be affected by changes in altitude or barometric pressure or the presence of certain gases such as ozone or carbon dioxide, which can damage the detector cell.

Combustible Gas Indicators

The second instrument to be used is a combustible gas indicator (CGI). The market for these instruments is large and so too is the variety. Keep in mind the considerations discussed in choosing which instrument is right for the task in mind. If the device is complicated, many people simply won't use it. Many CGIs are considered

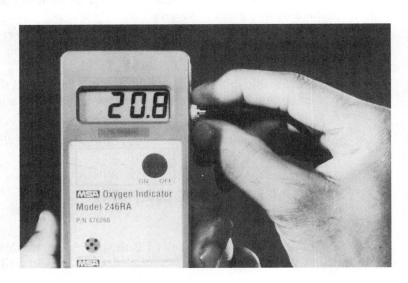

Figure 11-2 *Oxygen Meter.*

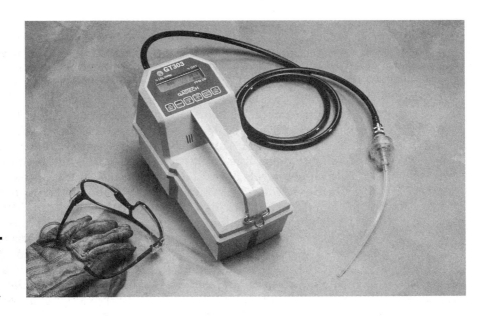

Figure 11-3
Combustible Gas Meter. Photo courtesy of Gastech.

■ **NOTE**
Never allow the intake port of a CGI to come in contact with any fluid including water.

multigas detectors and may have a built-in oxygen indicator. Others may detect specific gases such as carbon monoxide, methane, or hydrogen sulfide. If not designed for a specific gas, CGIs are designed to detect the presence of combustible gases, but not determine what combustible gas or gases are actually present.

Many CGI's operate on this principle: A fan draws in a sample of the test atmosphere over a heated catalytic filament. If a combustible gas is present, electrical resistance is created on one filament connected to a circuit causing a measurable imbalance. It is critical to remember to never allow the intake port of a CGI to come in contact with any fluid including water. Drawing a fluid into the machine compromises the device and requires extensive (and expensive) repairs. A typical device, as seen in Figure 11-3, is designed to give an visual and audible alarm usually at 10% of the lower explosive limit (LEL). This allows for a large safety factor due to variables in detecting the combustible gas. These variables include an oxygen-deficient atmosphere, the type of calibration gas (such as hexane or methane) used in the CGI itself, variation in temperature, and the presence of interfering materials such as lead, sulfur, silicones, hydrogen chloride, and hydrogen fluoride. When a CGI alarms, the machine is telling you of an oxygen-deficient atmosphere or that a combustible gas is present. The CGI does not determine what the combustible gas is unless the machine is a multigas detector calibrated for specific gases. In this situation the device can warn of the presence of gases such as carbon monoxide or hydrogen sulfide if it designed to do so. As with all monitors, a CGI requires proper calibration and zeroing before each use and recalibration with the test gas in a periodic manner.

Photoionization Detectors

Photoionization detectors (PIDs) are designed to detect relatively low concentrations of contaminants usually in the range of 0.1 to 2000 ppm. They are efficient in detecting aromatic hydrocarbons such as benzene, toluene, xylene, and vinyl chloride. They work by drawing a sample of the test atmosphere into a chamber with an ultraviolet (UV) lamp. The UV light breaks down the sample by displacing electrons and measures the energy required to do so. This process is called the *ionization potential* (IP) and each chemical compound has a unique IP. Thus, it is possible to determine an actual gas by reading the IP off the PID and comparing it to known levels of gases. One reference source is the NIOSH *Pocket Guide to Hazardous Materials,* which lists the IP for each compound. Another factor with the PID is that there is a variety of UV lamps with different electron volts (eV) capabilities. Ionization potential is measured in electron volts, so the correct lamp must be used to determine the gas. The contaminant gas must have an IP less than the eV capacity of the UV lamp. One of the key considerations for this device is that it must be properly calibrated and accurately **spanned** before use to ensure the proper IP is determined, which is best determined by consulting the manufacturer's data tables. Table 11-1 illustrates some ionization potentials.

Figure 11-4 displays a typical PID. The strap allows the body of the machine to be hung off the shoulder while the wand is used to collect the sample. PID's work well with small concentrations, but high concentrations may cause false low readings. Other variable factors include water vapor (humidity), nonionizing gases, diesel exhaust, smoke, soil, and a dirty UV lamp. Although a PID can detect many ionizing materials at low levels, they do not detect everything. For example, they cannot detect methane, therefore they are used in conjunction with the other instruments.

> ■ **NOTE**
> A PID must be properly calibrated and accurately spanned before use to ensure the proper IP is determined.

spanned
to have set an approximate calibration for a certain number of chemical compounds

Flame Ionization Detectors

Flame ionization detectors (FID) detect many organic gases and vapors by using charged particles or ions to detect chemicals in the air and a hydrogen flame to burn organic materials in air. As the test contaminant is burned, positively

Table 11-1 *Example ionization potentials.*

Chemical	IP (eV)	Chemical	IP (eV)
Acetone	9.69	Ethylene oxide	10.56
Ammonia	10.18	Hydrogen peroxide	10.54
Carbon dioxide	13.77	Toluene	8.82
Chlorine	11.48	Vinyl chloride	9.99

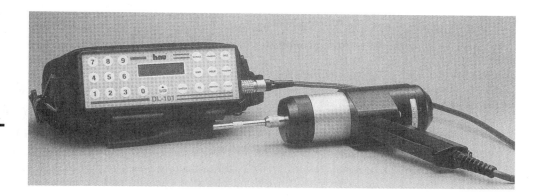

Figure 11-4
Photoionization Device. Photo courtesy of hnu.

charged ions are produced and a current is generated. This current is then measured on a scale relative to a calibrant gas. Because many of the capabilities of an FID are overlapped by those of a PID, and because of their high cost and maintenance, FIDs are not commonly used except by specialized hazardous materials teams. If they are used however, they do have an advantage over PIDs in that in the survey mode, they can read 0 to 1,000 ppm or up to 10,000 ppm. Many models of FIDs also have the configuration to work as a gas chromatograph, but this requires further training and experience to interpret the data correctly. Figure 11-5 shows one model of an FID. Other models may be more portable than this type including those that function solely as a gas chromatograph.

Figure 11-5 *Flame Ionization Device. Photo courtesy of Foxboro.*

Gas Chromatographs

If a device such as a PID or FID does incorporate a gas chromatograph (GC), then the device can be set in one of two modes: first, only in the survey mode, which allows the device to run as a detector only, and second, in the GC mode, which allows the device to run as a detector and a GC. However, in this latter mode, proper use of the device requires extensive training and correct readings require near ideal conditions. In principle, a GC draws a test sample into a tube. The tube contains a medium that adsorbs the test sample with a variation in bonding strength. Depending on the strength of the bond, the sample is retained in the tube and eventually passed out the other end. This retention time is measured and compared to known relations of gases. Under ideal conditions of temperature, medium, and flow rate, the GC can determine the quantity of a gas present based on retention time. It is possible to purchase a separate GC device for field use, but the cost (average $15,000) is a significant factor in most emergency response team's budget.

Colorimetric Tubes

The system that is used more often than PIDs and FIDs to determine unknown contaminants is colorimetric tubes as shown in Figure 11-6. Although colorimetric tubes also have limitations and their use may be time consuming, they are far easier to use and maintain than PIDs and FIDs. Colorimetric tubes work on the principle of detecting a specific gas by having the test atmosphere drawn into a tube that is filled with a chemical that will change to a specific color in response to the gas.

Figure 11-6
Colorimetric Tube System.

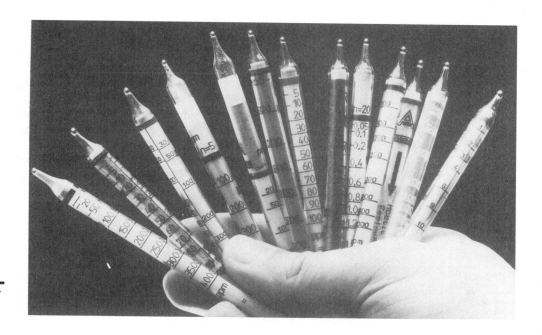

Figure 11-7 *Types of colorimetric tubes.*

■ NOTE
Colorimetric tubes work well in identifying a specific gas, but results in actual levels in parts per million may vary from what is actually present by 25–50%.

■ NOTE
Changes or extremes in temperature, high or low humidity, barometric pressure, sunlight, and other interfering gases may alter the reading.

Colorimetric tubes also can determine the quantity of gas present in parts per million. Observe in Figure 11-7 the graduated markings on the side of the tubes. These markings, or n scale can be read to determine the level of contaminant based on the number of pump strokes required. However, any concentration reading should be taken as general due to slight variation in color, irregular changes in color, or a "bleeding" effect in color change. Thus the prime use of colorimetric tubes is to determine what actual gas is present but not its actual concentration. Colorimetric tube systems work well in identifying a specific gas, but results in actual levels in parts per million may vary from what is actually present by 25–50%. It is not necessary to purchase a tube for every type of contaminant. This would not only be extremely expensive, but also unwieldy. Manufacturers such as Drager and Sensidyne offer qualitative tubes for acids, bases, organic amines, unsaturated hydrocarbons, halogenated hydrocarbons, and aromatic hydrocarbons. Also offered are polytest tubes that detect multiple gases such as ammonia, hydrogen chloride, hydrogen sulfide, and chlorine.

Colorimetric tubes have some other considerations that may affect their use. Changes or extremes in temperature, high or low humidity, barometric pressure, sunlight, and other interfering gases may alter the reading. When taking readings, the user should always take three tubes: one to perform monitoring, the second to monitor if the first tube becomes saturated, and the third as a blank to determine subtle color changes. This assists in determining if there are any cross-sensitivities to the gas being measured. Each box of tubes has directions that describe any cross-sensitivities that may exist.

Other Detection Devices

Other monitors may be used in some situations to determine unknown contaminants in an atmosphere. The prime example is radiation detection equipment as shown in Figure 11-8. As stated, if the presence of radioactive materials is known or suspected, then a device for measuring alpha, beta, and gamma radiation is used. There are also specialized monitors for materials such as lead, mercury, ozone, and solvent vapors that are used when a specific contaminant is suspected or known and verification is required.

Another type of air monitoring equipment that may sound rudimentary but can be essential is pH paper. If the presence of corrosive vapors is known or suspected, a simple procedure can protect sensitive equipment against corrosive

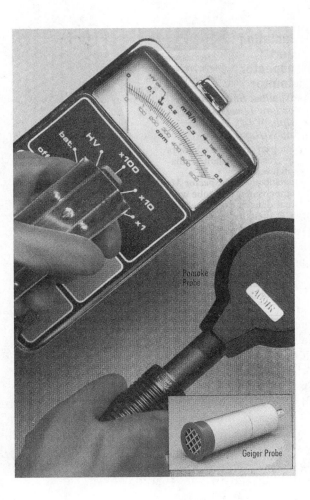

Figure 11-8
Radiation Detection Device. Photo courtesy of Lab Safety.

vapors. Simply attach a piece of pH paper wetted with deionized water to an extension device as simple as a broom handle and put it into the test atmosphere. If the paper detects the presence of a strong corrosive gas (pH less than 2.5 or greater than 12), then ventilating the area or using some other engineering control can assist in reducing the level of contamination. Caution must be exercised even with an extension because you may still be contaminated by the atmosphere.

> **Safety**
> Caution must be exercised even with an extension because you may still be contaminated by the atmosphere.

GENERAL PROCEDURES FOR AIR MONITORING

As with any piece of equipment, it is important to fully understand how it works before using it in the field. Becoming familiar with the monitor includes reading manufacturers' guidelines, practicing with other experienced personnel, and performing some basic drills before actual use. In addition, waste workers or responders shall be familiar with the procedures for calibrating and zeroing the monitor before use. Usually calibrating and zeroing requires energizing the monitor in a normal atmosphere before entering the test atmosphere. One of the prime reasons for equipment failure is not checking battery levels. Ensure all monitors to be used are properly charged and ready to go before use. Never go into the field with the faith that all equipment is in properly operating order. Involved personnel should know beforehand that all equipment is working properly. An ongoing maintenance program with a log that documents all work and tests performed is an excellent way to ensure air monitoring equipment can be counted on when the need arises.

> **NOTE**
> One of the prime reasons for equipment failure is not checking battery levels.

> **NOTE**
> Remember that gases have different vapor densities.

When testing for gases, it is important to remember that gases have different vapor densities. That is, some will rise and others will settle and possibly collect in low areas of the test atmosphere, therefore it is important to test high and low. Before entering a confined space, use an extension device to put the probe head of your oxygen meter or CGI into the space without making bodily entry. When using any sort of monitor, with or without an extension, allow time for the machine to draw the sample into the test chamber. The monitor pump may need a few moments before the sample reaches the detection device itself and is analyzed. In other words, do not keep walking into an area until you have a reading. Stop, wait a few moments, then proceed. Continue this process until the entire area has been tested. If dealing with a gas in a confined space or any enclosed area, think of the most logical place in which the gases may accumulate. Some gases stratify; they can occupy different layers dependent upon the vapor density of the associated atmosphere. Other gases such as hydrogen, acetylene, and ethylene may accumulate near the ceiling. Most hydrocarbon vapors and flammable gases such as propane and butane are heavier than air and lie low to the ground. These gases may follow the path of least resistance to the lowest area possible, therefore it is important to check areas such as sumps, subfloors and trenches. When checking gases and vapors over liquid spills such as in Figure 11-9, take care not to put a probe head into any standing liquid. Liquids, including water, that are drawn into the monitors, especially CGIs and PIDs, can alter readings and cause severe internal damage.

> **NOTE**
> If dealing with a gas in a confined space or any enclosed area, think of the most logical place in which the gases may accumulate.

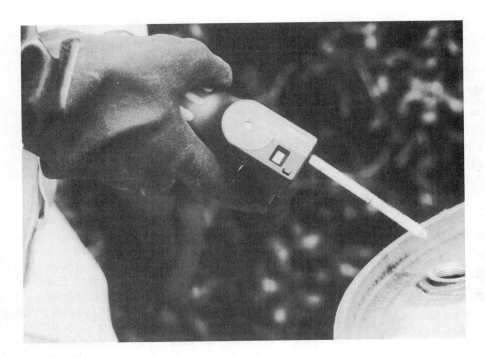

Figure 11-9
Sampling a gas over a liquid spill.

■ **NOTE**

In confined spaces, initial monitoring should be done without physical entry into the space.

MONITORING IN CONFINED SPACES

In confined spaces, it is important to remember that initial monitoring should be done without physical entry into the space. Attach extension devices to the probe head and wands if required. Keep in mind that any extension will increase the draw time. In the case of colorimetric tubes it may take the pump a long time to fully expand from each stroke. When monitoring confined spaces, ensure all corners are tested as well as any area that a suspect gas may pocket or accumulate. A good rule of thumb is that as much of the space as possible should be monitored. Monitoring in confined spaces, like all monitoring programs, shall continue as long as required; that is, several times during the workday, any time after workers have left the space and are returning, with any change in temperature or barometric pressure, if potential gases are produced or disturbed, or any other condition that may warrant monitoring the atmosphere. In many cases, monitors are kept energized from the beginning until the end of the task.

MONITORING OUTSIDE

When monitoring outside, it is best to approach the spill from uphill and upwind if possible. Again, take a reasonable amount of time to allow the monitor to cycle what it is sampling. When outside, the use of an oxygen indicator is not usually required unless the gas discharge is large enough to displace the normal atmospheric oxygen level and is continuing to discharge. If a visible cloud is present and relatively stable conditions are in effect (no significant wind), then detection for oxygen deficiency may be required. It is a good idea to understand the effects of variations in **topography**. For example, heavy gases will follow the path of least resistance into areas such as ravines, gullies, drainage ditches, and trenches. Remember too that many gases such as carbon monoxide are colorless and odorless and detection can be based only upon monitoring equipment.

Even after approaching from uphill and upwind, entry into the contamination area is necessary to detect what gas or vapor is present and the amount of contamination. Whenever this situation is encountered, proper personal protective equipment shall be used. Detection procedures shall include looking for a source or origin of the gas, such as a compromised cylinder, and then checking at various points downwind to determine the extent of the spread of the contamination. Monitoring should also be done on the cross-axis of the discharge area to determine the degree of dispersion. Finally, monitoring should be done upwind of the source to determine that there is no cross-contamination and to verify the source itself.

THE EFFECTS OF LOCAL WEATHER

Air monitoring procedures and the movement of gases and vapors can be significantly altered by changes in the weather. Changes in the temperature can increase vapor production (volatilization) and possibly lower vapor density, thus keeping gases near the ground instead of dissipating into the atmosphere. An increase in temperature may also affect the stability or reactivity of normally stable materials. With a change in temperature may also come a change in **relative humidity**. As the humidity level increases, so too may the rate of vaporization of water-based soluble material. The most significant weather event that may alter the situation is wind. In conditions of low wind (wind speed less than 10 knots), clouds of gases and vapors will be slow to dissipate and vapor concentration will remain high. When conditions are like this, the weather is considered stable and downwind dispersion remains small. When condition become unstable (high winds and/or variation in wind direction), gases and vapors tend to spread further, but do disperse more rapidly. An increase in winds can also make dusts and particulate-bound contaminants airborne. With any task involving air monitoring, these factors should be taken into account along with planning for expected daily changes, called **diurnal effects**. Examples of diurnal effects are changes in barometric pressure, wind direction, and temperature.

When dealing with a release of a hazardous material it is very important to know the effects of local weather. In many situations outside agencies such as the National Weather Service or local airports can assist with weather forecasts. **Land breezes** and **shore breezes** are prime examples of variation in weather. Due

---

topography
physical land features of an area

■ **NOTE**
Air monitoring procedures and the movement of gases and vapors can be significantly altered by changes in the weather.

relative humidity
expression of the amount of water vapor in the air

■ **NOTE**
The most significant weather event that may alter the situation is wind.

diurnal effects
expected daily weather patterns such as directional wind shifts, fog, or rain

land breezes
occur when the land mass cools faster than the adjacent body of water and creates an offshore breeze

to differences in the cooling ratios of land and water, wind tends to flow onshore or pick up velocity after about midday. Because this effect can happen almost every day in some areas, it is a diurnal change that can be forecast with high accuracy. Thus, air monitoring would have to be continued to ensure contaminants are not being spread in a new or larger area due to any change in wind direction. If involved personnel desire more direct on-site weather readings, then more sophisticated equipment can be purchased. However, equipment required for weather monitoring does not have to be complex. A simple wind sock or any type of flag or banner waving from a pole can determine wind speed and direction. Simple barometers and thermometers can be purchased and kept ready for use.

Summary

- Many types of gases and vapors are produced by both man-made and natural sources.
- A gas is a substance whose physical state is such that it always occupies the whole of the space in which it is contained or into which it is released from its primary container.
- The behavior of a gas can be radically altered by a change in temperature, pressure, and volume. If volume is forced to remain the same while the other factors are changed, then cylinder failure may occur (Amonton's Law).
- The only accepted manner to determine the presence (or absence) of gases is the use of properly calibrated and adjusted mechanical air monitoring equipment.
- OSHA defines any atmosphere with less than 19.5% oxygen as oxygen deficient and any atmosphere with more than 23.5% oxygen as oxygen enriched.
- Combustible gas indicators usually determine the presence of a combustible gas but not actually what it is. They should alarm at 10% concentration of the lower explosive limit.
- Photoionization detectors require accurate adjustment, spanning, and reading of results to determine the ionization potential (IP) for each respective material.
- Many flame ionization detectors have a built-in gas chromatograph but require extensive training in use and interpretation of results.
- Colorimetric tubes work well in verifying the presence of many gases but should not be counted on to accurately determine the actual amounts (in parts per million).
- If the presence of radioactive materials is known or suspected, then testing with radiation detectors shall be performed first.
- If corrosive gases are known or suspected to be present, then testing the atmosphere with a wet strip of pH paper first is recommended.

Review Questions

1. What three physical factors can be changed to alter the behavior of a gas?
2. What air monitoring device should be used first when evaluating the atmosphere of a confined space?
3. What are the six air monitoring devices that have been described. Give a brief description of how each one is used.
4. In your own words, describe some hazards that may pose a threat while testing the atmosphere in a confined space.
5. What factors should you take into account when conducting air monitoring outside?

Scenario

As part of an emergency response team, you are responsible for monitoring the atmosphere of an underground storage tank. Upon opening the access plate, you observe the tank is heavily rusted with what appears to be about six inches of a waterlike material on the bottom. The purpose of entry is to empty and clean the tank. You have access to any air monitoring device you require. List your actions and any specific hazards that may arise during this job.

Chapter 12

Physical Hazards

Learning Objectives

Upon completion of this chapter, you should be able to:

- Define confined space operations including:
  - Characteristics of confined spaces.
  - Hazards associated with confined spaces, particularly atmospheric hazards.
  - Management of confined space hazards.
  - The hazards posed by the release of dangerous materials and energy within a confined space.
  - Identifying different types of hazardous atmospheres including oxygen-deficient and oxygen-enriched atmospheres, environments with flammable vapors, and potentially toxic atmospheres.
  - Monitoring the atmosphere to detect hazards.
  - Respirators and other equipment used to reduce your exposure to hazards.
  - Complying with entry permit requirements.
  - Basic emergency procedures.
- Identify environments where a potential noise hazard exists.
- Understand the importance of controlling hazardous energy sources.

INTRODUCTION

In addition to chemical hazards found at waste sites and hazardous materials emergency scenes, a number of physical hazards may be present. These dangers include electrical hazards, trenches or other unstable workplaces, confined spaces, and exposure to environmental elements. Because these items can be quite hazardous on their own, the Occupational Safety and Health Administration (OSHA) directed Hazardous Waste Operations and Emergency Response (HAZWOPER) training programs to include some discussion on them as part of the curriculum. The range of these hazards is far too encompassing to cover in a study such as ours, but we focus our attention on a few of the most widely encountered problems found at work sites. These include confined space operations, control of hazardous energy, and noise exposure. In addition to these, you should always consider the potential for slips, trips, and falls, foot protection, ladder work, or any additional threat that complicates your chemical response or cleanup operation.

CONFINED SPACE

permit required confined space (PRCS)
a recognized confined space requiring permits and formal operations such as rescue, air monitoring, and training prior to entry

■ **NOTE**
OSHA estimates that the new rules and procedures (if followed) may prevent approximately 80% to 90% of the deaths in confined space accidents annually.

codified
adopted as a recognized piece of legislation and placed into the statutes in writing

In January of 1993, OSHA issued its final standard on **permit required confined spaces** (PRCS) as a result of numerous deaths and injuries that occur during confined space operations. Statistics showed that approximately 300 persons die each year in the United States as a result of confined space accidents while countless others are injured. OSHA estimates that the new rules and procedures (if followed) may prevent approximately 80% to 90% of those accidents annually.

Once promulgated, the confined space regulations were **codified** at the federal level in 29 CFR, Part 1910.146. In this standard, OSHA completely overhauled the previous confined space requirements. Additionally, within the regulation, OSHA also provided a working definition of exactly what constitutes a confined space. The regulation states that in order to be considered a confined space, the area must meet *all* of the following criteria:

The space must be large enough for someone to enter and,

Not be designed for continuous worker occupancy and,

Have limited openings for entry and exit.

The regulation went on to define a specific type of confined space, one that OSHA called a permit required confined space (PRCS). These PRCSs pose the highest hazard to workers and emergency response personnel. In order to be classified as a PRCS, the area must meet the definitions of a confined space as noted above, *and* have any of the following characteristics:

Have a known or potentially hazardous atmosphere

Contain a material that can engulf (bury) workers (e.g., soil, sand, sawdust, grain, coal, wood chips)

Have inwardly sloping walls or dangerously sloping floors that can entrap workers

Could have any other serious safety hazard

Obviously, anyone who works within a PRCS has a high potential for injury or death if safety procedures are not followed. For this reason, OSHA requires that the employer do the following for all PRCS:

- Assess the hazards within each confined space.
- Determine how to manage these hazards and develop entry procedures. If they can be mitigated, the employer is obligated to try to do so.
- Label all PRCS appropriately.
- Issue a permit authorizing entry into each of the PRCSs. The permits must be posted at the entrance and records kept documenting their existence.
- Train employees in safe operations regarding confined spaces.

In reviewing the definition of a PRCS, it is easy to understand that very common areas such as sewers, pipelines, tanks, silos, or vaults often present significant hazards to those who work within them. Given that some of these areas are found where hazardous waste operations and emergency response activities take place, OSHA mandates that working within such areas be covered as part of the HAZWOPER training programs.

Hazards of Confined Space Operations

One of the most common problems associated with working inside a confined space is the potential for a low level of oxygen to exist. As we saw in our study of respiratory protection in Chapter 6, any oxygen level below 19.5% is classified as oxygen deficient by OSHA. In most cases, when the oxygen content drops to this level, there are usually no noticeable physiological effects. It is once the oxygen level drops below 17% that the first effects on us are noted. These effects are usually difficult to recognize and respond to because we often are not aware of the oxygen-deficient condition.

At low levels, a person may become confused and lack the proper judgment to get out of the area. Asphyxiation is the most common cause of death within confined spaces. Such levels are common in spaces with little or no natural ventilation. In many cases, simply moving ambient air through a space is enough to ensure adequate oxygen concentration.

Low levels of oxygen within confined spaces can be caused by a number of factors, including:

Displacement. The introduction of other gases into the space may displace the oxygen. These gases might flush it out, force it up and out of the space, or cause it to settle to the bottom, depending on the density of the gases and the characteristics of the space.

■ **NOTE**
One of the most common problems associated with working inside a confined space is the potential for a low level of oxygen to exist.

Safety
Asphyxiation is the most common cause of death within confined spaces.

Bacterial action. The decomposition of organic matter such as sewage, leaves, grass, and wood can use available oxygen and make the levels within the spaces deficient.

Oxidation. Whenever metals oxidize (rust) they absorb atmospheric oxygen. While this process is not a huge cause of oxygen deficiency it is worth considering.

Combustion. Burning, welding, or cutting within a confined space uses great quantities of oxygen. The available oxygen may combine with carbon to form carbon dioxide, which can displace the oxygen levels within the area.

Absorption. Some materials such as wet activated carbon will absorb oxygen.

Painting. As paint dries, it absorbs oxygen in the drying process. Because many of the operations that take place within confined spaces involve painting, this common hazard should be considered.

It is wise to remember that the safest atmosphere for us to work in is one that is most similar to normal air. Recall from our discussion of respiratory protection that the composition of air is approximately:

78% nitrogen

20.9% oxygen

1% carbon dioxide and other gases

> **Safety**
> The safest atmosphere for us to work in is one that is most similar to normal air.

Toxic Atmospheres in Confined Spaces

In addition to the lack of oxygen within confined spaces, there is also the potential for other poisonous or toxic materials to accumulate. Confined spaces generally have poor air circulation and nonhomogenous atmospheres, which can result in pockets of toxic materials within the confined space. It is possible to have different atmospheres within the same confined space because of the vapor density of the materials, which cause some of the airborne contaminants to stratify and form pockets. These pockets can be quite stable and can persist for long periods of time.

A toxic or poisonous atmosphere contains excessive levels of toxic gases, vapors, mists, or dusts in concentrations that can cause illness or injury. These levels are generally those above the permissible exposure limit (PEL) for the substance or substances and require proper ventilation or respiratory protection. In some cases, the levels can be above the immediately dangerous to life and health (IDLH) value, causing some immediate effects on the person working in the area.

> **Safety**
> It is possible to have different atmospheres within the same confined space.

Sources of Atmospheric Toxins in Confined Spaces

Depending on the nature of the confined space, there are many possible sources of toxins as follows:

- Bacterial action of materials within the confined space can cause toxic or flammable materials to be produced.
- Products or chemicals that are currently stored or may have been stored in the confined space can emit harmful vapors.
- Substances (e.g., cleaners, solvents, paints) brought into the confined space may produce vapors or react with materials within the space to produce harmful materials.
- Work being performed, such as cleaning, welding, sandblasting, and painting can release toxins or other hazardous materials.
- Areas next to the confined space may release toxins and these materials may migrate into the confined space due to vapor density or other factors.
- Water blasting or steam cleaning a dry vessel can draw absorbed toxins from walls and floors.

While these are general causes of toxic atmospheres, two very specific materials are commonly found within confined spaces: hydrogen sulfide (H_2S) and carbon monoxide (CO). If we are to work safely within confined spaces, it is wise to have a good understanding of these materials.

Hydrogen Sulfide Hydrogen sulfide (H_2S) is a by-product of decomposing organic matter and can be produced by a variety of industrial processes. A poisonous and flammable gas, it is very irritating to the eyes, nose, and lungs at or above 50 ppm. Its very characteristic odor, which is similar to the odor of rotten eggs, is detectable by smell at as low as .004 ppm.

In fact, the odor is so pronounced and distinctive that many people believe they don't need a detector to identify its presence. This could be one of the biggest hazards of hydrogen sulfide because the odor is only detected by smell when the levels are low. In fact, when the level of the material is such that it can produce harm, H_2S has the ability to completely knock out our ability to smell by paralyzing our **olfactory** glands. The smell of H_2S is lost at concentrations at or above 150 ppm. In this state, a person can no longer smell the hydrogen sulfide, but also does not have the ability to smell many of the other harmful materials that may be present. Additionally, if this material is inhaled at this level, the paralytic effects soon spread to the respiratory system causing respiratory arrest. Some symptoms of mild overexposure include **photophobia,** tearing, blurred vision, and irritation of the eyes, nose, and lungs. Overexposures to high concentrations can cause intense anxiety, immediate fatigue, nausea, and unconsciousness. Rapid respiratory arrest may occur if exposed to over 1,000 ppm for greater than one minute. In addition to its toxicity, hydrogen sulfide is also extremely flammable with a wide flammable range of between 4% and 46%.

Carbon Monoxide Like hydrogen sulfide, carbon monoxide (CO) is commonly encountered in confined spaces. Detection is a problem with carbon monoxide as it is a

---

Safety The smell of hydrogen sulfide is lost at concentrations at or above 150 ppm.

olfactory relating to the sense of smell

photophobia intolerance of the eye to light

NOTE In addition to toxicity, hydrogen sulfide is also extremely flammable with a wide flammable range of between 4% and 46%.

Safety It is important to understand how hydrogen sulfide kills, because this type of asphyxiation can occur even when adequate levels of oxygen are present.

flammable range
the concentration of gas or vapor in air that will burn if ignited; expressed as a percentage that defines the range between a lower explosive limit and an upper explosive limit

> **Safety**
> Like hydrogen sulfide, carbon monoxide is also extremely flammable with a flammable range of 12% to 75%.

hemoglobin
iron-containing compound responsible for binding oxygen and transporting it to cells

ataxia
muscular incoordination manifested when voluntary muscular movements are attempted resulting, for instance, in an unsteady gait

syncope
transient loss of consciousness

carboxyhemoglobin
term for hemoglobin that is bound to carbon monoxide rather than oxygen

colorless, odorless, and tasteless gas. It is also a flammable gas with a **flammable range** of 12% to 74%. If inhaled, it interferes with the blood's ability to transport oxygen and leads to chemical asphyxiation at the cellular level. Unlike hydrogen sulfide, this material does not knock out the respiratory drive, but is toxic at low levels. The mechanism for this action is important to understand because this type of asphyxiation can occur even when adequate levels of oxygen are present.

Carbon monoxide is brought into the lungs during inspiration and travels to the lowest reaches of the respiratory tract. Along with oxygen and other gases, the carbon monoxide crosses the membranes of the alveoli and is picked up by the red blood cells within the capillaries. The **hemoglobin** of the red blood cells is the predominant oxygen transport system for the body and is responsible for carrying oxygen to the cells. Unfortunately, the hemoglobin has more of an affinity for the carbon monoxide than the oxygen, so the CO is bound up first. Additionally, the CO binding causes the oxygen that is bound by the hemoglobin molecule to be held more tightly. The result is that unoxygenated blood circulates around our bodies depriving the cells of the fuel they need for survival. Carbon monoxide has a great degree of influence on the brain and causes headache at low concentrations. Exposure to higher concentrations (above 400 ppm) may cause dizziness, nausea, vomiting, **ataxia,** and **syncope.** Once concentrations reach 2,000 ppm, immediate unconsciousness and death may occur. Ironically, the lung tissue itself only appears to be affected when CO levels are extremely high. For exposures that resulted in unconsciousness or respiratory or cardiac arrest, follow-up therapy may include high concentrations of oxygen inside a hyperbaric chamber. This combination of pressure and high oxygen concentration can often reduce the levels of **carboxyhemoglobin** in under a half an hour. For this reason, care should be taken to monitor the levels of both oxygen and carbon monoxide whenever there is the possibility of carbon monoxide to exist within the space. The hemoglobin prefers carbon monoxide 200 times more than it prefers oxygen!

A common situation where high carbon monoxide levels may exist is instances of incomplete combustion. Examples include the use of any internal combustion engine, such as a generator or power tool, or when open flames or welding occurs. Like hydrogen sulfide, carbon monoxide is also extremely flammable with a flammable range of 12% to 75%.

Combustible Atmospheres within Confined Spaces

A study of fire chemistry would show that in order to have fire, certain conditions are necessary. Obviously, a fire within a confined space would be a serious threat and would most likely kill or seriously injure anyone working within it. For this reason, the atmosphere within a confined space should be evaluated and monitored for the potential of fire.

In review, the conditions necessary for combustion include the following:

> **⚠ Safety**
> Hemoglobin prefers carbon monoxide 200 times more than it prefers oxygen!

- Sufficient oxygen. Generally, this needs to be above 15% by volume.
- A source of ignition. This can be an open flame, a spark, or a hot piece of metal.
- Sufficient fuel to burn and continue the combustion chain reaction. The amount of fuel needed depends on the level of oxygen, which is one of the reasons why you should never ventilate a confined space with pure oxygen.

The combustible material that may be present takes one of two basic forms within confined spaces: gases or vapors and particulates or dusts. Examples of commonly encountered gases include methane, hydrogen, carbon monoxide, hydrogen sulfide, and the vapor from the evaporation of a flammable liquid such as alcohol, gasoline, or other solvents used within the space. Remember that most of these gases or vapors are not readily visible and can go undetected. Additionally, the gases can **stratify** within various areas of the confined space.

> **stratify**
> relating to layers or strata

Particulates or dusts that occur within confined spaces are actually a suspension (cloud or mist) of flammable particles or liquids that are contained within the atmosphere. Examples of some of the flammable particulates include flour, pulverized coal, or any flammable dusts suspended in air. Generally, in order for particulates to ignite, the airborne concentration is thick enough to be seen and obscure vision at a distance of 5 feet or greater, depending on the flammability of the particles. Some substances, however, may pose a danger at concentrations lower than the 5-foot standard. A combination of gases with the particulates can act together to cause even more problems.

Other Confined Space Hazards

In addition to those previously discussed, several other hazards are worth discussion: These include engulfment, electrocution, heat disorders, being struck by objects, and entrapment due to the design of the space. Most of these hazards strike with little or no warning and can prevent self-rescue. Therefore, a basic understanding of them is necessary if we are to safely work within confined spaces. Accident studies show that most people get into trouble because they don't recognize the hazards or they ignore past training.

> **■ NOTE**
> Accident studies show that most people get into trouble because they don't recognize the hazards or they ignore past training.

Engulfment occurs when the person entering the space is buried or crushed by loose material present within the space. This material typically can include sand, sawdust, cement, or other particulate products. Once these materials are disturbed, their fluid nature often makes them move with considerable force and momentum, trapping the victim. In addition, the engulfing material may also be hot or corrosive leading to further complications.

Electrocution is a second of the relatively common causes of death and injury within confined spaces. The electrical hazards present within the spaces are often the cause of the electrocution, while in some cases, electrical tools or devices taken into the space may result in problems.

> **⚠ Safety**
> Always be aware of hazards above your head and wear a hard hat when necessary.

Objects that fall or strike workers can present significant hazards in certain types of confined spaces. Often the confined space contains augers or other moving parts that stir or move the materials into hoppers or conveyors. Getting a chemical suit entangled within a conveyor belt can pull a worker's body into the machinery. Always be aware of hazards above your head and wear a hard hat when necessary.

Confined Space Entrant's Responsibilities

> **■ NOTE**
> From a technical point of view you become an entrant when any part of your body breaks the plane of the permit required confined space.

Personnel who work in confined space operations fall into one of four categories. The first of these is the entrant, or the one who actually enters the space. From a technical point of view you become an entrant when any part of your body breaks the plane of the permit required confined space. This person runs a high risk of exposure once the entry plane to the space has been breached or entry has been made. Due to the potentially hazardous nature of confined space entry, OSHA requires that personnel engaged in such efforts be trained in the hazards and operations required for safe entry.

As part of this training program, the entrant must:

- Be familiar with all the hazards and potential hazards within a given space.
- Know the signs and symptoms of exposure to the hazards found in confined spaces.
- Be familiar with and be able to use all the personal protective equipment necessary for safe entry into the space.
- Understand the importance of maintaining contact with the attendant. Communication between the entrant and the attendant can be accomplished in a number of ways, and the entrant must be familiar with as many as may be appropriate.
- Understand the means and importance of alerting the attendant to any observed hazard or condition not allowed by the permit.
- Instantly obey any order to evacuate the permit space.

Confined Space Attendant's Responsibilities

A confined space attendant is someone who "attends" to the confined space entrant. The attendant is required to stay outside the confined space and monitor the activities that take place both within and outside the space. The attendant must:

- Be familiar with the hazards within the permit space and recognize the presence of hazards within the space.
- Monitor the entrant's behavior closely for signs and symptoms of exposure or other health problems.

- Keep count of the number of workers in the space and allow only authorized entrants access to the space.
- Keep in constant communication with the entrant.
- Protect the entrant from external hazards such as falling objects.
- Not leave the entrance unless relieved by a qualified attendant.
- Not enter the space to perform rescues and be able to instantly contact the rescue team.

Confined Space Entry Supervisor

The entry supervisor is the person (such as the employer, foreman, or crew chief) responsible for the overall operations and safety within the confined space. The supervisor is responsible for the following:

- Knowing the hazards that may be faced during the entry, including information on the mode of action of the chemical, signs or symptoms of overexposure, and consequences of the exposure.
- Verifying, by checking that the appropriate entries have been made on the permit, that all tests specified by the permit are in place before endorsing the permit and allowing entry to begin.
- Verifying that rescue services are available and that the means for summoning them are operable.
- Determining, whenever responsibility for a permit space entry operation is transferred and at intervals dictated by the hazards and operations performed within the space, that entry operations remain consistent with terms of the entry permit and that acceptable entry conditions are maintained.
- Authorizing entry into the confined space.
- Terminating the entry and canceling the permit as required by this section.
- Removing unauthorized individuals who enter or who attempt to enter the permit space during entry operations.

An entry supervisor may also serve as an attendant or as an authorized entrant, as long as that person is trained and equipped as required for each role he or she fills. Also, the duties of entry supervisor may be passed from one individual to another during the course of an entry.

PRCS Emergency Procedures

The easiest emergency to deal with is the one that never occurs. Given that there is a system for establishing procedures for the safe entry into a confined space, it is possible to prevent nearly all emergency situations that might occur. If we evaluate the hazards, properly deal with those that can be mitigated, train and equip the entrant, and have the proper safeguards in place, very few emergency situations should occur.

Despite proper planning, there might be cases where emergency procedures will need to be implemented during a confined space entry. When these situations are encountered, an order to evacuate and activities that cause evacuation to happen need to be implemented. Following is a list of some of the types of conditions that could lead to initiating an emergency response.

- The oxygen level drops below 19.5%. This condition would not always apply unless the personnel within the space were not equipped with supplied air breathing apparatus.
- The oxygen level rises above 23.5%.
- The level of combustible gases rises above 10% of the lower explosive limit (LEL).
- Concentrations of toxins within the confined space rise above the permissible limits.
- Conditions outside the space change to a point where they might endanger the entrants.
- Any entrant begins to exhibit signs of exposure to chemicals or other behavior indicating problems.
- Any other conditions arise that are not included on the entry permit.
- The attendant must leave his or her post.

At the first sign of changing conditions the attendant should begin emergency actions. These actions could include increasing the ventilation within the space, increasing the monitoring of the space or of the entrant, and increasing the communication with the entrant.

The prompt recognition of potential problems coupled with swift aggressive action is important in emergency response. It has been found that in confined space emergencies the situation can deteriorate rapidly further increasing the exiting time for the entrant. It is therefore important to remember that if there is a possibility of danger to the entrant, remove them immediately. When in doubt, get them out!

Safety When in doubt, get them out!

Rescue and Emergency Personnel

All rescue personnel must be provided with and trained to use the personal protective equipment and rescue tools necessary for making rescues from the space. They must be trained in the duties and responsibilities of working in a permit required confined space as outlined in this program. Rescue personnel shall also practice making a permit space rescue at least once every 12 months. Each member of the rescue team shall be trained in basic first aid and in CPR.

Initiating an Evacuation from a Confined Space

Either the entrant, attendant, or supervisor can order an evacuation of the confined space. When ordered to do so, all entrants must evacuate the confined space immediately. Self-rescue is the easiest and safest means of evacuating a confined space. Following this, **nonentry rescue** is indicated.

nonentry rescue
the process whereby the attendant initiates the rescue of the entrant who is unable to leave the space

Perhaps the most common means whereby rescue occurs is when the attendant is lifted from the space using one of the hauling devices that may be required. Using some type of mechanical advantage such as a pulley system, the attendant removes the entrant by means of a rope or cable that is attached to the entrant's body harness.

If self-rescue, and nonentry rescue procedures are not successful in removing the entrant, the last means of rescue is through the use of a specially trained confined space rescue team. The regulations require that such a team be available for rescue whenever entry into a PRCS occurs. What is not clear from the regulations is just how close the rescue team must be. For example, could the local fire department serve as the rescue team? Are the department members properly trained in accordance with the regulations? What if they have a delayed response time to your site? Obviously, a lot of questions need to be resolved for compliance with the regulation.

In any case, whether the rescue team is composed of personnel from the site or whether an outside agency is used, the team will need information from the attendant in order to effectively perform rescue. This information includes:

- The number of entrants working within the space.
- The condition of the entrants. This information may be unknown at the time of rescue.
- The types of hazards present within the space.
- The cause of the emergency and circumstances that caused it.

Equipment Used in Confined Space Operations

Various types of equipment are used and/or required for use within confined spaces. The equipment can do any of the following:

- Detect hazards before and during entry. Portable monitoring devices are especially suited for this application.
- Enable attendants to contact entrants and rescue team. Examples include radios, telephones, lifelines, and retrieval lines.
- Protects entrants from hazards. Examples include respirators, fall protection equipment, lifelines, and retrieval lines.

Using Atmospheric Monitors in Confined Space

As we know from Chapter 11, atmospheric monitoring means using an instrument to measure the concentrations and/or presence of airborne contaminants. Within confined spaces, harmful atmospheres for which monitoring should be initiated include:

> Oxygen. The level must be between 19.5% and 23.5%.
>
> Combustibles. Any level of combustibles above 10% of the LEL is classified as potentially flammable and hazardous for safe entry.
>
> Toxic gases or other materials. Permit spaces may contain one or more toxic gases that were already present or might be introduced through the work taking place.

Commonly encountered toxins include carbon monoxide (CO) and hydrogen sulfide (H_2S) as noted previously, but there are countless more depending on the type of space and the industrial process.

The monitors used in the confined space operation may be hand held by the entrants or attendants, worn by the entrant while working, or fixed and mounted within the confined space. In order to be most effective, the monitors should have both a visible readout and either a visual or audible alarm. Many of the monitors used in these operations are fitted with hoses or extensions so gas concentrations can be measured in remote locations, usually by the attendant. Whatever the case, all of the monitors used are sophisticated electronic devices but they are not foolproof. Unless used properly, they will give a false reading. The life of those entering the space may depend on the accuracy of monitoring devices.

Harnesses and Tripods

OSHA requires that all entrants wear some type of body harness whenever a permit required combined space is entered unless the use of the device will increase the hazards to the entrant. Of differing styles and configurations, body harnesses are designed to provide a higher degree of safety to the entrant by allowing the attendant to remotely retrieve the victim. As stated previously, this procedure is called nonentry rescue and is the first means used if the entrant is not able to perform self-rescue. Body harnesses are particularly useful for work in permit spaces that are at or above the IDLH level, have a vertical entry, or that pose a risk of engulfment.

Although useful in some situations, such retrieval devices may not work if the victim falls behind machinery or there are obstacles in the space. In this instance, a properly trained and certified confined space rescue team will need to make entry in order to retrieve the victim.

Confined Space Ventilation Equipment

Another set of tools well suited to purge hazardous atmospheres are blowers and ventilators. They have an effective capacity in CFM (cubic feet per minute) that

> **Safety**
> The life of those entering the space may depend on the accuracy of monitoring devices.

> **■ NOTE**
> Body harnesses are particularly useful for work in permit spaces that are at or above the IDLH level, have a vertical entry, or that pose a risk of engulfment.

determines purge times by extracting or injecting air into confined spaces. The actual amount of time necessary to effect this operation and render the atmosphere within the space safe should be listed on the permit.

While purging a confined space atmosphere, it is important to follow certain procedures to ensure the maximum efficiency for the operation. The dropped end of the hose should hang vertically within the space to ensure proper air movement. It should be positioned not more than 1 foot below the ceiling and/or not more than 2 feet above the floor of the space to ensure the maximum air movement.

When using a blower to inject fresh air into the space for the entrant to breathe, it is important to make sure that the blower intake is away from vehicle exhaust or other airborne contaminants to avoid the introduction of harmful air contaminants into the space. If a gas-powered blower is used, proper care should be taken to ensure that the exhaust is not immediately sucked into the blower air intake. The blower and the air intake should be at least 5 feet away from the confined space entry to ensure that air is not simply recycled from within the space, that the noise does not impede communication, and that the device does not impede the ingress or egress from the space.

> **Safety**
> If a gas-powered blower is used, proper care should be taken to ensure that the exhaust is not immediately sucked into the blower air intake.

Assessing Air Quality within Confined Spaces

Assessing the air quality within a confined space is similar to that done in other monitoring operations as discussed in Chapter 11. However, because of the configuration and nature of confined space operations, some procedures are noteworthy. If possible, the following procedures should be considered:

- Whenever possible, the initial air sampling should be done before opening up the space. It may be possible in some cases to insert the gas detector probe into a sample hole. If this is not possible, the next best procedure is to gently crack the opening just enough to test the atmosphere.
- Test for oxygen first, because a significantly low oxygen level (less than 10%) will result in an artificially low combustible vapor level readout.
- Once the oxygen level is established, determine the combustibles and toxins that might be present in the space.
- Test all levels within the space including the lowest level, **breathing zone,** above the breathing zone, and just below the ceiling to account for the various vapor densities, temperature inversions, and other gas stratification that might occur.

breathing zone
an imaginary zone encompassing a horizontal plane from the middle of the chest to the top of the head

Isolation of Confined Spaces

Although not a specific type of equipment, the use of isolation procedures is an important element of a confined space safety procedure. Many accidents are

blanking
a procedure to physically seal pipes (e.g., hydraulic, pneumatic, chemical, fuel, water) by inserting and bolting a piece of metal between flanges in the piping

bleeding
releasing pressure in hydraulic or pneumatic lines

■ **NOTE**
The permit specifies measures that must be taken. If the permit is complete it will not only identify the hazards but also the means that need to be taken to control them.

caused when hazardous materials or energies enter the permit space and endanger or contact the entrant. The actual procedures involved in isolation of a confined space vary, but generally have the following elements:

Lockout and tag-out of all electrical sources.

Blanking and **bleeding** of hydraulic/pneumatic power sources.

Disconnecting all mechanical linkages that can activate machinery or equipment within the confined space.

Securing all mechanical moving parts within the permit space.

Alerting others who may be in the area of the work to be done.

The permit specifies measures that must be taken. If the permit is complete, it will not only identify the hazards but also the means that need to be taken to control them.

Confined Space Entry Permit Compliance Checklist

The following list of items denote those required to ensure a high degree of safety when working within a PRCS. The list is not all encompassing, but it provides a summary of the basics of safe confined space operations. You may also encounter many different types and styles of permits. There are some standard formats in the industry, but most facilities develop site-specific permits and forms. It is incumbent on you to be familiar with any permits you may encounter as part of your job.

- Read the permit and note the hazards. Do they agree with what you know about the space?
- Note the safety equipment required. Does it address all the hazards mentioned? Is the equipment available and have you inspected it?
- Look at the list of required tools. Are they available and have they been inspected?
- Check the list of authorized entrants and attendants. Do they have the training required?
- Go through the preparation checklist. Do you have all the equipment to perform the required steps? Have all the steps been taken?
- Read the atmospheric tests specified. Do you have the monitors needed to make the measurements? Is the equipment operating properly? Have the instruments been calibrated and zeroed?
- Does the space need to be isolated? Has that been done?
- Does the space need to be ventilated? Has that been done? Ventilation should include five changes of air (not less than 5 minutes).
- During entry you should continuously ventilate and monitor.

Making the Entry into a PRCS

Part of a successful entry is dependent on thorough planning prior to going into the space. The following procedures can be used as a guide for your site-specific operations.

Prior to the Entry

Authorize entry after all acceptable entry conditions have been verified.

Sign the permit and post it outside the space.

During the Operation

Oversee entry.

Ensure compliance with work procedures.

Ensure communication between entrants, attendants, and rescue service.

Terminating the Entry

Debrief.

Cancel permit.

Review for accuracy.

Recommend changes.

Keep records.

Contractors Entering Permit Required Confined Spaces

When contractors are required to perform work that involves entry into a permit required confined space, the following conditions should be addressed:

- The contractor shall be informed that the work location contains a permit required confined space and that entry is allowed through compliance with the written confined space program.
- The contractor shall be told of any possible hazards that may arise that requires this space to be a permit required confined space.
- The contractor shall be told of the precautions that have been taken to protect employees from the hazards associated with the confined space.
- The contractor must be informed that the site safety official must be notified prior to entry operations and at the completion of the confined space work.
- Upon notification from the contractor that the work has been completed, the site safety official or his or her designee shall debrief the contractor regarding any hazards confronted or created in the permit space during entry operations.

Any contractor working in a confined space shall obtain any available information regarding permit space hazards and entry operations from the Environmental Health and Safety Department (EH&S) or site safety official prior to entry operations.

Any contractor who will be performing work in a confined space must show proof to the EH&S department or site safety official that the contractor's employees have completed training in confined space. This must be done prior to entry operations in any permit required confined space.

NOISE HAZARDS

Noise exposures can cause serious hearing damage and should be treated like any other workplace hazard. Because the areas that we are discussing may be confined, noises can be intensified within these spaces. These hazards may include intermittent and continuous sources as well as impact or pulsating type sounds. A commonly accepted noise benchmark for the workplace is an exposure of up to 85 dB (decibels) for 8 hours. Above that exposure level hearing protection may need to be used. There are tables in the standard that factor in high impact noise over minutes or seconds and list acceptable exposure limits.

Note that the hazard posed by noise depends on several factors, not just the intensity of the sound. These factors include the length of exposure time, the force of the sound waves (loudness), and the frequency (pitch) of the sound. Specific components of noise hazards and hearing conservation programs can be found in CFR 29, Part 1910.95. The regulation is quite comprehensive and lists such items as hearing conservation programs, audiometric testing programs, and audiometric testing criteria. Just because noise exists at a work site does not mean that it exceeds the accepted OSHA levels. Specific testing of the area must be performed to provide adequate data from which to base hearing conservation programs. The bottom line is that if a potential noise hazard exists, it must be dealt with using the same regard for safety as any other workplace hazard.

> ■ **NOTE**
> Specific testing of the area must be performed to provide adequate data from which to base hearing conservation programs.

> ■ **NOTE**
> Approximately 80% of the workers killed never attempted to shut off energy sources before working on or maintaining machinery. Another 20% lost their lives when power was inadvertently turned on as they were engaged in completing repairs or other activities relating to the machinery.

CONTROL OF HAZARDOUS ENERGY (LOCKOUT/TAG-OUT)

Uncontrolled energy sources account for approximately 6% of all worker deaths nationwide. Although this may not seem to be a large number, many of these deaths could have been prevented. If there had been proper procedures, increased awareness of the potential dangers, and qualified training, many of these workers could be alive today. Approximately 80% of the workers killed never attempted to shut off energy sources before working on or maintaining machinery. Another 20% lost their lives when power was inadvertently turned on as they were engaged in completing repairs or other activities relating to the machinery. Reflecting on the statistics should illustrate that most deaths and/or accidents could easily have been avoided. Some of the most common mistakes leading to injury or death include:

- Failure to stop equipment prior to maintenance. Working on running equipment has an enormous potential for disaster.
- Failure to disconnect power sources or accidentally restarting of equipment. Accidentally hitting switches or buttons may unexpectedly energize the equipment.
- Failure to dissipate residual power, including bleeding off pressure from pipes and allowing electrical energy to dissipate from capacitors.
- Failure to clear work areas prior to restarting machinery, which can cause clothing or gloves to get snagged in gears or conveyors, paths of conduction to be created by metallic objects contacting electrical wires, or shards of metal or glass becoming projectiles when machines are activated.

> **■ NOTE**
> Potential energy sources may be hard to identify because it is stored within the process and becomes active only upon release.

> **Safety**
> Any piece of equipment that requires an input of energy from external sources in order to operate should be identified as a possible hazardous energy source.

kinetic energy
actual energy being produced by the action of the equipment or process

potential energy
energy stored within a source and available when released

First of all, it is necessary to look at the potential sources of uncontrolled energy. The various sources may originate from electrical, mechanical, pneumatic, chemical, thermal, hydraulic, and or radioactive processes. Any piece of equipment that requires an input of energy from external sources in order to operate should be identified as a possible hazardous energy source. Any of the various sources have two types of energy you should be aware of. The first is called **kinetic energy.** This energy is usually easy to identify because the process it is powering gives some sort of indicator such as noise of operation, pressure in gauges, temperature readings and so on.

A second form, called **potential energy,** may also be present in machinery, springs, electrical capacitors, and other working parts of the item in question. A coiled spring has potential energy; if it is allowed to uncoil, the potential energy is released thereby converting to kinetic energy. Potential energy sources may be harder to identify because it is stored within the process and becomes active only upon release.

As with most worker safety issues, OSHA has promulgated a regulation for working safely around uncontrolled energy sources. This standard is commonly referred to as The Control of Hazardous Energy (lockout/tag-out) and can be found in 29 CFR, Part 1920.147. As with the noise standard, lockout/tag-out is quite complex and warrants a full reading if it applies to your operation. Essentially, however, it is designed to cover service and maintenance of equipment in which unexpected start up or release of stored energy is possible. This release, of course, may have the potential to injure or kill workers who are unprepared for the release. Some operations like oil and gas drilling, maritime employment, construction or agriculture are exceptions to the standard but all workers should follow good practices and use common sense.

Lockout/tag-out programs are key in ensuring safe operations when working around energized electrical equipment. These programs should be implemented when employees are to bypass safety guards or actually place a body part inside a machine or process. Common sense should play a big role when deciding when to actually lock out an energy source. When cord and plug devices such as minor hand tools are used, it may not warrant the full implementation of lockout/tag-out. If, however, there are extenuating circumstances or some extraordinary hazard, take all precautions and lockout/tag-out the system. Lockout is a process whereby energy sources are physically controlled via locks on breakers or other sources of energy. Simply turning off switches may not be enough to completely secure a machine for maintenance or repair. Time after time, these switches have been accidentally bumped or turned on, causing a machine to start up and injure a worker. Lockout devices should be standardized within the facility, assigned to a worker by name, strong enough to require considerable force to remove, and easily recognized. After a power source is locked, it should be tagged by the same worker. Tag-out is another process for securing equipment and includes identifying the time, duration, and reason for disallowing power to a particular piece of equipment. When used together, lockout/tag-out is an effective safety control measure. Tag-out devices should stand up to environmental hazards such as rain, heat, and sunshine and not be easily removed from the locking device. They may be reusable. A tag should identify the worker and the time and date it was applied along with some warning like *DO NOT CLOSE, DO NOT ENERGIZE,* or *DO NOT OPERATE.*

Lockout/tag-out procedures may apply to cleanup or emergency activities as well as day-to-day operations. In the event flammable vapors are present and ignition sources need to be secured, it may be necessary to completely control machinery or other processes. Rather than leaving personnel in harm's way, the use of a lockout/tag-out system achieves the goal of safety while reducing exposures. The following steps should be incorporated when implementing lockout/tag-out procedures.

> *Assess the system.* This requires a knowledge of the system and the identification of the type of energy the system uses. Additionally, you must identify how much energy is present and how it is best controlled. You may also be required to inform other persons or divisions that may be affected by the shutdown. Consult manufacturer's notes or specifications if necessary.
>
> *Shut down the system.* The system should be turned off by regular means. This may include the normal on–off switches or closing valves.
>
> *Secure the main power source for the equipment.* This step may include isolating electrical subpanels or major valve junctions. Do not remove fuses—they are too easily replaced! Review all potential sources of energy including backup supplies and remember to consider multiple sources. It is also necessary to determine if your own work may energize or start the machinery.
>
> *Lockout/tag-out the system.* Use a standardized system that is known and understood by all. It is important to ensure that all lockout devices function properly. Valve covers should be properly sized, locks for circuit breakers should be appropriate or adjusted using lock adapters. It is also acceptable to use multiple locks with multiple workers.

> **!Safety**
> When used together, lockout/tag-out is an effective safety measure.

> **!Safety**
> Do not remove fuses—they are too easily replaced!

If more than one person is working on the system, all involved should place a lock.

Release all energy in system. This phase includes releasing energy stored in springs, compressed air or gaseous systems, electrical capacitors, and so forth. Essentially it is necessary to *bleed off* whatever energy exists after the shutdown. You may also need to install ground wires or some other method in order to control sources of static electricity.

Verify that all has been done. This is a final check. Walk through the procedure one last time to ensure that all potential energy sources are identified and secured. It is better to check twice than be wrong once!

> **NOTE**
> It is better to check twice than be wrong once!

Test the system. Turn on the normal operating switches and valves, and ensure that the machinery will not operate. Volt meters or pressure gauges should assist you in the verification of a *dead* system.

Perform the work. Once the system is secured, normal work can be completed.

Remove the system. Once the work is finished and it is safe to return power, the lockout/tag-out devices may be removed. Guards should be replaced, tools should be removed from the machine, junction boxes should be covered, valves replaced, and a final safety walk through should be performed to ensure that the equipment is indeed ready to be returned to normal operating status. Lockout/tag-out is an individualized system, which means that if you put it on, you are the only one who should remove it. Additionally, notices should be sent to persons who may have been affected by the shutdown. These notices formally notify everyone that the systems are operating normally and that there are no restrictions to using appropriate power sources.

> **Safety**
> Lockout/tag-out is an individualized system, which means that if you put it on, you are the only one who should remove it.

Summary

- In January of 1993, OSHA issued its final standard on Permit Required Confined Spaces (PRCS) which can be found in 29 CFR, Part 1910.146.
- The confined space regulation states that in order to be considered a confined space, the area must meet *all* of the following criteria:
  - The space must be large enough for someone to enter and,
  - Not be designed for continuous worker occupancy and,
  - Have limited openings for entry and exit.

  And have any of one of the following characteristics:
  - Have a known or potentially hazardous atmosphere.
  - May contain a material that can engulf (bury) workers (e.g., soil, sand, sawdust, grain, coal, wood chips).
  - Have inwardly sloping walls or a dangerously sloping floors that can entrap workers.
  - Could have any other serious safety hazard.
- OSHA requires that the employer do the following for all PRCS:
  - Assess the hazards within each confined space and determine if they are PRCS.
  - Label all PRCS in an appropriate manner.
  - Determine how to manage these hazards and develop entry procedures. If they can be mitigated, the employer is obligated to try to do so.
  - Issue a permit authorizing entry into each of the PRCS. The permits must be posted at the entrance and records kept documenting their existence.
  - Train employees in safe operations regarding confined spaces.
- Confined spaces generally have poor air circulation and nonhomogenous atmospheres, which can result in pockets of toxic materials within the confined space. It is possible to have different atmospheres within the same confined space. Asphyxiation is the most common cause of death within confined spaces.
- While these are general causes of toxic atmospheres encountered, two very specific materials are commonly found within confined spaces: hydrogen sulfide (H_2S) and carbon monoxide (CO). Hydrogen sulfide is also a flammable gas with a flammable range of between 4% and 46%. Carbon monoxide is also extremely flammable with a flammable range of 12% to 75%.
- Hemoglobin prefers carbon monoxide 200 times more than it prefers oxygen.
- In review, the conditions necessary for combustion include the following:
  - Sufficient oxygen, generally above 15% by volume.

- A source of ignition. This can be an open flame, a spark, or a hot piece of metal.
- Sufficient fuel to burn and continue the combustion chain reaction. The amount of fuel needed depends on the level of oxygen, which is one of the reasons why you should never ventilate a confined space with pure oxygen.
- Engulfment occurs when the person entering the space is buried or crushed by loose material present within the space.
- The duties of a confined space entrant include:
  - Be familiar with all the hazards and potential hazards within a given space.
  - Know the signs and symptoms of exposure to the hazards found in confined spaces.
  - Be familiar with and be able to use all the personal protective equipment necessary for safe entry into the space.
  - Understand the importance of maintaining contact with the attendant. Communication between the entrant and the attendant can be accomplished in a number of ways, and the entrant must be familiar with as many as may be appropriate.
  - Understand the means and importance of alerting the attendant to any observed hazard or condition not allowed by the permit.
  - Instantly obey any order to evacuate the permit space.
- The duties of a confined space attendant include:
  - Be familiar with the hazards within the permit space and recognize the presence of hazards within the space.
  - Monitor the entrant's behavior closely for signs and symptoms of exposure or other health problems.
  - Keep count of the number of workers in the space and allow only authorized entrants access to the space.
  - Keep in constant communication with the entrant.
  - Protect the entrant from external hazards such as falling objects.
  - Not to leave the entrance unless relieved by a qualified attendant.
  - Not to enter the space to perform rescues and to be able to instantly contact the rescue team.
- The duties of the supervisor of a confined space operation include the following:
  - Knowing the hazards that may be faced during the entry, including information on the mode of action of the chemical, signs or symptoms of overexposure, and consequences of the exposure.
  - Verifying by checking that the appropriate entries have been made on the permit, that all tests specified by the permit are in place before endorsing the permit and allowing entry to begin.

- Verifying that rescue services are available and that the means for summoning them are operable.
- Determining, whenever responsibility for a permit space entry operation is transferred and at intervals dictated by the hazards and operations performed within the space, that entry operations remain consistent with terms of the entry permit and that acceptable entry conditions are maintained.
- Authorizing entry into the confined space.
- Terminating the entry and canceling the permit as required by this section.
- Removing unauthorized individuals who enter or who attempt to enter the permit space during entry operations.

- At the first sign of changing conditions the attendant should begin emergency actions.
- When using a blower to inject fresh air into the space for the entrant to breathe, it is important to make sure that the blower intake is away from vehicle exhaust or other airborne contaminants to avoid the introduction of harmful air contaminants into the space.
- Assessing the air quality within a confined space should include procedures:
  - Whenever possible, the initial air sampling should be done before opening up the space.
  - Test for oxygen first.
  - Once the oxygen level is established, determine the combustibles and toxins that might be present in the space.
  - Test all levels within the space including the lowest level, **breathing zone**, above the breathing zone, and just below the ceiling.
- Noise exposures can cause serious hearing damage and should be treated like any other workplace hazard.
- A commonly accepted noise benchmark for the workplace is an exposure of up to 85 dB (decibels) for 8 hours. Above that exposure level hearing protection may need to be used.
- Specific components of noise hazards and hearing conservation programs can be found in CFR 29, Part 1910.95.
- Some of the most common mistakes leading to the injury or death of those working with energized electrical equipment include:
  - Failure to stop equipment prior to maintenance.
  - Failure to disconnect power sources or accidental restarting of equipment.
  - Failure to dissipate residual power.
  - Failure to clear work areas prior to restarting machinery
- This lockout/tag-out standard can be found in 29 CFR, Part 1920.147.

- Lockout/tag-out programs are key in ensuring safe operations when working around energized electrical equipment.
- The following steps should be incorporated when implementing lockout/tag-out procedures.
  - Prepare for shutdown
  - Shut down the system
  - Secure the main power source for the equipment
  - Lockout/tag-out the system
  - Release all energy in system
  - Verify that all has been done
  - Test the system
  - Perform the work
  - Remove lockout/tag-out system
- Lockout/tag-out is an individualized system, which means that if you put it on, you are the only one who should remove it.

Review Questions

1. Which of the following pieces of legislation contains the information relating to confined space?
   A. 29 CFR, Part 1910.120
   B. 29 CFR, Part 1910.140
   C. 29 CFR, Part 1910.115
   D. 29 CFR, Part 1910.146

2. In order to be considered a confined space, an area *must* meet which of the following?
   A. Must be large enough for a person to enter
   B. Must not be designed for continuous worker occupancy
   C. Must have limited openings for entry and exit
   D. All of the above

3. True or False? Employers must label all recognized PRCS.

4. True or False? Asphyxiation is the most common cause of death in confined spaces.

5. List three potential causes of low oxygen atmospheres inside a confined space.

6. True or False? Water blasting or steam cleaning inside a dry storage vessel can draw out absorbed chemicals from walls, floors, and ceilings.

7. Which of the following most accurately defines hemoglobin?
   A. An iron rich compound in the red blood cells responsible for transporting oxygen to the cells.
   B. An iron poor protein that binds carbon monoxide so that it is not harmful to the body.
   C. A compound that produces feratin.
   D. An iron rich compound that improves the absorption of carbon dioxide.

8. Which of the following chemical compounds can deaden the sense of smell at low concentration levels?
   A. Hydrogen bromide
   B. Carbon monoxide
   C. Carbon dioxide
   D. Hydrogen sulfide
9. True or False? In addition to its toxicity, carbon monoxide is also a flammable gas.
10. Define syncope.
11. List the three conditions required for combustion to occur.
12. Engulfment occurs when
    A. A person is buried or crushed by loose material.
    B. A person is in a space with sides that are too smooth to climb.
    C. A person is left unattended within a confined space.
    D. A person is rendered unconscious by falling debris.
13. List four responsibilities of a confined space entrant.
14. True or False? Technically speaking, you become an entrant anytime you are working near a confined space.
15. Describe the typical duties of a confined space attendant.
16. A commonly accepted benchmark for workplace exposures to noise is
    A. 80 decibels for 10 hours
    B. 85 decibels for 8 hours.
    C. 85 decibels for 10 hours.
    D. 80 decibels for 8 hours.
17. True or False? Approximately 80% of the workers killed in electrically related incidents never attempted to shut off energy sources before working on machinery.
18. List three common mistakes workers make that lead to injury or death when dealing with energized electrical equipment.
19. Define lockout and tag-out.

Appendix A

State Plan States

Alaska
Alaska Department of Labor
1111 West 8th Street, Room 306
Juneau, AK 99801
907-465-2700

Arizona
Industrial Commission of Arizona
800 W. Washington
Phoenix, AZ 85007
602-542-5795

California
California Department of Industrial Relations
455 Golden Gate Avenue, 4th Floor
San Francisco, CA 94102

Connecticut
Connecticut Department of Labor
200 Folly Brook Blvd.
Wethersfield, CT 06109
203-566-5123

Hawaii
Hawaii Department of Labor and Industrial Relations
830 Punchbowl Street
Honolulu, HI 96813
808-586-8844

Indiana
Indiana Department of Labor
State Office Building, Room W195
402 West Washington Street
Indianapolis, IN 46204
317-232-2693

Iowa
Iowa Division of Labor Services
1000 E. Grand Avenue
Des Moines, IA 50319
515-281-3447

Kentucky
Kentucky Labor Cabinet
1049 U.S. Highway 127 South
Frankfort, KY 40601
502-564-3070

Maryland
Maryland Division of Labor Industry
Department of Licensing and Regulation
501 St. Paul Place, 2nd Floor
Baltimore, MD 21202-2272
410-333-4179

Michigan
Michigan Department of Labor
Victor Office Center
201 N. Washington Square
P. O. Box 30015
Lansing, MI 48933
517-373-9600

Minnesota
Minnesota Department of Labor and Industry
443 Lafayette Road
St. Paul, MN 55155
612-296-2342

Nevada
Division of Industrial Relations
400 West King Street
Carson City, NV 89710
702-687-3032

New Mexico
New Mexico Environmental Dept.
Occupational Health and Safety Bureau
1190 St. Francis Drive
P. O. Box 26110
Santa Fe, NM 87502
505-827-2850

New York
New York Department of Labor
State Office Building, Campus 12, Room 457
Albany, NY 12240
518-457-2741

North Carolina
North Carolina Department of Labor
4 West Edenton Street
Raleigh, NC 27601
919-733-7166

Oregon
Oregon Occupational Safety and Health Division
Department of Consumer and Business Services
Labor and Industries Building, Room 430
350 Winter Street, NE
Salem, OR 97310
503-378-3272

Puerto Rico
Puerto Rico Department of Labor and Human Resources
Prudencio Rivera Martinez Building
505 Munoz Rivera Avenue
Hato Rey, PR 00918
809-754-2119

South Carolina
South Carolina Department of Labor
3600 Forest Drive
P. O. Box 11329
Columbia, SC 29211-1329
803-734-9594

Tennessee
Tennessee Department of Labor
Gateway Plaza - Suite A - 2nd floor
701 James Robertson Parkway
Nashville, TN 37243-0655

Utah
Industrial Commission of Utah
160 East 300 South, 3rd floor
P. O. Box 146600
Salt Lake City, UT 84114-6600
801-530-2288

Appendix A State Plan States

Vermont
Vermont Department of Labor and Industry
120 State Street
Montpelier, VT 05620
802-828-2288

Virgin Islands
Virgin Islands Department of Labor
2131 Hospital Street
Christiansted
St. Croix, Virgin Islands 00840-4666
809-773-1994

Virginia
Virginia Department of Labor and Industry
Powers-Taylor Building
13 South 13th Street
Richmond, VA 23219
804-786-2377

Washington
Washington Department of Labor and Industries
General Administration Building
P. O. Box 44001
Olympia, WA 98504-4001
206-956-4213

Wyoming
Occupational Safety and Health Administration
Herschler Building, 2nd Floor East,
112 West 25th Street
Cheyenne, WY 82002
307-777-7786

Appendix B

Human Carcinogens

The following agents are listed by the International Agency for Research on Cancer as Group 1 carcinogens, substances shown to cause cancer in humans. This list is current as of April, 1997.

Aflatoxin B1
Aflatoxins
Aluminum production
Aminobiphenyl, 4- (4-Aminodiphenyl)
Ammonium dichromate (VI)
Analgesic mixtures containing phenacetin
Arsenic acid, calcium salt
Arsenic acid, calcium salt (2:3)
Arsenic trioxide
Arsenic, and certain arsenic compounds

Arsenious acid, monosodium salt
Asbestos
Asbestos, Actinolite
Asbestos, Amosite
Asbestos, Anthophyllite
Asbestos, Chrysotile
Asbestos, Crocidolite
Asbestos, Tremolite
Auramine, manufacture of
Azathioprine

Barium chromate (VI)

Benzene

Benzidine

Beryllium aluminum alloy

Beryllium aluminum silicate

Beryllium and certain beryllium compounds

Beryllium chloride

Beryllium compounds, n.o.s.

Beryllium fluoride

Beryllium hydrogen phosphate (1:1)

Beryllium hydroxide

Beryllium oxide

Beryllium oxide carbonate

Beryllium sulfate

Beryllium sulfate, tetrahydrate (1:1:4)

Beryllium zinc silicate

Betel quid with tobacco

Butadiene, 1,3-

Butanedioldimethylsulfonate, 1,4- (Busuphan; Myleran)

Cadmium chloride

Cadmium fume

Cadmium sulfate (1:1)

Cadmium sulfide

Calcium chromate (VI)

Chlorambucil

Chloromethyl ether, bis-

Chromate (1-), Hydroxyoctaoxodizincatedi-, potassium

Chromic acid, disodium salt

Chromic acid, lead (2+) salt (1:1)

Chromium (VI) chloride

Chromium (VI compounds, certain water insoluble

Chromium (VI) dioxychloride

Chromium (VI) oxide (1:3)

Ciclosporin

Coal-tar pitches

Coal-tars

Coke production (Coke oven emissions)

Cyclophosphamide

Cyclosporin A

Dichromic acid, Diammonium salt

Diethylstilbestrol (DES)

Erionite

Fowler's solution

Furniture and cabinet making

Haematite mining, underground, with exposure to radon gas

Iron and steel founding

Isopropyl alcohol manufacture, strong-acid process

Lead chromate (VI) oxide

Magenta, manufacture of

Melphalen

Methoxypsoralen, 8-, plus ultraviolet A radiation (Methoxsalen)

Methoxypsoralen, 8-, plus UV radiation (Methoxsalen)

Mineral oil, petroleum condensates, vacuum tower

Mineral oil, petroleum distillates, acid-treated heavy naphthenic

Mineral oil, petroleum distillates, acid-treated heavy paraffinic

Mineral oil, petroleum distillates, acid-treated light naphthenic

Mineral oil, petroleum distillates, acid-treated light paraffinic

Mineral oil, petroleum distillates, heavy naphthenic

Mineral oil, petroleum distillates, heavy paraffinic

Mineral oil, petroleum distillates, hydrotreated (mild) heavy naphthenic

Mineral oil, petroleum distillates, hydrotreated (mild) heavy paraffinic

Mineral oil, petroleum distillates, hydrotreated (mild) light naphthenic

Mineral oil, petroleum distillates, hydrotreated (mild) light paraffinic

Mineral oil, petroleum distillates, light naphthenic

Mineral oil, petroleum distillates, light paraffinic

Mineral oil, petroleum distillates, solvent-dewaxed heavy naphthenic

Mineral oil, petroleum distillates, solvent-dewaxed heavy paraffinic

Mineral oil, petroleum distillates, solvent-dewaxed light naphthenic

Mineral oil, petroleum distillates, solvent-dewaxed light paraffinic

Mineral oil, petroleum distillates, solvent-refined (mild) heavy naphthenic

Mineral oil, petroleum distillates, solvent-refined (mild) heavy paraffinic

Mineral oil, petroleum distillates, solvent-refined (mild) light naphthenic

Mineral oil, petroleum distillates, solvent-refined (mild) light paraffinic

Mineral oil, petroleum extracts, heavy naphthenic distillate solvent

Mineral oil, petroleum extracts, heavy paraffinic distillate solvent

Mineral oil, petroleum extracts, light naphthenic distillate solvent

Mineral oil, petroleum extracts, light paraffinic distillate solvent

Mineral oil, petroleum extracts, residual oil solvent

Mineral oil, petroleum naphthenic oils, catalytic dewaxed heavy

Mineral oil, petroleum naphthenic oils, catalytic dewaxed light

Mineral oil, petroleum paraffin oils, catalytic dewaxed heavy

Mineral oil, petroleum paraffin oils, catalytic dewaxed light

Mineral oil, petroleum residual oils, acid treated

Mineral oils, untreated and mildly-treated

Molybdate orange

MOPP and other combined chemotherapy including alkylating agents

Mustard gas

Naphthylamine, 2-

Nickel and certain nickel compounds

Nitrosourea, 1-(2-Chloroethyl)-3-(4-methylcyclohexyl)-1-(Methyl-CCNU; Semustine)

Oestrogen replacement therapy

Oestrogen, nonsteroidal

Oestrogen, steroidal
Oral contraceptives, combined
Oral contraceptives, sequential
Radon and its decay products
Rubber industry
Shale-oils
Silica, Crystalline
Silicic acid, beryllium salt
Sodium dichromate (VI)
Soots
Strontium chromate (VI)
Sulfur trioxide
Sulfuric acid
Talc (containing asbestos fibers)
Talc containing asbestiform fibres
Tetrachlorodibenzo-para-dioxin, 2,3,7,8- (TCDD)
Tobacco products, smokeless
Tobacco smoke
Treosulphan
Vinyl chloride
Wood dust (certain hard woods)
Zinc chromate (VI) hydroxide

Suggested Reading

Bergeron, J. David. 1995. *Managing the Incident.* Fire Protection Publications, Stillwater, OK.

Code of Federal Regulations 29, 40, and 49. 1993. Office of the Federal Register, Washington, DC.

International Fire Service Training Association, Hazardous Materials for First Responders. 1988. Stillwater, OK.

NFPA 471. 1992. *Recommended Practice for Responding to Hazardous Materials Incidents.* NFPA, Quincy, MA.

NFPA 472. 1992. *Professional Competence of Responders to Hazardous Materials Incidents.* NFPA, Quincy, MA.

NFPA 1991. 1994. *Vapor Protective Suits for Hazardous Chemical Emergencies.* NFPA, Quincy, MA.

NFPA 1992. 1994. *Liquid Splash Protective Suits for Hazardous Chemical Emergencies.* NFPA, Quincy, MA.

Noll, Gregory G., Michael S. Hildebrand, and James G. Yvorra. 1988. *Hazardous Materials: Managing the Incident,* Peake Productions, Annapolis, MD.

SITE SAFETY PLAN REFERENCES

California State Fire Marshal, California Specialized Training Institute, *Technical/Specialist Instructors Guide,* Module 1C—Chapter 15, September 1995.

Environmental Compliance—A Simplified National Guide, STP Specialty Publishers, Inc. North Vancouver, BC, 1992.

Hazardous Materials Emergency Planning Guide, (NRT-1) National Response Team, Washington, DC, March 1987.

NIOSH/OSHA/USCG/EPA. *Occupational Safety and Health Guidance Manual for Hazardous*

Waste Site Activities, U.S. Department of Health and Human Services, Public Health Services, October 1985, Publication #85-115.

Technical Guidance for Hazardous Analysis—Emergency Planning for Extremely Hazardous Substances, US EPA, FEMA, US DOT, Washington, DC, December 1989.

NFPA 1600 *Recommended Practice for Disaster Management*

Emergency Response-Related Internet Directory

Emergency Management and Planning Sites

US Environmental Protection Agency: Chemical Emergency Preparedness—http://www.epa.gov/swercepp/

US Department of Transportation Office of Hazardous Materials—http://ohm.volpe.dot.gov/ohm

OSHA—http://osha/gov

FEMA—http://www.fema.gov/homepage.html

Disaster Management Center—http://epd_hp9k.caenn.wisc.edu/dmc/>

Emergency Management Associates—http://www.coastside.net/USERS/bkinsman/ema

EPIX Emergency Preparedness Information—eXchangegopher://hosi.cic.sfu.ca:5555/11/epix

Haz Net by IDNDR—http://hpshi.cic.sfu/ca/~hazard/

The Alliance for Fire & Emergency Management—http://beep.road.com:80/afem

Weather Sites

Weather Channel	http://www.weather.com
Weather Visualizer	http://covis.atmos.uiuc.edu
WeatherNet	http://cirrus.sprl.umich.edu
US InfraRed Radar	http://clunix.msu.edu
National Hurricane Center	http://nhc.noaa.gov
National Severe Storms Lab	http:// www.nssl.uoknor.edu
NOAA	http://www.noaa.gov
Severe Weather Information	http://aspi.sbs.ohio.state.edu

News Sites

CNN News	http://www.cnn.com
Reuters News	http://www.reuters. com
ABC News	http://abcradio.ccabc.com
AP News	http://www.trib.com
NBC News	http://www.nbc.com
NY Times	http://nytimes.com
USA Today	http://usatoday.com

Acronyms

ACGIH	American Conference of Governmental Industrial Hygienists	CPC	Chemical protective clothing
AMU	Atomic mass unit	CPSC	Consumer Product Safety Commission
ANFO	Ammonium nitrate and fuel oil	CRZ	Contamination reduction zone
ANSI	American National Standards Institute	CWA	Clean Water Act
APR	Air purifying respirator	DASHO	Designated Agency Safety and Health Official
ASTM	American Society for Testing and Materials	DHHS	Department of Health and Human Services
BLEVE	Boiling liquid expanding vapor explosion	DMSO	Dimethyl sulfoxide
		DOL	Department of Labor
BOM	Bureau of Mines	DOT	Department of Transportation
CAA	Clean Air Act	EAP	Emergency action plan
CAS	Chemical Abstracts Service	EH & S	Environmental Health and Safety
CERCLA	Comprehensive Environmental Response, Compensation, and Liability Act	EPA	Environmental Protection Agency
		EPCRA	Emergency Planning and Community Right to Know
CFR	Code of Federal Regulations	ETO	Ethylene oxide
CGI	Combustible gas indicator	FACOSH	Federal Advisory Council for Occupational Safety and Health
CHEMTREC	Chemical Emergency Transportation Center	FDA	Food and Drug Administration
CMA	Chemical Manufacturers Association	FFSHC	Field Federal Safety and Health Council
CNS	Central nervous system	FGAN	Fertilizer grade ammonium nitrate
COC	Cleveland Open Cup	FID	Flame ionization detector

FIFRA	Federal Insecticide, Fungicide, and Rodenticide Act
FRA	First Responder Awareness
FRO	First Responder Operations
GC	Gas chromatograph
HAZWOPER	Hazardous Waste Operations and Emergency Response
HEPA	High efficiency particulate air
HF	Hydrofluoric acid
HMIS	Hazardous Materials Information System
HMTA	Hazardous Materials Transportation Act
IARC	International Agency for Research on Cancer
IC	Incident commander
ICS	Incident command system
IDLH	Immediately dangerous to life and health
IP	Ionization potential
IPA	Isopropyl alcohol
IUPAC	International Union of Pure and Applied Chemistry
LDL	Lethal dose low
LEL	Lower explosive limit
LFL	Lower flammable limit
MEK	Methyl ethyl ketone
MIBK	Methyl isobutyl ketone
MSDS	Material safety data sheet
MSHA	Mine Safety and Health Administration
MW	Molecular weight
NCI	National Cancer Institute
NFPA	National Fire Protection Association
NIOSH	National Institute for Occupational Safety and Health Administration
NOx	Oxides of nitrogen
NPIRS	National Pesticide Information Retrieval System
NPL	National Priority Site List
NRC	National Response Center
NTP	National Toxicology Program
OFAP	Office of Federal Agency Programs
OSHA	Occupational Safety and Health Administration
PCB	Polychlorinated biphenyl
PEL	Permissible exposure limit
PID	Photoionization detector
PIO	Public information officer
PMCC	Pensky Martens Closed Cup
PPE	Personal protective equipment
PRCS	Permit required confined space
PSM	Process safety management
PVC	Polyvinyl chloride
RBC	Red blood cell
RCRA	Resource Conservation and Recovery Act
SAC	Site access control officer
SAR	Supplied air respirator
SARA	Superfund Amendment and Reauthorization Act
SCBA	Self-contained breathing apparatus

SETA	Setaflash Closed Tester	TLV/C	Threshold limit value—ceiling
SLUD	Salivation, lacrimation, urination, defecation	TLV/TWA	Threshold limit value—time weighted average
SOP	Standard operation procedure	TNT	Trinitrotoluene
SPCC	Spill Prevention Control and Countermeasure Plan	TSCA	Toxic Substances Control Act
		TSD	Treatment storage and disposal
STEL	Short-term exposure limit	TSDF	Treatment storage and disposal facilities
STP	Standard temperature and pressure		
SUS	Saybolt Universal Seconds	UEL	Upper explosive limit
TCC	Tag (Tagliabue) Closed Cup	USDA	U.S. Department of Agriculture
TDL	Toxic dose low	UV	Ultraviolet
TLV	Threshold limit value		

Glossary

The following glossary explains terms commonly used to describe chemical properties and hazards. Some terms are used extensively on chemical labels and material safety data sheets (MSDSs). In most cases, the terms are explained from a work site standpoint based on health and safety. These definitions may differ from the more technical ones used in a chemistry or physics course, but provide working information necessary for field use.

Absorb A process whereby one substance penetrates into the structure of another.

Absorbents Any material capable of soaking up a spilled liquid. Commonly known as 3M pads or spill socks, absorbents come in many shapes and sizes. For spills on a flat surface such as a concrete pad, the best type is the sausage type absorbent boom that can be placed around the spill. Absorbents are also useful for stopping liquids from entering a storm drain or sewer. All absorbents that are used to absorb hazardous materials must be disposed of as would the hazardous material itself.

Absorption Passage of toxic materials through some body surface into body fluids and tissue. Generally speaking, this refers to passage through the skin, eyes, or mucous membranes.

Acceptable entry conditions The conditions that must exist in a permit required confined space to ensure that employees involved in the operation can safely enter into and work within the space.

Accountability log A written record of information such as products involved in the incident, times of significant changes in the incident, names and entry/exit times of personnel in the exclusion zone, level of personal protective equipment used, and function of personnel in the exclusion.

Acid An inorganic or organic compound that (1) reacts with metals to yield hydrogen; (2) reacts with a base to form a salt; (3) dissociates in water to yield hydrogen ions; (4) has a pH of less than 7.0; and (5) neutralizes bases or alkalis. All acids contain hydrogen and turn litmus paper red. They are corrosive to human tissue and should be handled with care. Examples include sulfuric and hydrochloric acid.

Action level The exposure level (concentration in air) at which OSHA regulations to protect employees take effect (29 CFR 1910.1001–1047); e.g., workplace air analysis, employee training, medical monitoring, and record keeping. Exposure at or above action level is termed *occupational exposure*. Exposure below this level can also be harmful. The action level is generally half the PEL.

Acute exposure Exposure of short duration, usually to relatively high concentrations or amounts of material.

Acute health effect Adverse effect on a human or animal, which has severe symptoms developing rapidly and coming quickly to a crisis. Acute effects are usually immediate and can be severe or life threatening.

Acute toxicity The ability of a substance to do damage (generally systemic) as a result of a one-time exposure from a single dose of or exposure to a material. This exposure is generally brief.

Adhesion A union of two surfaces that are normally separate.

Adiabatic heat The technical definition is descriptive of a system in which no net heat loss or gain is allowed.

Adsorption A process or reaction that causes molecules to adhere to the surface of another substance.

Aerosol A fine aerial suspension of particles sufficiently small in size to confer some degree of stability from sedimentation (e.g., smoke or fog).

Afterdamp Carbon monoxide

Air bill A shipping paper, prepared from a bill of lading, that accompanies each piece in an air shipment.

Air line respirator A respirator that is connected to a compressed breathing air source by a hose of small inside diameter. The air is delivered continuously or intermittently in a sufficient volume to meet the wearer's breathing requirements from a tank or compressor in a remote location.

Air monitoring The process of evaluating the air in a given work environment. This process can be accomplished using a variety of instrumentation depending on the anticipated hazards. In relation to confined spaces, this function must be carried out by the confined space attendant during entry. Air monitoring includes the use of instrumentation to determine if the confined space is oxygen deficient, has a combustible gas present, or a known hazardous gas such as hydrogen chloride or ammonia. Air monitoring must be conducted before entry is made, any time conditions change (temperature, humidity, hot work), and on a regular basis while entry personnel are working in areas of airborne contamination in the confined space.

Air purifying respirator A respirator that uses chemicals to remove specific gases and vapors from the air or that uses a mechanical filter to remove particulate matter. An air purifying respirator must be used only when there is sufficient oxygen to sustain life and the air contaminant level is below the concentration limits of the device. This type of respirator is effective for concentrations of substances that are generally no more than ten times the threshold limit value (TLV) of the contaminant but never more than the immediately dangerous to life and health (IDLH) value, and if the contaminant has warning properties (odor or irritation) below the TLV.

Air reactive materials Substances that ignite when exposed to air at normal temperatures. Also called *pyrophoric*.

Alkali Any chemical substance that forms soluble soaps with fatty acids. Alkalis are also referred to as *bases*. They may cause severe burns to the skin. Alkalis turn litmus paper blue and have pH values from 7.1 to 14. Sodium hydroxide, sodium bicarbonate, and ammonium hydroxide are all examples of alkalis.

Allergic reaction An abnormal physiological response to chemical or physical stimuli by a sensitive person.

Alpha particle A small, charged particle emitted from the nucleus of an unstable atom. The small particle is essentially a helium nucleus consisting of two neutrons and two protons. This particle is of high energy and is thrown off by many radioactive elements.

American Council of Governmental Industrial Hygienists (ACGIH) Founded in 1938 for the purpose of determining standards of exposure to toxic and otherwise harmful materials in workroom air, this organization is comprised of persons employed by official government units responsible for programs of industrial hygiene, education, and research. The standards are revised annually. *See* Threshold limit value.

American National Standards Institute (ANSI) ANSI is a privately funded, voluntary membership organization that identifies industrial and public needs for national consensus standards and coordinates development of such standards.

American Society for Testing and Materials (ASTM) ASTM is the world's largest source of voluntary consensus standards for materials, products, systems, and services. ASTM is a resource for sampling and testing methods, health and safety aspects of materials, safe performance guidelines, and effects of physical and biological agents and chemicals.

AMU Atomic Mass Unit.

Anesthetic A chemical that causes a total or partial loss of sensation. Exposure to anesthetic can cause impaired judgment, dizziness, drowsiness, headache, unconsciousness, and even death. Examples include alcohol, paint remover, and degreasers.

ANFO Ammonium nitrate fuel oil.

Anhydrous Containing no free water. A chemical compound such as anhydrous ammonia would be virtually pure ammonia with only trace amounts of water.

Anion A negatively charged ion created when an atom gains an electron in its orbit. The addition of a negatively charged electron results in an imbalance between the protons and electrons. Because there are more electrons in this case, it shifts the electrical charge to the negatively charged state.

Antidote A remedy to relieve, prevent, or counteract the effects of a poison.

Aquatic toxicity The adverse effects to marine life that result from exposure to a toxic substance.

Asbestos A fibrous form of silicate minerals that has many uses in society. Although found naturally in California, it is mainly used in fireproof fabrics, brake linings, gaskets, roofing compositions, electrical and mechanical insulation, paint filler, chemical filters, and as a reinforcing agent in rubber, plastics, and tile flooring. Its prime hazard is being a carcinogen (*see* carcinogen). It is highly toxic by inhalation of dust particles.

Asphyxiate To smother. Asphyxiation is one of the principal potential hazards of working in confined and enclosed spaces.

Asymptomatic Showing no symptoms.

Ataxia Muscular incoordination especially that manifested when voluntary muscular movements are attempted, as when one walks with an unsteady gait.

Atm Atmosphere, a unit of pressure equal to 760 mm Hg (mercury) at sea level.

Atomic number The number of protons in the nucleus of a particular atom. This number determines the elements position on the periodic table. *See also* Proton.

Atomic symbol A one- or two-letter designation for a given element. These symbols are found on the periodic table and pertain to only one element. Symbols are also used in chemical formulas to identify the components of a given substance. LiF for example, is called lithium fluoride and is made up of the elements lithium and fluorine.

Atomic weight The total weight of any atom is the sum of all its subatomic parts—the protons, neutrons, and electrons.

Atomizing The process of rapidly compressing a liquid into mist or vapor.

Authorized attendant *See* Confined space attendant.

Autoignition temperature The lowest temperature to which a material will ignite spontaneously or burn. Sometimes referred to as ignition temperature. Almost all materials have an autoignition temperature because most materials must be heated to this level in order to begin to burn.

Base *See* Alkali.

Benign A term expressing that a disease (usually tumor) is not recurrent, not progressing, and/or not malignant.

Beta particle A by-product of radioactive decay from an unstable nucleus. The process changes a neutron into a proton and subsequently emits an electron from its orbit. The released electron can take the form of a positron (positive charge) or a negatron (negative charge). This type of decay involves a change in the atomic number but no change in the mass of the atom. Beta particles are of high velocity and in some cases exceed 98% of the speed of light.

Biodegradable Capable of being broken down into innocuous products by the action of living things. If the material is biodegradable, it is generally less hazardous to the environment.

Biohazard *See* Pathogen.

Blanking A procedure to physically seal pipes (e.g., hydraulic, pneumatic, chemical, fuel, water) by inserting and bolting a piece of metal between flanges in the piping. Some pipes might also be isolated with double valves that can be locked out.

Blasting agent A material designed for blasting that has been tested in accordance with Sec. 173.114a(b) of 49 CFR and found to be so insensitive that there is very little probability of accidental initiation to explosion or of transition from deflagration to detonation. (Sec. 173.114a(a))

Bleeding Releasing pressure in hydraulic or pneumatic lines.

Boiling liquid expanding vapor explosion (BLEVE) A major failure of a closed liquid container into two or more pieces. It is usually caused when the temperature of the liquid is well above its boiling point at normal atmospheric pressure.

Boiling point The temperature at which a liquid changes to a vapor state at a given pressure. The boiling point is usually expressed in degrees Fahrenheit at sea level pressure (760 mm Hg, or one atmosphere). For mixtures, the initial boiling point or the boiling range may be given. Some examples of flammable materials with low boiling points include:

Propane	−44°F
Anhydrous ammonia	−28°F
Butane	−31°F
Gasoline	100°F
Ethylene glycol	387°F

Bonding The interconnecting of two objects by means of a clamp and bare wire. Its purpose is to equalize the electrical potential between the objects to prevent a static discharge when transferring a flammable liquid from one container to another. The conductive path is provided by clamps that make contact with the charged object and a low resistance flexible cable that allows the charge to equalize. *See* Grounding.

Bradycardia A slow heart beat, usually slower than sixty beats per minute.

Brass An alloy consisting of copper and zinc.

Breakthrough A term that denotes the passage of a substance from one side of a barrier to the other.

Breakthrough time The elapsed time from initial contact of the outside surface of the personal protective equipment with chemical to the first detection of chemical on the inside surface.

Breathing zone An imaginary zone encompassing a horizontal plane from the middle of the chest to the top of the head.

Brisance An expression of the shattering effect of a particular explosive material.

Buddy system A concept of teamwork among responders and site workers. The rule of "No one works alone" requires a minimum team of two to operate in hazardous areas.

BuMines Bureau of Mines, U.S. Department of Interior.

C Centigrade, a unit of temperature that has a scale ranging from 0 to 100 with 0 being the temperature at which water freezes and 100 the temperature at which water boils.

C, or Ceiling The maximum allowable human exposure limit, usually for an airborne substance, that is not to be exceeded even momentarily. *Also see* Permissible exposure limit and Threshold limit value.

Cancer, Carcinoma An abnormal multiplication of cells that tends to infiltrate other tissues and metastasize (spread). Each cancer is believed to originate from a single "transformed" cell that grows (splits) at a fast, abnormally regulated pace, no matter where it occurs in the body. Cancer is the second most common cause of death in the United States and is expected to be the number one cause by the year 2000. Most cancers are caused by our lifestyle, i.e., smoking and diet.

Carbon dioxide (CO_2) A heavy, colorless gas produced by the combustion and decomposition of organic substances and as a by-product of many chemical processes. CO_2 does not burn and is relatively nontoxic (although high concentrations, especially in confined spaces, can create hazardous oxygen-deficient environments).

Carbon monoxide (CO) A colorless, odorless, flammable, and very toxic gas produced by the incomplete combustion of carbon. It is a by-product of many chemical processes.

Carboxyhemoglobin The term for hemoglobin which is bound to carbon monoxide rather that oxygen.

Carcinogen A material that either causes cancer in humans, or, because it causes cancer in animals, is considered capable of causing cancer in humans. Findings are based on the feeding of large quantities of a material to test animals or by the application of concentrated solutions to the animals' skin. A material is considered a carcinogen if: (1) the International Agency for Research on Cancer (IARC) has evaluated it and found it a carcinogen or potential carcinogen; (2) the National Toxicology Program's (NTP) Annual Report on Carcinogens lists it as a carcinogen or potential carcinogen; (3) OSHA regulates it as a carcinogen; or (4) one positive study has been published. Following is a listing of one breakdown of carcinogens. It is the system used by the ACGIH in its classification of carcinogens.

A1—Confirmed Human Carcinogen: The agent is carcinogenic to humans based on the weight of evidence from epidemiological studies of or convincing clinical evidence in exposed humans.

A2—Suspected Human Carcinogen: The agent is carcinogenic in experimental animals at dose levels, by route(s) of administrations, at site(s), of histologic type(s), or by mechanism(s) that are considered relevant to worker exposure. Available epidemiological studies are conflicting or insufficient to confirm an increased risk of cancer in exposed humans.

A3—Animal Carcinogen: The agent is carcinogenic in experimental animals at relatively high dose, by route(s) of administration, at site(s), of histologic types(s), or by mechanisms(s) that are not considered relevant to worker exposure. Available epidemiological studies do not confirm an increased risk of cancer in exposed humans. Available evidence suggest that the agent is not likely to cause cancer in humans except under uncommon or unlikely routes or levels of exposure.

A4—Not Classifiable as a Human Carcinogen: There are inadequate data on which to classify the agent in terms of its carcinogenicity in humans and/or animals.

A5—Not Suspected as a Human Carcinogen: The agent is not suspected to be a human carcinogen on the basis of properly conducted epidemiological studies in humans. These studies have sufficiently long follow-up, reliable exposure histories, sufficiently high dose, and adequate statistical power to conclude that exposure to the agent does not convey a significant risk of cancer to humans. Evidence suggesting a lack of carcinogenicity in experimental animals will be considered if it is supported by other relevant data.

Carcinogenicity The ability to produce cancer.

CAS number (CAS registration number) An assigned number used to identify a chemical. CAS stands for Chemical Abstract Service, an organization that indexes information published in Chemical Abstracts by the American Chemical Society and that provides index guides by which information about particular substances may be located in the abstracts. The CAS number is a concise, unique means of material identification.

Catalyst A substance that modifies (slows, or more often quickens) a chemical reaction without being consumed in the reaction.

Cation A positively charged ion, created when an atom loses an electron from its orbit. The reduction of a negatively charged electron results in an imbalance between the protons and electrons. Because there are more protons in this case, the electrical charge shifts to the positively charged state.

Caustic See Alkali.

cc Cubic centimeter is a volume measurement in the metric system that is equal in capacity to one milliliter (ml). One quart is about 946 cubic centimeters.

Central nervous system (CNS) The term that generally refers to the brain and spinal cord. These organs supervise and coordinate the activity of the entire nervous system. Sensory impulses are transmitted into the central nervous system and motor impulses are transmitted out. The remainder of the nervous system that is not part of the CNS is generally classified as the peripheral nervous system.

Chemical Any element, chemical compound, or mixture of elements and/or compounds where chemical(s) are distributed.

Chemical cartridge respirator See Air Purifying Respirator.

Chemical family A group of single elements or compounds with a common general name. Example: acetone, methyl ethyl ketone (MEK), and methyl isobutyl ketone (MIBK) are of the ketone family: acrolein, furfural, and acetaldehyde are of the aldehyde family.

Chemical hygiene plan A written program developed and implemented by the employer setting forth procedures, equipment, personal protective equipment, and work practices that (1) are capable of protecting employees from the health hazards presented by hazardous chemicals used in the particular workplace.

Chemical name The name given to a chemical in the nomenclature system developed by the International Union of Pure and Applied Chemistry (IUPAC) or the Chemical Abstracts Service (CAS).

Chemical structure The arrangement within the molecule of atoms and their chemical bonds.

Chemical Transportation Emergency Center (CHEMTREC) CHEMTREC is a national center established by the Chemical Manufacturers Association (CMA) to relay pertinent emergency information concerning specific chemicals on

Glossary

requests from individuals and response agencies. CHEMTREC has a 24-hour toll-free telephone number (800-424-9300) to help agencies who respond to chemical transportation emergencies.

Chemistry The branch of study concerning the composition of chemical substances and their effects and interactions with one another. This study includes atoms and their behavior, the makeup of chemical compounds, the reactions that occur between these compounds, and the energies resulting from those reactions. Chemistry is not an isolated discipline however, and it includes theories from thermodynamics, physics, and biology.

Chronic exposure Long-term contact with a substance. Such exposures often result in long-term health effects such as the development of cancers or organ damage.

Chronic toxicity Harmful systemic effects produced by long-term, low-level exposure to chemicals. Typically, this period of time is considered to be more than 3 months.

Class A Flammable Materials These are fires involving solid, organic materials including wood, cloth, paper, and many plastics.

Class A Explosive A mass of explosive substance capable of detonating or otherwise reaching maximum explosion hazard. The various types of class A explosives are defined in Sec 173.53 of 49 CFR.

Class B Explosive In general, the type of explosive that functions by rapid combustion rather than detonation and includes some explosive devices such as special fireworks or flash powders. Flammable hazard. (Sec 173.88 of 49 CFR)

Class C Explosive Certain types of materials that are manufactured articles containing Class A or Class B explosives, or both, as components but in restricted quantities, and certain types of fireworks. These present a minimum hazard.

Class IA Flammable Liquids A class of flammable liquids with a flash point below 73°F and a boiling point below 100°F.

Class IB Flammable Liquids A class of flammable liquids with a flash point below 73°F and a boiling point at or above 100°F.

Class IC Flammable Liquids A class of flammable liquids with a flash point at or above 73°F and below 100°F.

Class II Liquids A class of combustible liquids with a flash point at or above 100°F and below 140°F.

Class III Liquids A class of flammable liquids with a flash point above 140°F.

Clean Air Act Federal law enacted to regulate and reduce air pollution and administered by the Environmental Protection Agency.

Clean Water Act Federal law enacted to regulate and reduce water pollution. The CWA is administered by the EPA.

Clinical toxicology Designates within the realm of medical science an area of professional emphasis concerned with diseases caused by, or uniquely associated with toxic substances. This area of toxicology generally deals with drug overdoses and its treatment.

COC Cleveland Open Cup is a flash point test method.

Code of Federal Regulations (CFR) Codification of the various federal regulations. The most important example for workplace safety is book number 29 (29 CFR), which contains the OSHA regulations. Other examples include 40 CFR, which contains EPA regulations, and 49 CFR, which contains Department of Transportation regulations.

Codified Adopted as a recognized piece of legislation and placed into the statutes in writing.

Colorimetric detector tubes These are sealed glass tubes, that, when the tips are broken off and an air sample is drawn through with a bellows device, will turn reagents in the tube a different color when exposed to a specific air contaminant.

Combustible A term used by NFPA, DOT, and others to classify on the basis of flash points certain liquids that will burn. NFPA generally defines combustible liquids as having a flash point of 100°F (38.7°C) or higher. In 1992, DOT modified its definition to a liquid with a flash point greater than 140°F. *Also see* Flammable. Another use for the term is with nonliquid substances such as wood and paper. In this case, materials capable of burning are often referred to as combustible or as "ordinary combustibles."

Combustible gas indicator (CGI) A mechanical device that can detect any gas or vapor with a defined flash point (*see* Flash point) and lower explosive limit (which see), if the concentration is high enough. This includes both flammable and combustible materials. Most CGIs also have the capability to monitor oxygen levels in the atmosphere, which is especially important for monitoring areas that have the potential to be oxygen deficient (below 19.5% oxygen) or oxygen enriched (above 23% oxygen).

Combustible liquid NFPA classifies combustible liquid as any liquid having a flash point at or above 100°F (37.8°C). Combustible liquids are also referred to as either Class II or Class III liquids depending on their flash point. Note that DOT has a definition that differs. *See* Combustible.

Command post The geographic location of the IC and the control center for the response.

Command staff These job functions work for the IC separately from the general staff. These functions include Safety Officer, Public Information Officer, and Liaison Officer.

Compound The chemical combination of two or more elements, which results in the creation of unique properties and a definite, identifiable composition of the substance.

Comprehensive Environmental Response, Compensation, and Liability Act of 1980 (CERCLA) This act requires that the Coast Guard provide a national response capability that can be used in the event of a hazardous substance release. The act also provides for a fund (the Superfund) to be used for the cleanup of abandoned hazardous waste disposal sites.

Compressed gas (1) A gas or mixture of gases having, in a container, an absolute pressure exceeding 40 psi at 70°F (21.1°C); or (2) a gas or mixture of gases having, in a container, an absolute pressure exceeding 104 psi at 130°F (54.4°C) regardless of the pressure at 70°F (21.1°C); or (3) a liquid having a vapor pressure exceeding 40 psi at 100°F (37.8°C) as determined by ASTM D-323-72.

Concentration The relative amount of a substance when combined or mixed with other substances. Examples: 2 ppm hydrogen sulfide in air, or a 50% caustic solution.

Confined space Any area as defined by OSHA that has the following characteristics: (1) is large enough so that a person can bodily enter; (2) is not designed for continuous occupancy; and (3) has limited ingress and egress. To be classified as a permit required confined space, the area must also possess, or have the potential to possess one of the following: (1) an oxygen-deficient atmosphere; (2) a potentially hazardous atmosphere (toxicity); (3) a flammable atmosphere; (4) the risk of entrapment due to converging walls or sloping floors; (5) risk of engulfment, (6) any other recognized serious health or safety hazard. Examples of confined spaces include storage tanks, pits, vaults, or chambers.

Confined space attendant Whenever an entry is made into a confined space, an attendant is required. Entry can be as simple as sticking one's head into an opening or as complex as entering an underground storage tank. The attendant is responsible for ensuring entry personnel have the correct respiratory protection and chemical protective clothing. The attendant also conducts monitoring when required, ensures communications are in place, and ensures the entry team act in a safe and competent manner. At no time is the attendant allowed to leave the confined space area.

Confined space permit This form is filled out by a competent authority such as a qualified confined space attendant, industrial hygienist, safety professional, or marine chemist. The form states what the space was tested for, what specific (if any) equipment is required, reports that it is safe for entry, and is signed.

Consist A rail shipping paper containing a list of cars in the train by order. Those containing hazardous materials are indicated. Some railroads include information on emergency operations for the hazardous materials on the train with the consist.

Consumer commodity A material that is packaged or distributed in a form intended and suitable for sale through retail sales agencies or for consumption by individuals for personal care or household use. This term also includes drugs and medicines. (See ORM-D)

Consumer Product Safety Commission (CPSC) Has responsibility for regulating hazardous materials when they appear in consumer goods. For CPSC purposes, hazards are defined by the Hazardous Substances Act and the Poison Prevention Packaging Act of 1970.

Container Any bag, barrel, bottle, box, can, cylinder, drum, reaction vessel, storage tank, or the like that contains a hazardous chemical. For the purposes of an MSDS or the Hazard Communication Standard, pipes or piping systems are not considered to be containers.

Contamination reduction zone (CRZ) Also known as the warm zone. The area where the decon corridor is placed.

Control point The monitored area where the decon corridor enters the hot zone.

Corrosive A chemical that causes visible destruction of, or irreversible alterations in, living tissue by chemical action at the site of contact. For example, a chemical is considered to be corrosive if, when tested on the intact skin of albino rabbits by the method described by the U.S. Department of Transportation in Appendix A to 49 CFR, Part 173, it destroys or changes irreversibly the structure of the tissue at the site of contact following an exposure period of 4 hours.

CPC Chemical protective clothing.

Cryogen Gases that are cooled to a very low temperature, usually below $-150°F$ ($-101°C$), to change to a liquid. Also called *refrigerated liquids*. Some common cryogens include liquid oxygen, nitrogen and helium.

Cutaneous toxicity See Dermal toxicity.

Cutting, welding, and burning Any action involving the use of oxyacetylene or heliarc welding on metal that creates sparks, heat, or fumes that may collect in one space or effect the atmosphere in a nearby space. Welding also applies to using Weld-on type glues in affixing PVC piping and joints. All of the actions may change the hazards and conditions, especially in a confined space.

Dangerous cargo manifest A cargo manifest listing the hazardous materials on board a ship and their location.

Dangerous goods (Canada) Any product, substance, or organism included by its nature or by the regulation in any of the classes listed in the schedule (UN 9 Classes of Hazardous Materials).

Decomposition Breakdown of a material or substance (by heat, chemical reaction, electrolysis, decay, or other processes) into parts or elements or simpler compounds.

Decon corridor The designated area for decontamination to take place. Located at the exit of the exclusion zone (hot zone) and bridging across into the support or cold zone.

Decontamination (decon) The complete and systematic removal of contaminants from people, property, and equipment.

Decon team leader This management position is responsible for the operations and support of the decontamination team.

Deflagration A rapid combustion of a material occurring in the explosive mass at subsonic speeds. The event is usually caused by contact with a flame source but may also be caused by mechanical heat or friction.

Degradation Decomposition of a substance at a molecular level.

Delaney clause of the 1958 Food Additives Amendments Act The legislation that requires that chemicals used as food additives be considered as human carcinogens if they produce cancer in any animal species at any level of exposure.

Density The mass (weight) per unit volume of a substance. For example, lead is much more dense than aluminum.

Dermal Relating to the skin.

Dermal toxicity Adverse effects resulting from skin exposure to a substance. The term was ordinarily used to denote effects in experimental animals.

Dermis The "true skin" that covers the body. The dermis contains such structures as the blood vessels and nerves that supply the skin. The dermis supports the epidermis, which is the outermost covering of the body.

Descriptive toxicology The branch of toxicology concerned directly with toxicity testing.

Designated Agency Safety and Health Official (DASHO) The executive official of a federal department or agency who is responsible for safety and occupational health matters within a federal agency and who is so designated or appointed by the head of the agency.

Designated area An area that may be used for work with select carcinogens, reproductive toxins, and substances with a high degree of acute toxicity. A designated area may be the entire laboratory, or a device such as a laboratory hood.

Detonation An extremely rapid decomposition of an explosive material. This decomposition propagates throughout the explosive agent at supersonic speeds and is accompanied by pressure and temperature waves. This detonation could be initiated by mechanical friction, impact, or heat.

Diatomic gas A gaseous element requiring two atoms to achieve stability. Examples of diatomic gases are oxygen and nitrogen.

Diffusion The action of a gas to move from an area of higher concentration to an area of lower concentration.

Dike A barrier constructed to control or confine hazardous substances and prevent them from entering sewers, ditches, streams, or other flowing waters.

Diurnal effects Expected daily weather patterns such as directional wind shifts, fog, or rain.

Dose The amount of a given pollutant or chemical that enters the body of an exposed organism in a given period of time. The time can be as short as a few seconds (to inject a substance) or as long as a lifetime (in the case of chronic exposures). *Internal dose* refers to the amount of a chemical absorbed by the body. *Biologically effective dose* refers to the amount that interacts with a particular target tissue or organ.

$$\text{Dose} = \text{Concentration} \times \text{Time} \times \text{Exchange Rate}$$

where exchange rate refers to the breathing rate (e.g., m^3/hr) or ingestion rate (liters or grams/day).

Note that if the internal dose is to be calculated, then another factor must be included, namely, the fraction of the chemical taken up. This fraction is the retention factor, which ranges from 0 to 1.0.

Ductile A material that will not return to its original dimension when stress is removed.

Dynamite An industrial high explosive that is moderately sensitive to shock and heat. The main ingredient is nitroglycerin or sensitized ammonium nitrate. This material is then distributed in diatomaceous earth or a mass of hydrated silica.

Edema An abnormal accumulation of clear watery fluid in the tissues.

Electrolytes Substances such as sodium, potassium, or calcium that dissociate into charged particles in water. Rapid loss of these electrolytes can cause acute medical problems.

Electromagnetic radiation The propagation of waves of energy of varying electric and magnetic fields through space. The waves move at the speed of light from matter in the form of photons or energy packets. The strength of the waves depends on their individual frequency. *See also* Gamma radiation.

Electron A subatomic particle having a negative electric charge. The electron has mass, which is approximately 1/1837th of a proton. Electrons surround the nucleus of atoms and are equal in number to the protons for the given element they are orbiting. Electrons that have been separated from an atomic orbit are said to be *free electrons*. Electrons are the subatomic particle responsible for bonding and are pivotal in the formation of compounds.

Element One of the 109 recognized substances that comprise all matter at the atomic level. Elements are the building blocks for the compounds formed by chemical reactions. A listing of the elements can be found on the periodic table. Some examples include sodium, oxygen, neon, and carbon.

Emergency Any occurrence such as, but not limited to, equipment failure, rupture of containers, or failure of controls, that results in an uncontrolled release of a hazardous chemical into the workplace.

Encapsulating A description of a one-piece chemical resistant garment that completely covers the entire body of the wearer.

Endothermic A description of a process that ultimately absorbs heat and requires large amounts of energy for initiation and maintenance.

Engulfment The surrounding and effective capture of a person by a liquid or finely divided solid substance that can be aspirated to cause death by filling or plugging the respiratory system or that can exert enough force on the body to cause death by strangulation, constriction, or crushing.

Entry The action by which a person passes through an opening into a permit required confined space. Entry includes ensuing work activities in that space and is considered to have occurred as soon as any part of the entrant's body breaks the plane of an opening into the space.

Environmental Protection Agency (EPA) Established in 1970, the federal EPA is required to ensure the safe manufacture, use, and transportation of hazardous chemicals. The State of California has also established California EPA, which follows the same guidelines on the state level.

Environmental toxicity Information obtained as a result of conducting environmental testing designed to study the effects on aquatic and plant life.

Environmental toxicology The branch of toxicology dedicated to developing an understanding of "chemicals in the environment and their effect on man and other organisms." Environmental media may include air, groundwater, surface water, and soil. Impact may be from fish to philosopher. This area of toxicology is (or should be) the most significant for the development of risk assessments.

Epidemiology Science concerned with the study of disease in a general population. Determination of the incidence (rate of occurrence) and distribution of a particular disease (as by age, sex, or occupation) that may provide information about the cause of the disease.

Epithelium The covering of internal and external surfaces of the body.

Etiologic agent A material or substance capable of causing disease. The material could contain a viable microorganism or its toxin, which causes or may cause human disease. A biohazard or biologic agent.

Evaporation A physical change of a solid or liquid into its gaseous or vapor phase. Temperature greatly influences the evaporation rate of a given substance. The change of solids into gases without going into a liquid phase is called *sublimation*.

Evaporation rate The rate at which a material vaporizes (evaporates) when compared to the known rate of vaporization of a standard material. The evaporation rate can be useful in evaluating the health and fire hazards of a material. The designated standard material is usually normal butyl acetate with a vaporization rate designated as 1.0. Vaporization rates of other solvents or materials are then classified as:

> Fast evaporating if greater than 3.0. Examples: methyl ethyl ketone = 3.8, acetone = 5.6, hexane = 8.3.
>
> Medium evaporating if 0.8 to 3.0. Examples: 190 proof (95%) ethyl alcohol = 1.4, naphtha = 1.4.
>
> Slow evaporating if less than 0.9. Examples: xylene = 0.6, isobutyl alcohol = 0.6, normal butyl alcohol = 0.4, water = 0.3, mineral spirits = 0.1.

Exclusion zone Also known as the hot zone. The area of highest contamination and hazard to be entered only by properly trained and protected individuals.

Exothermic An expression of a reaction or process that evolves energy in the form of heat. The process of neutralization evolves heat. This reaction is considered to be exothermic.

Explosive A chemical that causes a sudden, almost instantaneous release of pressure, gas, and heat when subjected to sudden shock, pressure, or high temperature. Explosives are broken down into various classifications by the Department of Transportation.

Exposure An event in which a pollutant (chemical toxicant in the present context) has contact with an organism, such as a person, over a certain interval of time. Exposure is thus the product of concentration and time (e.g., mg/m^3 × hr).

Exposure factor(s) Conditions of a chemical exposure that determine how damaging the event may be. These factors include the concentration of the chemical, the duration of the exposure, the uptake rate, and the possible complications of chemical interactions.

Exposure limit Values established by scientific testing. These values denote safe levels that enable workers to maintain a margin of safety when functioning in contaminated atmospheres.

Extremely hazardous substance Chemicals determined by the Environmental Protection Agency (EPA) to be extremely hazardous to a community during an emergency spill or release as a result of their toxicities and physical/chemical properties.

Eye protection Recommended safety glasses, chemical splash goggles, face shields, and so forth to be utilized when handling a hazardous material.

F Fahrenheit is a scale for measuring temperature. On the Fahrenheit scale, water boils at 212°F and freezes at 32°F.

Federal Advisory Council for Occupational Safety and Health (FACOSH) A joint management–labor council that advises the secretary of labor on matters relating to the occupational safety and health of federal employees.

Federal Insecticide, Fungicide, and Rodenticide Act (FIFRA) Regulates poisons, such as chemical pesticides, sold to the public and requires labels that carry health hazard warnings to protect users. It is administered by EPA.

Fetus The developing young in the uterus from the seventh week of gestation until birth.

FGAN Fertilizer grade ammonium nitrate.

Field Federal Safety and Health Councils (FFSHC) Organized throughout the country to improve federal safety and health programs at the field level and within a geographic location.

Firedamp Methane gas.

First Responder Awareness Level (FRA) The first level of training under the HAZWOPER standard for persons who would likely come across a hazardous materials release and whose job it would be to report the spill to the correct agency.

First Responder Operations Level (FRO) The second level of training under the HAZWOPER standards. This level is designed for those personnel whose role it is to respond to a hazardous materials release and initiate defensive control operations.

Flammable liquid Any liquid having a flash point below 100°F (37.8°C), except any mixture having components with flash points of 100°F (37.8°C) or higher, the total of which make up 99% or more of the total volume of mixture.

Class IA: A class of flammable liquids with a flash point below 73°F and a boiling point below 100°F.

Class IB: A class of flammable liquids with a flash point below 73°F and a boiling point at or above 100°F.

Class IC: A class of flammable liquids with a flash point at or above 73°F and below 100°F.

Flammable range The concentration of gas or vapor in air that will burn if ignited. It is expressed as a percentage that defines the range between a lower explosive limit (LEL) and an upper explosive limit (UEL). A mixture below the LEL is too "lean" to burn; a mixture above the UEL is to "rich" to burn.

Flammable solid A solid, other than a blasting agent or explosive as defined in 24 CFR 1910.109 (A), that is likely to cause fire through friction, absorption of moisture, spontaneous chemical change, or retained heat from manufacturing or processing, or which can be ignited readily and when ignited burns so vigorously and persistently as to create a serious hazard. A substance is a flammable solid if, when tested by the method described in 16 CFR 1500.44, it ignites and burns with a self-sustained flame at a rate greater than one-tenth of an inch per second along its major axis.

Flash point The minimum temperature at which a liquid gives off a vapor in sufficient concentration to ignite when tested by the following methods:

1. Tagliabue Closed Tester (see American National Standard Method of Test for Flash Point by Tag Closed Tester, Z11.24 1979 [ASTM D5-79]) for liquids with a viscosity of less than 45 Saybolt Universal Seconds (SUS) at 100°F (37.8°C), that do not have a tendency to form a surface film under test; or

2. **Pensky-Martens Closed Tester** (see American National Standard Method of Test for Flash Point by Pensky-Martens Closed Tester, Z11.771979 [ASTM D9-79]) for liquids with a viscosity equal to or greater than 45 SUS at 100°F (37.8°C), or that contain suspended solids, or that have a tendency to form a surface film under test; or

3. **Setaflash Closed Tester** (see American National Standard Method of Test for Flash Point by Setaflash Closed Tester [ASTM D 3278-78]).

Organic peroxides, which undergo autoaccelerating thermal decomposition, are excluded from any of the flash point determination methods specified above.

For practical purposes, this temperature denotes the point at which vapors are emitted from a liquid and are ignitable in the presence of a flame or ignition source. It is probably the single most important fire term used in our study.

Forbidden A hazardous material that must not be offered or accepted for transportation.

Forensic toxicology The branch of toxicology concerned with medico-legal aspects of the harmful effects of chemicals on humans and animals.

Formula The scientific expression of the chemical composition of a material (e.g., water is H_2O, sulfuric acid is H_2SO_4, sulfur dioxide is SO_2).

Freelancing A term used to denote the actions of a person or group of individuals operating outside the organized plan of action.

Fume A solid condensation particle of extremely small diameter, commonly generated from molten metal as metal fume.

Gamma radiation This form of radiation is essentially electromagnetic radiation. The wavelength of gamma rays are shorter than x-rays and are emitted as photons. Photons are massless and are considered to be pure energy. *See also* Electromagnetic radiation.

Gas A substance whose physical state is such that it always occupies the whole of the space in which it is contained.

General exhaust A system for exhausting air containing contaminants from a general work area. *Also see* Local exhaust.

General staff The group of leaders heading the major divisions under the Incident Commander. The four general staff leaders are Operations, Logistics, Planning, and Finance.

Genetic Pertaining to or carried by genes; hereditary.

Gestation The development of the fetus from conception to birth.

g/kg Grams per kilogram is an expression of dose used in oral and dermal toxicology testing to denote grams of a substance dosed per kilogram of animal body weight. *Also see* kg (kilogram).

Grounding The procedure used to carry an electrical charge to ground through a conductive path. A typical ground may be connected directly to a conductive water pipe or to a grounding bus and ground rod. *See* Bonding.

Half-life The time required for the decay process to reduce the energy production to one half of its original value. This means that half the atoms are present, therefore, half the radioactive energy is occurring. This time period varies from isotope to isotope. Half-lives can range from millionths of a second to more than a million years.

Hand protection Specific types of gloves or other hand coverings required to prevent harmful exposure to hazardous materials.

Hazard warning Words, pictures, symbols, or combination thereof presented on a label or other appropriate form to inform of the presence of various materials.

Hazardous atmosphere An atmosphere that may expose employees to the risk of death, incapacitation, impairment of ability to self-rescue (for example, escape unaided from a permit required confined space), injury, or acute illness from one or more of the following causes:

1. Flammable gas, vapor, or mist in excess of 10% of its lower flammable limit (LFL);

2. Airborne combustible dust at a concentration that meets or exceeds its LFL;

Note: This concentration may be approximated as a condition in which the dust obscures vision at a distance of 5 feet (1.52m) or less.

3. Atmospheric oxygen concentration below 19.5% or above 23.5%;

4. Atmospheric concentration of any substance for which a dose or a permissible exposure limit is published in Subpart G, Occupational Health and Environmental Control, or in Subpart Z, Toxic and Hazardous Substances, of this part and which could result in employee exposure in excess of its dose or permissible limit;

Note: An atmospheric concentration of any substance that is not capable of causing death, incapacitation, impairment of ability to self-rescue, injury, or acute illness due to its health effects is not covered by this provision.

5. Any other atmospheric condition that is immediately dangerous to life or health.

Note: For air contaminants for which OSHA has not determined a dose or permissible exposure limits, other sources of information, such as material safety data sheets that comply with the Hazard Communication Standard 1910.1200 of this part, published information, and internal documents can provide guidance in establishing acceptable atmospheric conditions.

Hazardous chemical Any chemical whose presence or use is a physical hazard or a health hazard. For purposes of the OSHA Laboratory Standard, a chemical for which there is significant evidence, based on at least one study conducted in accordance with established scientific principles, that acute or chronic health effects may occur in exposed employees. The term *health hazard* includes chemicals that are carcinogens, toxic or highly toxic agents, reproductive toxins, irritants, corrosives, sensitizers, hepatotoxins, nephrotoxins, neurotoxins, agents which act on the hematopoietic system, and agents that damage the lungs, skin, eyes, or mucous membranes.

Hazardous substance Any substance designated under the Clean Water Act and the Comprehensive Environmental Response, Compensation, and Liability Act (CERCLA) as posing a threat to waterways and the environment when released.

Hazardous wastes Discarded materials regulated by the Environmental Protection Agency because of public health and safety concerns. Regulatory authority is granted under the Resource Conservation and Recovery Act.

HAZWOPER Hazardous Waste Operations and Emergency Response.

Heat stroke A serious medical condition caused by the inability of the body to regulate its temperature. Usually brought on by high temperature environmental conditions. Characterized by the cessation of sweating, hot dry skin, and the ultimate collapse of the victim. Heat stroke can be fatal.

Hemoglobin An iron-containing compound responsible for binding oxygen and transporting it to the cells. Hemoglobin is found in the red blood cells.

HEPA High efficiency particulate air.

Hepatotoxin A substance that causes injury to the liver.

Heroic dose An expression of exposure levels used in animal testing. Heroic doses are referred to as the maximum tolerated doses of the test chemical. Essentially, it is the highest non lethal dose the animal can tolerate during the testing process.

Heterogeneous A substance that has parts that are unlike or without interrelation; having composition that differs from sample point to sample point.

HF Hydrofluoric acid.

Highly toxic Defined by OSHA as a chemical within any of the following categories:

1. A chemical with a median lethal dose (LD_{50}) of 50 milligrams or less per kilogram of body weight when administered orally to albino rats weighing between 200 and 300 grams each.
2. A chemical with a median lethal dose (LD_{50}) of 200 milligrams or less per kilogram of body weight when administered by continuous contact for 24 hours (or less if death occurs within 24 hours) with the bare skin of albino rabbits weighing between 2 and 3 kilograms each.
3. A chemical that has a median lethal concentration (LC_{50}) in air of 200 parts per million by volume or less of gas or vapor, or 2 milligrams per liter or less of mist, fume, or dust, when administered by continuous inhalation for 1 hour (or less if death occurs within 1 hour) to albino rats weighing between 200 and 300 grams each.

Homogeneous A substance that displays a uniform composition throughout.

Hot work permit The employer's written authorization to perform operations (e.g., riveting, welding, cutting, burning, and heating) capable of providing a source of ignition.

Hydrolysis A chemical decomposition of a substance by water. The products of the decomposition results in the formation of two or more new substances.

Hygroscopic The ability of a substance to absorb moisture from the air.

Hyperbaric chamber A pressure chamber that places the victim in a high oxygen environment under pressure. Pressure inside the chamber is usually in the range of 2–3 atmospheres absolute.

Hypergolic Substances that spontaneously ignite on contact with another. Many hypergolic materials are used as rocket fuels.

Hyperkalemia A medical condition characterized by an abnormally high level of potassium in the body.

Hyperthermia An increase in the body temperature caused by heat transfer from the external environment.

Hypocalcemia A condition characterized by abnormally low levels of calcium in the blood.

Hypotension A blood pressure with a systolic value below 100 mm Hg, as measured with a blood pressure cuff and stethoscope.

Hypothalamus A gland located in the brain that regulates many bodily functions such as metabolic rate and other functions of the endocrine system.

Ignition temperature The minimum temperature to which a fuel in air must be heated in order to start self-sustained combustion independent of the heating source.

Immediately Dangerous to Life and Health (IDLH) The term is usually expressed in parts per million and reflects the atmospheric level of any toxin, corrosive, or asphyxiant that poses a danger to life or would cause irreversible or delayed adverse health effects or would impair the ability of an individual to escape the area. In many cases this value is based on an exposure of 30 minutes.

Impervious A material that does not allow another substance to pass through or penetrate it.

Incident commander The person responsible for all operations at a hazardous materials emergency.

Incident command system (ICS) A management tool used by responders to facilitate an incident. The system includes such features as common terminology, standardized communications, and modular structure. The incident command system (ICS) is flexible and can be utilized on incidents of any size.

Incompatible Materials that could cause dangerous reactions by direct contact with one another.

Inerting The displacement of the atmosphere in an area such as a permit required confined space by a noncombustible gas (such as nitrogen) to such an extent that the resulting atmosphere is noncombustible. *Note:* This procedure often produces an IDLH oxygen-deficient atmosphere.

Ingestion The act of taking in by the mouth.

Inhalation Breathing in of a substance in the form of a gas, vapor, fume, mist, or dust.

Inhibitor A chemical added to another substance to prevent an unwanted chemical change.

Injection A route of entry that a chemical may take to enter the body. Injection can happen when chemicals enter through cuts or other breaches of the skin. Injection also occurs when chemicals are forced through the skin by air pressure or other forceful means.

Insoluble Incapable of being dissolved in a liquid.

International Agency for Research on Cancer (IARC): One of the leading agencies that list and identify carcinogens and suspected carcinogens.

Ion An atom or molecule that has acquired a positive or negative charge by gaining or losing an electron. The atom or molecule is no longer considered neutral in this state.

Ionization The formation of ions. This ion formation occurs when a neutral molecule of an inorganic solid, liquid, or gas undergoes a chemical change. These highly energetic, short wavelength rays are capable of causing mutations in cell nuclei and DNA. These changes in the cell structures of the body may cause cancer or other long-term disease processes.

Ionizing radiation Radiation capable of causing the ionization of solids, liquids, or gases either directly or indirectly. This process can be caused by alpha and beta or gamma radiation.

IPA Isopropyl alcohol.

Irritant A chemical, that is not corrosive but that causes a reversible inflammatory effect on living tissue by chemical action at the site of contact. A chemical is a skin irritant if, when tested on the intact skin of albino rabbits by the methods of 16 CFR 1500.41 for 4 hours exposure or by other appropriate techniques, it results in an empirical score of 5 or more. A chemical is an eye irritant if so determined under the procedure listed in 16 CFR 1500.42 or other appropriate techniques.

Irritating An irritating material, as defined by DOT, is a liquid or solid substance which, upon contact with fire or when exposed to air, gives off dangerous or intensely irritating fumes (not including poisonous materials). *See* Poison, Class A and Poison, Class B.

Isolation The process by which an area such as a permit required confined space is removed from service and completely protected against the release of energy and material into the area by such means as blanking or blinding, or removing sections of lines, pipes, or ducts; a double block and bleed system; lockout or tag-out of all sources of energy; or blocking or disconnecting all mechanical linkages.

Isotonic Of equal tension as in a solution

containing a like concentration of electrolytes or nonelectrolytes exerting the same osmotic pressure as the solution with which it is compared.

Isotope One of two or more types of atoms of an element. These atoms have the same atomic number but a differing number of neutrons. Uranium 238 as compared to uranium 235 is an example of an isotope. U-238 has 92 protons and 146 neutrons. U-235 has 92 protons and 143 neutrons. *See* Radioactive isotope.

kg Kilogram, a metric unit of weight, about 2.2 U.S. pounds. *Also see* g/kg, and mg.

Kinetic energy The actual energy produced by the action of the equipment or process.

L Liter is a metric unit of capacity. A U.S. quart is about 9/10 of a liter.

Lab packing A packing procedure performed when several smaller containers such as 1-gallon glass bottles, or cans, need to be stabilized for transport.

Label Notice attached to a container, bearing information concerning its contents.

Laboratory A facility where the "laboratory use" of hazardous chemicals occurs. It is a workplace where relatively small quantities of hazardous chemicals are used on a nonproduction basis.

Laboratory-type hood A device located in a laboratory, enclosed on five sides with a moveable sash or fixed partial enclosure on the remaining side, constructed and maintained to draw air from the laboratory and to prevent or minimize the escape of air contaminants into the laboratory. A hood allows a worker to conduct chemical manipulations in the enclosure without inserting any portion of the body other than hands and arms. Walk-in hoods with adjustable sashes meet the above definition provided that the sashes are adjusted during use so that the airflow and the exhaust of air contaminants are not compromised and employees do not work inside the enclosure during the release of airborne hazardous chemicals.

Lactic acid An acid formed in the muscles during activity by the breakdown of glycogen.

Land breeze Occurs when the land mass cools faster than the adjacent body of water and creates an offshore breeze.

LC Lethal concentration, the concentration of a substance being tested that is expected to kill. It is a measurement of the material in the air and is expressed in either ppm or ppb.

LClo Lethal concentration low; lowest concentration of a gas or vapor as measured in the air that is capable of killing a specified species over a specified time.

LC_{50} The concentration of a material in air that is expected to kill 50% of a group of test animals with a single exposure (usually 1 to 4 hours). The LC_{50} is expressed as parts of material per million parts of air, by volume (ppm) for gases and vapors, or as micrograms of material per liter of air (mcg/l) or milligrams of material per cubic meter of air (mg/m^3) for dusts and mists, as well as for gases and vapors.

LD Lethal dose is the quantity of a substance being tested that will kill. It is a measurement of the actual dose taken in via ingestion, injection, or dermal exposure and is expressed in mg/kg.

LDhi The concentration of a chemical by dermal contact or absorption that will be fatal to 100% of a test population

LDlo Lethal dose low; lowest administered dose of a material capable of killing a specified test species.

LD_{50} A single dose of a material expected to kill 50% of a group of test animals. The LD_{50} dose is usually expressed as milligrams or grams of material per kilogram of animal body weight (mg/kg or g/kg). The material may be administered by mouth or applied to the skin.

Limited quantity The maximum amount of hazardous material, as specified in those sections applicable to the particular hazard class, for which there are specific exceptions from the requirements of this subchapter. See Sec. 173.118, 173.118(a), 173.153. 173.244, 173.306, 173.345, and 173.364 of 49 CFR.

Local An effect of a chemical exposure limited to the area of contact.

Local exhaust A system for capturing and exhausting contaminants from the air at the point where the contaminants are produced (welding, grinding, sanding, or other processes or operations). *Also see* General exhaust.

Lower explosive limit or lower flammable limit (LEL or LFL) The lowest concentration (lowest percentage of the substance in air) that will produce a flash of fire when an ignition source (heat, arc, or flame) is present. At concentrations lower than the LEL, the mixture is too "lean" to burn. *Also see* Upper explosive limit.

M Meter, a unit of length in the metric system. One meter is about 39 inches.

m^3 Cubic meter, a metric measure of volume, approximately 35.3 cubic feet or 1.3 cubic yards.

Malignant Tending to become progressively worse and to result in death.

Malleable Substances exhibiting the properties of flexibility; the ability to bend or be hammered into thin sheets. Most metallic elements are considered to be malleable.

Matter Anything that has mass and occupies space. Matter is generally found in three basic forms: solid, liquid, and gas. Most of the subatomic particles that make up matter are invisible to the naked eye.

Mechanical exhaust A powered device, such as a motor-driven fan or air steam venturi tube, for exhausting contaminants from a workplace, vessel, or enclosure.

Mechanical filter respirator A respirator used to protect against airborne particulate matter like dusts, mists, metal fume, and smoke. Mechanical filter respirators do not provide protection against gases, vapors, or oxygen-deficient atmospheres.

Melting point The temperature at which a solid substance changes to a liquid state.

Metabolic acidosis A medical condition characterized by the production of lactic acid within the body.

Metabolism Physical and chemical processes taking place among the ions, atoms, and molecules of the body. Metabolism is an important element in the study of how the toxic materials that we take into our bodies are processed by it.

mg Milligram, a metric unit of weight that is one-thousandth of a gram.

mg/kg Milligrams of substance per kilogram of body weight is an expression of toxicological dose.

mg/m^3 Milligrams per cubic meter is a unit for expressing concentrations of dusts, gases, or mists in air.

Micron Micrometer is a unit of length equal to one-millionth of a meter. A micron is approximately 1/23,000 of an inch.

Miscibility *See* Solubility in water.

Mist Suspended liquid droplets generated by condensation from the gaseous to the liquid state, or by breaking up a liquid into a dispersed state, such as splashing, foaming, or atomizing. Mist is formed when a finely divided liquid is suspended in air.

Mitigation An operational term used in the industry to describe the process of bringing a chemical release or emergency under control.

Mixture Any combination of two or more chemicals if the combination is not, in whole or part, the result of a chemical reaction.

ml Milliliter is a metric unit of capacity, equal in volume to one cubic centimeter (cc), or approximately one-sixteenth of a cubic inch; one-thousandth of a liter.

mm Hg Millimeters (mm) of mercury (Hg) is a unit of measurement for gas pressure.

Molecular weight Weight (mass) of a molecule based on the sum of the atomic weights of the atoms that make up the molecule.

mppcf Million particles per cubic foot is a unit for expressing concentration of particles of a substance suspended in air. Exposure limits for mineral dusts (silica, graphite, Portland cement, nuisance dusts, and others), formerly expressed as mppcf, are now more commonly expressed in mg/m.

MSDS Material safety data sheet. The written information on a specific chemical compound that expresses such items as physical hazards, signs and symptoms of exposure, toxicology information, and other pertinent data.

Mutagen A substance or agent capable of altering the genetic material in a living cell.

Mutual aid Predetermined agreements between jurisdictions that allow for shared resources in the event of a large-scale incident.

National Cancer Institute (NCI) That part of the National Institutes of Health that studies cancer causes and prevention as well as diagnosis, treatment, and rehabilitation of cancer patients.

National Fire Protection Association (NFPA) An international membership organization that promotes and improves fire protection and prevention and establishes safeguards against loss of life and property by fire. Best known on the industrial scene for the National Fire Codes, thirteen volumes of codes, standards, recommended practices, and manuals developed (and regularly updated) by NFPA technical committees. Among these are NFPA 704, the code for showing hazards of materials as they might be encountered under fire or related emergency conditions, using the familiar diamond-shaped label or placard with appropriate numbers or symbols, and NFPA 471 and 472 that cover practices for hazardous materials incidents and procedures for responding to hazardous materials incidents.

National Institute for Occupational Safety and Health (NIOSH) A government agency under the Department of Health and Human Services that is responsible for investigating the toxicity of workroom environments and all other matters relating to safe industrial practice. NIOSH publishes the *Pocket Guide to Chemical Hazards,* which is an excellent source of health hazards relating to hazardous materials. Other activities that NIOSH is involved in includes testing and certifying respiratory protective devices and air sampling detector tubes.

National Pesticide Information Retrieval System (NPIRS) An automated database operated by Purdue University containing information on EPA registered pesticides, including reference file MSDSs.

National Response Center (NRC) A notification center that must be called when significant oil or chemical spills or other environment-related accidents occur. The toll-free telephone number is 1-800-424-8802.

National Toxicology Program (NTP) The NTP publishes an Annual Report on Carcinogens.

Nausea Tendency to vomit, feeling of sickness at the stomach.

Neonatal The first four weeks after birth.

Nephrotoxin A substance that causes damage or injury to the kidneys.

Neurotoxin A material that affects the nerve cells and may produce emotional or behavioral abnormalities.

Neutralization A neutralization reaction happens when a mutual reaction occurs between an acid and a base. The products of the reaction include a salt compound, water, and heat.

Neutron An elementary subatomic particle existing

in the nucleus having a mass of 1.009 amu's (atomic mass units). The neutron has no electrical charge and exists in the nucleus of all atoms except hydrogen. This atom is comprised of one proton and one electron. To find the amount of neutrons in an atom, subtract the atomic number from the atomic weight. The remainder will be the number of neutrons.

ng Nanogram, one billionth of a gram.

Nonentry rescue The process whereby the attendant remotely initiates the rescue of the entrant who is unable to leave the space.

Nonflammable Not easily ignited, or if ignited, not burning rapidly.

Nonflammable gas Any compressed gas other than a flammable compressed gas.

Non liquefied gases A gas other than a gas in solution that under the charging pressure is entirely gaseous at 70°F (21°C).

Non-permit confined space A confined space that does not contain or, with respect to atmospheric hazards, have the potential to contain any hazard capable of causing death or serious physical harm.

Nonsparking tools Tools made from beryllium-copper or aluminum-bronze greatly reduce the possibility of igniting dusts, gases, or flammable vapors. Although these tools may emit some sparks when striking metal, the sparks have a low heat content and are not likely to ignite most flammable liquids.

NOx Oxides of nitrogen, which are undesirable air pollutants. NOx emissions are regulated by the EPA under the Clean Air Act.

NPL National Priority Site List.

Occupational Safety and Health Administration (OSHA) On both the federal and state levels, OSHA is responsible for establishing and enforcing standards for exposure of workers to harmful materials in industrial atmospheres and other matter affecting the health and well-being of industrial workers. Federal OSHA is part of the U.S. Department of Labor.

Odor A description of the smell of the substance.

Odor threshold The lowest concentration of a substance's vapor, in air, that can be smelled.

Office of Federal Agency Programs (OFAP) The organizational unit of OSHA that provides federal agencies with guidance to develop and implement occupational safety and health programs for federal employees.

Olfactory Relating to the sense of smell.

Operational goal(s) The end point of the incident as identified by the IC. Strategy and tactics are based on the desired operational goal. Everyone on the incident should understand the goal in order to reduce freelancing.

Oral Used in or taken into the body through the mouth.

Oral toxicity Adverse effects resulting from taking a substance into the body by mouth. Ordinarily used to denote effects in experimental animals.

Organic peroxide An organic compound that contains the bivalent —O—O— structure and may be considered a structural derivative of hydrogen peroxide where one or both of the hydrogen atoms have been replaced by an organic radical.

Organogenesis The development of tissues into different organs in embryonic development. This period begins approximately 18 days into the pregnancy in humans.

ORMS A class of materials used by the Department of Transportation that does not meet the definition of a hazardous material but poses some risk when transported in commerce. The materials are broken into classifications called ORM-A, ORM-B, ORM-C, ORM-D, and ORM-E. These terms are used less with the new DOT regulations.

ORM-A: A material that has an anesthetic,

irritating, noxious, toxic, or other similar property and that can cause extreme annoyance or discomfort to passengers and crew in the event of leakage during transportation. 49 CFR, Sec 173.500(b)(1))

ORM-B: A material (including a solid when wet with water) capable of causing significant damage to a transport vehicle from leakage during transportation. Materials meeting one or both of the following criteria are ORM-B materials: (1) A liquid substance that has a corrosion rate exceeding 0.250 inch per year (IPY) on aluminum (nonclad 7075-T6) at a test temperature of 130°F.

ORM-C: A material that has other inherent characteristics not described as an ORM-A or ORM-B but that make it unsuitable for shipment, unless properly identified and prepared for transportation. Each ORM-C material is specifically named in Sec. 172.101. (Sec. 173.500(b)(4)) of 49 CFR.

ORM-D: A material such as a consumer commodity which, though otherwise subject to the regulations of this subchapter, presents a limited hazard during transportation due to its form, quantity, and packaging. They must be materials for which exceptions are provided in Sec. 172.101. A shipping description applicable to each ORM-D material or category of ORM-D materials is found in Sec. 172.101 (Sec. 173.500(b)(4)) of 49 CFR.

ORM-E: A material that is not included in any other hazard class, but is subject to the requirements of this subchapter. Materials in this class include hazardous wastes and hazardous substances as defined in Sec. 171.8 of 49 CFR.

Orthostatic hypotension A decrease in blood pressure when quickly changing from a supine to a sitting or to a standing position.

overboots Disposable covers slipped over heavier rubber boots.

Oxidation A chemical reaction that brings about an oxidation reaction. An oxidizing agent may (1) provide the oxygen to the substance being oxidized (in which case the agent has to be oxygen or contain oxygen), or (2) it may receive electrons being transferred from the substance undergoing oxidation. Chlorine is a good oxidizing agent for electron-transfer purposes, even though it contains no oxygen.

Oxidizer A chemical other than a blasting agent or explosive that initiates or promotes combustion in other materials, causing fire either by itself or through the release of oxygen or other gases.

Oxygen-deficient atmosphere An atmosphere containing less than 19.5% oxygen by volume.

Oxygen-enriched atmosphere An atmosphere containing more than 23.5% oxygen by volume.

Particulate matter A solid or liquid matter that is dispersed in a gas, or insoluble solid matter dispersed in a liquid. The prime hazard of particulate matter is inhalation, along with the possibility of the matter lodging in the lung tissue. Asbestos fibers are especially dangerous when captured by lung tissues.

Pathogen A disease-causing agent.

PCB An abbreviation for polychlorinated biphenyl. This chemical compound was commonly used as a cooling agent in electrical transformers. It is an aromatic hydrocarbon compound consisting of two benzene nuclei combined with two chlorine atoms. This compound is highly toxic.

Pensky Martens Closed Cup (PMCC) *See* Flash point.

Periodic table An arrangement of the elements in such a form as to emphasize the similarities of their physical and chemical properties.

Permanent gas A gas that cannot be liquefied by pressure alone.

Glossary

Permeation The slow movement of a chemical from the outside to the inside of the particular item of PPE.

Permissible exposure limit (PEL) Permissible exposure limit is an exposure limit established by OSHA's regulatory authority. It is generally the 8-hour time-weighted average (TWA) limit, but also could be expressed as the maximum concentration exposure limit. The PEL is usually established for airborne hazards.

Permit required confined space A recognized confined space requiring permits and formal operations such as rescue, air monitoring and training prior to entry.

Personal protective equipment (PPE) The correct clothing and respiratory equipment needed to perform a job involving hazardous materials and protect the worker. PPE includes proper boots, gloves, splash protective clothing, gas protective clothing, Tyvek suits, eye protection, hearing protection, air purifying respirators (which *see*) and air supplying respirators (*see* Self-contained breathing apparatus). It is important that all PPE be used properly and when required.

pH The symbol relating the hydrogen ion concentration to that of a given standard solution. A pH of 7 is neutral. Numbers increasing from 7 to 14 indicate greater alkalinity. Numbers decreasing from 7 to 0 indicate greater acidity.

Photophobia Intolerance of the eye to light.

Physical hazard Describes a chemical for which there is scientifically valid evidence that is it a combustible liquid, a compressed gas, explosive, flammable, an organic peroxide, an oxidizer, pyrophoric, unstable (reactive), or water reactive.

Placards 10 3/4-inch square diamond markers required on transporting vehicles, such as trucks, rail cars, or freight containers 640 cubic feet or larger.

Pneumoconiosis A condition of the lung in which there is permanent deposition of particulate matter and tissue reaction to its presence. It may range from relatively harmless forms of iron oxide deposition to destructive forms of silicosis.

Poison A material falling into one of the following two categories.

> Class A: A DOT term for extremely dangerous poisons, poisonous gases, or liquids that, in very small amounts, either as gas or as vapor of the liquid, mixed with air, are dangerous to life. Examples: phosgene, cyanogen, hydrocyanic acid, nitrogen peroxide.
>
> Class B: A DOT term for liquid, solid, paste, or semisolid substances, other than Class A poisons or irritating materials, that are known (or presumed on the basis of animal tests) to be so toxic to humans that they are hazardous to health during transportation.

Polar The description of a molecule where the positive and negative charges are permanently separated. This differs from nonpolar substances in which the electrical charges may coincide. Polar molecules will ionize in water and conduct an electrical current. Water is the most common polar substance. Nonpolar substances include gasoline, diesel fuel, and most hydrocarbons. If a substance is polar, it will mix with water.

Polyethylene A thermoset plastic that is lightweight, strong, and highly chemical resistant.

Polymerization A chemical reaction in which one or more small molecules combine to form larger molecules. A hazardous polymerization is such a reaction that takes place at a rate that releases large amounts of energy. If hazardous polymerization can occur with a given material, the MSDS usually lists conditions that could start the reaction and, since the material usually contains polymerization inhibitors, the length of time during which the inhibitor will be effective.

Polyvinyl chloride (PVC) A synthetic thermoplastic polymer resistant to most acids, fats, and oils. A prime hazard is the decomposition of PVC at temperatures above 148°C (298°F) which can produce hydrogen chloride fumes.

Positive pressure A characteristic of supplied air systems to maintain a few PSI of air pressure inside the face piece even after the wearer has inhaled. The advantage of positive pressure is that it helps to prevent airborne contaminants from entering through ineffectively sealed masks.

Potential energy The energy stored within the source and available if released.

Potentiate The ability of one substance to interact with, and make more hazardous, another substance.

Pounds per square inch (PSI) Pounds per square inch (for MSDS purposes) is the pressure a material exerts on the walls of a confining vessel or enclosure. For technical accuracy, pressure must be expressed as psig (pounds per square inch gauge) of psia (pounds per square inch absolute). Absolute pressure is gauge pressure plus sea level atmospheric pressure, or psig plus approximately 14.7 pounds per square inch. *Also see* mm Hg.

ppb Parts per billion is the concentration of a gas or vapor in air, parts (by volume) of the gas or vapor in a billion parts of air. Usually used to express extremely low concentrations of unusually toxic gases or vapors; also the concentration of a particular substance in a liquid or solid.

ppm Parts per million is the concentration of a gas or vapor in air, parts (by volume) of the gas or vapor in air, parts (by volume) of the gas or vapor in a million parts of air; also the concentration of a particular substance in a liquid or solid.

Pressure *See* Pounds per square inch.

Prohibited condition Any condition in a permit required confined space that is not allowed by the permit during the period when entry is authorized.

Proton A basic subatomic particle existing in the nucleus of all atoms. A proton has mass and an atomic weight of 1 amu (atomic mass unit). The number of protons is also expressed as the atomic number for a given element. Carbon has an atomic number of 6 which means it has 6 protons in the nucleus.

Proximity suit A type of firefighting gear used when operating in areas of extreme heat.

Pulmonary Relating to, or associated with, the lungs.

Pulmonary edema Fluid in the lungs.

Pyrophoric A chemical that ignites spontaneously in air at a temperature of 130°F (54.4°C) or below.

Radiation safety officer This person is in charge of the program that handles radiation-related issues at fixed facilities. This person develops training programs, safe handling procedures, and emergency response information.

Radiation sickness A group of symptoms associated with a radiation exposure. These symptoms run from mild to extreme and include nausea, vomiting, and malaise.

Radioactive isotope Also referred to as *radioisotope*. A radioactive isotope of any element. These isotopes can be naturally occurring or artificially created by bombarding an atom with neutrons. *See* Isotope.

Radioactive materials Materials that spontaneously emit ionizing radiation from the nucleus of an atom. *See* Ionizing radiation.

RBC Red blood cell.

Reaction A chemical transformation or change.

The interaction of two or more substances to form new substances.

Reactive *See* Unstable.

Reactivity Chemical reaction with the release of energy. Undesirable effects, such as pressure buildup, temperature increase, formation of noxious, toxic, or corrosive by-products, may occur because of the reactivity of a substance to heating, burning, direct contact with other materials, or other conditions in use or in storage.

Reducing agent In a reduction reaction (which always occurs simultaneously with an oxidation reaction) the reducing agent is the chemical or substance that (1) combines with oxygen or (2) loses electrons in the reactions. *See* Oxidation.

Relative humidity An expression of the amount of water vapor in the air.

Reproductive toxin A substance that affects either male or female reproductive systems and may impair the ability to have children; the term includes chromosomal damage (mutagenesis) and effects on fetuses (teratogenesis).

Resource Conservation and Recovery Act (RCRA) Environmental legislation aimed at controlling the generation, treatment, storage, transportation, and disposal of hazardous wastes. It is administered by EPA.

Respiratory protection Devices that protect the wearer's respiratory system from exposure to airborne contaminants by inhalation. Respiratory protection is used when a worker must work in an area where he or she might be exposed to concentrations in excess of the allowable exposure limit.

Respiratory system The breathing system that includes the lungs and the air passages (trachea, larynx, mouth, and nose) to the air outside the body, plus the associated nervous and circulatory supply.

Retention factor(s) Conditions of a chemical exposure that largely determine how the body will respond. These factors include the inherent toxicity of the chemical agent as well as the physical well-being and metabolism of the person exposed.

Retrieval system The equipment (including a retrieval line, chest or full body harness, wristlets if appropriate, and a lifting device or anchor) used for nonentry rescue of persons from permit spaces.

Routes of entry The means by which materials may gain access to the body, for example, inhalation, ingestion, injection, and absorption.

Saponification A reaction between an alkali and a fatty substance that forms water soluble byproducts and a soap.

Sea breeze Same as onshore breeze.

Select carcinogen A chemical that meets one of the following criteria:

(1) it is regulated by OSHA as a carcinogen; (2) it is listed under the category "known to be carcinogens" in the National Toxicology Program's latest Annual Report on Carcinogens; (3) it is listed under Group 1 ("carcinogenic to humans") by the International Agency for Research on Cancer Monographs; or (4) it is listed in Group 2A or 2B by IARC or under the category "reasonably anticipated to be carcinogens" by NTP, and causes statistically significant tumor incidence in experimental animals after inhalation exposure of 6–7 hours per day, 5 days per week, for a significant portion of a lifetime to dosages of less than 10 mg/m^3; or after repeated skin application of less than 300 mg/kg of body weight per week; or after oral dosages of less than 50 mg/kg of body weight per day.

Self-contained breathing apparatus (SCBA) A respiratory protection device that consists of a supply or a means of usable air, oxygen, or oxygen-generating material, carried by the wearer. If an air purifying respirator cannot be used due to one of the conditions listed under APR, then a worker must be protected with a SCBA. Sometimes known as Scott Packs (a brand name), SCBA's and air line respirators deliver air to the user from a tank (SCBA) or an air line. Although they negate all of the problems of an APR, they still have problems of their own. Air line systems are bulky and more difficult to work with due to the attached air line. SCBA's are heavy, more expensive, require more training, and reduce the mobility of the worker. Both, however, have the advantage of having positive pressure in the face mask which greatly reduces the risk of exposure if there is a leak in the mask.

Sensitizer A chemical that causes a substantial proportion of exposed people or animals to develop an allergic reaction in normal tissue after repeated exposure to the chemical.

Setaflash Closed Tester (SETA) *See* Flash point.

Shipping papers A shipping order, bill of lading, manifest, waybill, or other shipping document issued by the carrier.

shore breeze Occurs when the adjacent body of water cools faster than the surrounding land mass creating an onshore breeze.

Silicosis A disease of the lungs caused by the inhalation of silica dust.

Site access control officer (SAC) This function controls the movement of personnel and equipment into and out of the contaminated areas. In many cases, the SAC reports to the Hazardous Materials Group Supervisor.

Size-up A mental process used to evaluate the incident. Size-up involves gathering as much information as quickly as you can, but realizing that the incident is not going to be put on "hold" until you complete this.

Skasol Trade name for a corrosive etch. Primary component is hydrochloric acid typically in a 17% concentration.

Skin absorption Ability of some hazardous chemicals to pass directly through the skin and enter the bloodstream.

Skin toxicity *See* Dermal toxicity.

SLUD A mnemonic denoting the effects of exposure to certain types of pesticides (mostly organophosphates and carbamates). It stands for salivation, lacrimation, urination, and defecation.

Solubility in water A term expressing the percentage of a material (by weight) that will dissolve in water at ambient temperature. Solubility information can be useful in determining spill cleanup methods and reextinguishing agents and methods for a material.

Solvent A substance, usually a liquid, in which other substances are dissolved. The most common solvent is water.

SOx Oxides of sulfur.

Span of control The maximum number of people that one can effectively supervise at a given time. In emergency response a 5:1 ratio is acceptable.

Spanning This action allows the user to set an approximate calibration for a certain number of chemical compounds.

Specific chemical identity The chemical name, Chemical Abstracts Service (CAS) Registry Number, or any precise chemical designation of a substance.

Specific gravity The weight of a material compared to the weight of an equal volume of water is an expression of the density (or heaviness) of a material. Insoluble materials with specific gravity of less than 1.0 float on water. Insoluble materials with specific gravity greater than 1.0 sink in water. Most (but not all) flammable liquids have specific gravity less than 1.0 and, if not soluble, will float on water, an important consideration for fire suppression.

Spontaneously combustible A material that ignites as a result of retained heat from processing, or oxidizes to generate heat and ignite, or absorbs moisture to generate heat and ignite.

Stability The ability of a material to remain unchanged. For MSDS purposes, a material is stable if it remains in the same form under expected and reasonable conditions or storage or use. Conditions that may cause instability (dangerous change) are stated on the MSDS; for example, temperatures above 150°F; shock from dropping.

Stable atom An atom that is not in the process of radioactive decay or the formation of an ion. Stable atoms have equal numbers of protons and electrons without an unusual imbalance of protons to neutrons.

stage/staging The act of identifying and controlling a specific area for the computation of resources.

state plan states The concept of the state plan states is that the various levels of government work together to enact regulations that are appropriate for every area of the country.

STEL A term that denotes one of the occupational exposure limits for workers. It stands for Short-term exposure limit and was developed by the ACGIH. The term represents the maximum exposure limit for workers based on a 15-minute exposure. This level would presumably be harmful if it were exceeded for more than this 15-minute exposure.

Strategy The overall theme or goal of the incident. Usually the strategy is either offensive or defensive.

Stratify Relating to layers or strata.

Stratum corneum A thin, clear layer of tissue that covers the pupil and iris.

Strong acid An acid with a pH of 2.5 or less.

Strong base A base with a pH of 12.5 or greater.

Subcutaneous Beneath the layers of skin.

Superfund Amendments and Reauthorization Act of 1986 (SARA) SARA was the actual forerunner to the HAZWOPER regulation in that it mandated the Occupational Safety and Health Administration (OSHA) to develop a worker protection regulation for those persons who work with hazardous wastes.

Supine A description of a person lying flat on their back; face up

Supplied air respirators (SAR) Air line respirators or self-contained breathing apparatus. This type of respirator differs from the air purifying respirator in that the air that is breathed by the individual does not come from the atmosphere in the area where work is performed. Air that is breathed with a SAR system comes either in an air bottle carried by the individual or an air line system that either uses bottles or compressors that supply air from an outside clean air source.

Support zone Also referred to as the cold zone. The area of no contamination where logistic and support functions are performed. No PPE above level D should ever be required in the cold zone.

Surface area The amount of area exposed from a given system. For example, the surface area of the skin is approximately 25 square feet. Surface area also greatly influences vapor production in spilled liquids. The larger the surface area, the greater the vapor production.

Syncope A transient loss of consciousness.

Synonym Another name or names by which a material is known. Methyl alcohol, for example, also is known as methanol or wood alcohol.

Systemic poison A poison that spreads throughout the body, affecting all or some of the body systems and organs. Its adverse effect is not localized in one spot or area. Carbon monoxide has systemic effects upon exposure.

Systemic toxicity Adverse effects caused by a

substance that affects the body in a general rather than local manner.

Tachycardia A pulse rate above 100 beats per minute.

Tactics Methods or procedures used to deploy various tactical units (resources) to achieve objectives.

Tag (Tagliabue) Closed Cup (TCC) *See* Flash point.

Target organ effects The following is a target organ categorization of effects that may occur, including examples of signs and symptoms and chemicals that have been found to cause such effects. These examples are presented to illustrate the range and diversity of effects and hazards found in the workplace and the broad scope employers must consider in this area, but are not intended to be all inclusive.

Hepatotoxins: Chemicals that produce liver damage. Signs and Symptoms: Jaundice; liver enlargement. Chemicals: Carbon tetrachloride; nitrosamines.

Nephrotoxins: Chemicals that produce kidney damage. Signs and Symptoms: Edema; proteinuria. Chemicals: Halogenated hydrocarbons; uranium.

Neurotoxins: Chemicals that produce their primary toxic effects on the nervous system. Signs and Symptoms: Narcosis; behavioral changes; decrease in motor functions. Chemicals: Mercury, carbon disulfide.

Agents that act on blood hematopoietic system: Chemicals that decrease hemoglobin function; deprive the body tissues of oxygen. Signs and Symptoms: Cyanosis; loss of consciousness. Chemicals: Carbon monoxide; cyanides.

Agents that damage the lung: Chemicals that irritate or damage the pulmonary tissue. Signs and Symptoms: Cough, tightness in chest, shortness of breath. Chemicals: Silica; asbestos.

Reproductive toxins: Chemicals that affect the reproductive capabilities, including chromosomal damage (mutations) and effects on fetuses (teratogenesis). Signs and Symptoms: Birth defects; sterility. Chemicals: Lead.

Cutaneous hazards: Chemicals that affect the dermal layer of the body. Signs and Symptoms: Defatting of the skin; rashes; irritation. Chemicals: Ketones; chlorinated compounds.

Eye Hazards: Chemicals that affect the eye or visual capacity. Signs and Symptoms: Conjunctivitis; corneal damage. Chemicals: Organic solvents; acids.

Target Organ Toxin: A toxic substance that attacks a specific organ of the body. For example, exposure to carbon tetrachloride can cause liver damage.

Teratogen A substance or agent that can cause malformations in the fetus of a pregnant female exposed to it.

Testing The process by which the hazards that may confront personnel such as entrants of a confined space are identified and evaluated. Testing includes specifying the tests that are to be performed in the permit space.

Threshold The dividing line between effect and no-effect levels of exposure.

Threshold Limit Value (TLV) A set of standards established by the American Conference of Governmental Industrial Hygienists for concentrations of airborne substances in workroom air. They are time-weighted averages based on conditions that it is believed workers may be repeatedly exposed to day after day without adverse effects. The TLV values are revised annually and provide the basis for the safety regulations of OSHA.

Glossary

Threshold Limit Value/Ceiling (TLV/C) Another term denoting an occupational exposure limit. This term represents the maximum level or amount of a substance that a worker can be exposed to at any given time. It is usually given in either ppm, ppb, or mg/m^3.

Threshold Limit Value/Time Weighed Average (TLV/TWA) The TLV/TWA is a term used to denote one of the occupational exposure limits for types of hazardous materials. Developed by the ACGIH, the term denotes the average amount of a material that the average worker can be exposed to in the course of a typical 8-hour day, 5-day work week. At or below this level, it is presumed that a worker would suffer no ill effects from exposure to a given substance. It is usually given in either ppm, ppb, or mg/m^3 of air and is for airborne substances.

Time Weighted Average (TWA) TWA exposure is the airborne concentration of a material to which a person is exposed, averaged over the total exposure time, generally the total workday (8 to 12 hours). *Also see* Threshold Limit Value/Time Weighted Average.

TNT Trinitrotoluene.

Tool drop An area designated inside the exclusion zone, just outside the decon corridor. Grossly contaminated tools and other equipment are placed here for later cleaning.

Topography The physical land features of an area.

Toxic A substance as defined by OSHA as falling within any of the following categories:

1. A substance that has median lethal dose (LD$_{50}$) of 50 milligrams or more per kilogram of body weight but not more than 500 milligrams per kilogram of body weight when administered orally to albino rats weighing between 200 and 300 grams each.

2. A substance with a median lethal dose (LD$_{50}$) of more than 200 milligrams or less per kilogram of body weight but not more than 1,000 milligrams per kilogram of body weight when administered by continuous contact (dermal) for 24 hours (or less if death occurs within 24 hours) with the bare skin of albino rabbits weighing between 2 and 3 kilograms each.

3. A chemical that has a median lethal concentration (LC$_{50}$) in air of more than 200 parts per million but not more than 2,000 parts per million by volume or vapor, or 2 milligrams per liter but not more than 20 milligrams per liter of mist, fume, or dust, when administered by continuous inhalation for 1 hour (or less if death occurs within 1 hour) to albino rats weighing between 200 and 300 grams each.

Toxic concentration low (TCL) The lowest concentration of a gas or vapor capable of producing a defined toxic effect in a specified test species over a specified time.

Toxic dose low (TDL) Lowest administered dose of a material capable of producing a defined toxic effect in a specified test species.

Toxic Substances Control Act (TSCA) Federal environmental legislation (administered by EPA) that regulates the manufacture, handling, and use of materials classified as "toxic substances."

Toxicologist A person trained to examine the nature of adverse health effects and to assess the probability of their occurrence.

Toxicology The study of the adverse effects of chemicals on living organisms.

Translucent A barrier that allows light to pass but is not clearly seen through.

TSDF Treatment Storage and Disposal Facility.

Tyvek Suit fabric made of lightweight paperlike substance.

U.S. Department of Health and Human Services (DHHS) Replaced U.S. Department of Health,

Education, and Welfare. NIOSH and the Public Health Service are part of DHHS.

The U.S. Department of Labor (DOL) OSHA and MSHA are part of DOL.

U.S. Department of Transportation (DOT) regulates transportation of chemicals and other substances.

Unstable Tending toward decomposition or other unwanted chemical change during normal handling or storage.

Unstable reactive A chemical that, in the pure state or as produced or transported, will vigorously polymerize, decompose, condense, or become self-reactive under conditions of shock, pressure, or temperature.

Upper explosive limit or upper flammable limit (UEL or UFL) The highest concentration (highest percentage of the substance in air) that will produce a flash of fire when an ignition source (heat, arc, or flame) is present. At higher concentrations, the mixture is too "rich" to burn. *Also see* Lower explosive limit or lower flammable limit.

Vapor The gaseous form of a solid or liquid substance as it evaporates.

Vapor density The weight of a vapor or gas compared to the weight of an equal volume of air is an expression of the density of the vapor or gas. Materials lighter than air have vapor densities less than 1.0. (For example, propane, hydrogen sulfide, butane, chlorine, and sulfur dioxide have vapor densities greater than 1.0.) All vapors and gases mix with air, but the lighter materials tend to rise and dissipate (unless confined). Heavier vapors and gases are likely to concentrate in low places, along or under floors, in sumps, sewers and manholes, in trenches and ditches, where they may create fire or health hazards.

Vapor pressure The pressure exerted by a saturated vapor above its own liquid in a closed container. When quality control tests are performed on products, the test temperature is usually 100°F, and the vapor pressure is expressed as pounds per square inch (psig or psia), but vapor pressures reported on MSDSs are in millimeters of mercury (mm Hg) at 68°F (20°C), unless stated otherwise.

Three facts are important to remember:

1. Vapor pressure of a substance at 100°F will always be higher than the vapor pressure of the substance at 68°F (20°C).
2. Vapor pressures reported on MSDSs in mm Hg are usually very low pressures; 760 mm Hg is equivalent to 14.7 pounds per square inch.
3. The lower the boiling point of a substance, the higher its vapor pressure.

Vascular Pertaining to the circulatory system; an area of the body with high blood flow or a large amount of vasculature. The vasculature of the body includes such structures as capillaries, vessels, arteries, and arterioles.

Ventilation *See* General exhaust; Local exhaust; Mechanical exhaust.

Viscosity The ability or the resistance of a liquid material to flow. The higher the viscosity, the less the material will flow. Maple syrup is a substance with a high viscosity. Water is less viscous and tends to flow easier. Viscosity will assist you in determining if a liquid has a high risk of spreading out and potentially finding its way to a storm drain or sewer before adequate containment or cleanup procedures can be initiated.

Vital signs A set of evaluations reflective of a patient's general medical status including a measure of pulse rate, respiratory rate, blood pressure, pupil status, and skin signs.

Warm zone *See* Contamination reduction zone.

Warning property A physical characteristic of a chemical compound that can be detected in some fashion by humans; could include smell, taste, eye irritation.

Water-reactive Any solid substance (including sludges and pastes) which, by interaction with water, is likely to become spontaneously flammable or to give off flammable or toxic gases in dangerous quantities.

Water solubility The ability of a liquid or solid to mix with or dissolve in water.

Waybill The shipping paper used by the railroads indicating origin, destination, route, and product. There is a waybill for each car and they are carried by the conductor.

Zeroing A term used to express the action of bringing a monitoring device to a baseline or zero readout on gauges or digital readouts. It is much like ensuring that a scale is balanced before anything is weighed.

Zones Although zoning is usually applied to hazardous materials spills and isolation, setting up work zones such as hot or exclusionary zone assists any task involving hazardous materials. By isolating the work area and letting other personnel know of the hazards, an employer can reduce the possibility of other workers being injured or exposed.

Index

Absorbents, 277
Absorption, 152–155
Accountability log, 285
ACGIH (American Conference of Governmental Industrial Hygienists), 161
Acids, 124–128
   acidic concentration, 126–128
   acidic strength, 125–126
Acronyms, 399–401
Acute exposure, 158–159
Additive chemical interactions, 146
Adiabatic heat of compression, 77
Afterdamp, 344
Air line respirators, 229–230
Air monitoring, 343–362, 373–374
   detection of gases, 347–349
   effects of local weather, 359–360
   gas laws, 345–347
   general procedures, 357
   monitoring in confined spaces, 358, 373–374
   monitoring outside, 358–359
   sequence of air monitoring, 349
   use of air monitoring equipment, 349–357
Air purifying respirator (APR), 210–222
   advantages of, 215
   full-face respirators, 218
   half-face respirators, 215–217
   maintenance, storage and recordkeeping, 230–231
   respirator selection, 218–222
   types of cartridges, 212
Air quality, assessing, 375
Air supplying respirators, 222–230
   advantages of, 223–224
   air line respirators, 229–230
   limiting factors, 224
   maintenance, storage, and recordkeeping, 230–231
   self-contained breathing apparatus (SCBA), 224–229
Alkali, 128
Alpha particles, 118–119
American Conference of Governmental Industrial Hygienists (ACGIH), 161
Ammonia, 77, 152, 181
ANFO (ammonium nitrate and fuel oil), 69
Anhydrous substance, 128, 154
Anion, 120
Antagonistic interactions, 147
Aqueous, 155
Arsenic, 110
Arsine, 76
Asphyxiate, 157, 365
Ataxia, 368
Atmospheric toxins in confined space, 366–367
Atom, 37
   signature, 38
   stable, 37
Atomic mass units (AMUs), 45
Atomic number, 48–50

Atomic symbols, 47–48, 49
Atomic weight, 43, 45–47
Atomizing a liquid, 206

Bases, 128–129
Beta particles, 119
Biohazards, 160
Black powder, 68
Blanking pipes, 376
Blasting agents, 68–69
Bleeding hydraulic or pneumatic lines, 376
BLEVE, 79–81
Blowers, 374–375
Boiling point, 79
Boyle's law, 57–58, 346–347
Breakthrough, 212
Breakthrough time, 256
Breathing zone, 375
Brisance, 66
Buddy system, 282

Carbon, 38, 45
  atomic configuration, 38
Carbon dioxide (CO_2), 157
Carbon monoxide (CO), 41, 108, 367–368
Carboxyhemoglobin, 368
Carcinogens, 112, 142, 163–164, 391–394
  carcinogenic substances, 113
  health hazards of, 114
  safe handling, use, and storage, 114–115
Cation, 120
Caustics, 128
Central nervous system (CNS), 159
CERCLA (Comprehensive Environmental Response, Compensation, and Liability Act), 3
Charles's law, 58–59, 346–347
Chemical change/reaction, 54
Chemical compatibility charts, 260–261
Chemical degradation, 278

Chemical Emergency Transportation Center (CHEMTREC), 189, 329
Chemical exposure, effects of, 158–160
  chemical effects, 159
  modes of action, 158–159
  time frames, 159–160
Chemical interactions, 145–147
  additive, 146
  antagonistic, 147
  potentiation, 146–147
  synergistic, 146
Chemical pneumonitis, 92, 115
Chemical protective clothing, 241–242, 255–260
  health considerations and, 261–265
  heat cramps, 262–263
  heat exhaustion, 263–264
  heat stroke, 261, 264–265
Chemistry, 35–61
  basic gas laws, 55–59
  defined, 37
  compounds, 51–53
  exothermic and endothermic reactions, 54–55
  periodic table, 43–50
  physical and chemical change, 53–54
CHEMTREC (Chemical Emergency Transportation Center), 189, 329
Chlorine, 38, 77, 181
  atomic configuration, 38
Chronic exposure, 158–159
Clean Air Act, 330–331
Clean Water Act, 330
Closed head polyethylene drum, 199, 201
Closed head stainless steel drum, 201
Closed head steel drum, 197–198
Code of Federal Regulations, 2
Codified regulation, 364
Cold zone, 312
Colorimetric tubes, 354–355
Combustible atmospheres in confined spaces, 368–369

Combustible gas indicators (CGI), 350–351
Combustible liquids, 88–89
Command post, 306
Command staff, 304–305
   liaison officer, 305
   public information officer (PIO), 304–305
   safety officer, 304
Compound, 38, 39, 51–53
   mixtures, 51–52
Comprehensive Environmental Response, Compensation and Liability Act of 1980 (CERCLA), 3
Compressed gases, 77–78
Concentration
   acidic, 126
   chemical, 145
Confined space, 364–378
   assessing air quality within, 375
   attendant's responsibilities, 370–371
   combustible atmospheres within, 368–369
   contractors entering PRCS, 377–378
   defined, 364
   emergency procedures, 371–372
   entrant's responsibilities, 370
   entry permit compliance checklist, 376
   entry supervisor, 371
   equipment used in, 373
   harnesses and tripods, 374
   hazards of, 365–366, 369–370
   initiating an evacuation from, 373
   isolation of, 375–376
   making entry into PRCS, 377
   permit required confined space (PRCS), 364–365
   rescue and emergency personnel, 372
   sources of atmospheric toxins in, 366–368
      carbon monoxide, 367–368
      hydrogen sulfide, 367
   toxic atmospheres in, 366
   using atmospheric monitors in, 373–374
   ventilation equipment, 374–375
Containers, handling, 21, 22
Control of Hazardous Energy (lockout/tag-out), 378–381
Control point, 284
Corrosives, 122–134
   acidic concentration, 126–128
   acidic strength, 125–126
   acids, 124–125
   bases, 128–129
   corrosive substances, 131
   defined, 124
   emergency procedures, 133–134
   health hazards of, 130–132
   neutralization, 129–130
   safe handling, use, and storage, 132–133
Cryogenics, 81–83

Decon corridor, 274–276, 278–290
Decontamination, 21, 22, 273–292
   defined, 274
   emergency decontamination procedures, 276
   establishing control zones, 274–276
   methods of, 276–278
   post incident analysis, 287
   procedures, 283–290
   site selection and management, 278–283
      managing decon area, 279–280
      planning extent of decontamination, 280–283
   six-step Level B decontamination, 287–290
Decontamination leader, 315
Decon team leader, 278, 279, 280
Deflagration, 68
Degradation, 238, 258–259
Delayed effects, 160
Department of Transportation (DOT), 66, 172, 329
   hazardous materials definition, 173
   identification system, 173–180
   labels and placards, 176–180

Dermis, 153
Detection devices, 356–357
Detonate, 67
Detonation velocity, 71
Diffusion, 74
Diurnal effects, 359
DOT (Department of Transportation), 66, 172–180, 329
Drums, 197–202
   handling, 21, 22
Ductile, 49
Dynamite, 67

Electrocution, 369
Electrolytes, 262
Electromagnetic radiation, 117
Electron, 37
   cloud, 37
Element, 38–39
Emergency action plan (EAP), 326
Emergency response
   determining response levels, 31–32
   to hazardous substance releases, 24–31
   at uncontrolled hazardous waste sites, 23
Emergency response plan, 333–334
Encapsulating suit, 241, 250
Endothermic reaction, 54
Engineering controls, 20
Engulfment, 369
Entry leader, 314–315
Environmental Protection Agency (EPA), 3, 176
EPA (Environmental Protection Agency), 3, 176
Evaporation, 207
Event review, 318–319
Exothermic reaction, 55
Explosives, 65–73
   additional explosive situations, 69–72
   blast effects, 66
   blasting agents, 68–69
   categories, 66–67

   defined, 66
   emergency procedures, 73
   health hazards of, 72–73
   high explosives, 67–68
   low explosives, 68
   safe handling, use, and storage, 73
Exposure limits, 161–163
   PEL, 162–163
   STEL, 162
   TLV/C, 162
   TLV-TWA, 162
Exposure mechanisms, 144–158
   exposure factors, 144–147
      chemical interactions, 145–147
      concentration of chemical, 145
      duration of exposure, 145
      uptake rate, 145
   route of entry, 148–158
      absorption, 152–155
      ingestion, 155, 156
      inhalation, 149–152
      injection, 155, 157
      protection for, 157–158
   toxicity factors, 144
   You factors, 147–148
      metabolism, 147–148
      retention factors, 147–148

FGAN (fertilizer grade ammonium nitrate), 69
Finance section, 304
Firedamp, 344
First Responder Awareness level (FRA), 25–26
First Responder Operational level (FRO), 26–27
Fission, 119
Flame ionization detectors (FID), 352–353
Flammable liquids, 68, 87–95
   defined, 88
   emergency procedures, 94–95
   flash point and flammability, 88–91
   health hazards of, 93

Index

safe handling, use, and storage, 93–94
solvents, 91–93
Flammable range, 71–72, 368
   lower explosive limit (LEL), 71
   upper explosive limit (UEL), 71
Flammable solids, 96–99
   defined, 96–97
   emergency procedures, 99
   examples, 97
   health hazards of, 97–98
   safe handling, use, and storage, 98–99
Flash point, 88–91
Flash suit, 251, 253, 254
Freelancing, 295
Full-face respirators, 218

Gamma radiation, 119
Gas chromatographs (GCs), 354
Gases, 40–43, 73–87, 344
   basic laws, 55–59, 345–347
   compressed, 77–78
   cryogenics, 81–83
   defined, 74
   detection of, 347–349
   emergency procedures, 87
   flammable, 76
   general hazards of, 75
   health hazards of, 83–86
   liquefied, 78–81
   poisonous, 76
   safe handling, use, and storage, 86–87
General site worker, 17–18
General staff, 303

Half-face respirators, 215–217
Half-life, 118
Harnesses, 374
Hazard classes, 63–139
   corrosives, 122–134
   explosives, 65–73

flammable liquids, 87–95
flammable solids, 96–99
gases, 73–87
oxidizers and organic peroxides, 99–105
poisons, pesticides, and carcinogens, 105–116
radioactives, 116–122
Hazard Communication Standard, OSHA, 5–12, 72, 327
Hazardous materials group supervisor, 313–314
Hazardous Materials Scene Commander, 30–31
Hazardous Materials Specialist level, 29–30
Hazardous Materials Technician level, 28–29
Hazardous Materials Transportation Act of 1974, 173
Hazardous waste manifest, 186, 188–190
Hazardous Waste Operations and Emergency Response (HAZWOPER), 1, 2, 5, 12, 294, 364
   planning requirements, 324, 325–326
   regulations, 12–32
Hazard wastes, 176
HAZWOPER (Hazardous Waste Operations and Emergency Response), 2, 5, 12, 294, 364
   planning requirements, 324, 325–326
   regulations, 12–32
Heat cramps, 262–263
Heat exhaustion, 263–264
Heat stroke, 261, 264–265
Hemoglobin, 368
Heroic doses, 113
Heterogeneous mixture, 52
High explosives, 67–68
Highly toxic materials, 107
High-temperature protective clothing, 240–241
HMIS (hazardous materials identification system), 185–186
Homogeneous mixture, 51–52
Hot zone, 274, 312
   access to, 312–313
Hydrofluoric acid (HF), 110

Hydrogen sulfide (H_2S), 367
Hygroscopic, 69
Hyperkalemia, 110, 264–265
Hyperthermia, 261
Hypocalcemia, 110
Hypotension, 264
Hypothalamus, 264

Identification systems, 171–204
   DOT *North American Emergency Response Guidebook*, 190–197
   DOT system, 173–180
   drum profiles, 197–202
   HMIS, 185–186
   NFPA 704 system, 180–185
   penalties, 190
   shipping papers and hazardous waste manifests, 186–190
IDLH (Immediately Dangerous to Life or Health), 163, 211, 213, 221, 366
Immediate effects, 160
Incident commander (IC), 294–296
   conducting event review, 318–319
   dealing with small incidents, 306–308
   mistakes made by, 317–318
   responsibilities of, 296–300, 303, 312, 313
Incident command system (ICS), 24, 294–319
   characteristics of ICS, 301–308
      command staff, 304–305
      functional sections, 303–304
      managing small incidents, 306–308
      modular structure, 301
      span of control, 302
      standardized terminology, 302
      unity of command, 301–302
   and decontamination, 279
   establishing safe working areas/zones, 310, 312–313
      hot zone, 312–313
   incident commander, 294–300

incident termination, 317–319
   event review, 318–319
   mistakes made by incident commanders, 317–318
initial steps in incident response, 308–310, 311
   sample command procedures, 308–310, 311
job titles and descriptions, 313–317
Informational program, 21
Ingestion, 155, 156
Inhalation, 149–152
Injection, 155, 157
International Maritime Dangerous Goods (IMDG), 178
Ion, 120
Ionization, 125
Ionization potential (IP), 352
Ionizing radiation, 119–120
Irritants, 107
Isolation of the problem, 308
Isopropyl alcohol (IPA), 91
Isotonic solution, 263

Kinetic energy, 379

Labels, 176–180
Lab packing, 198
Lactic acid, 264
Land breezes, 359
LChi, 164, 165
LClo, 164, 165
LC_{50}, 164
LDhi, 164, 165
LDlo, 164, 165
LD_{50}, 164
Liaison officer, 305
Lighting, 23
Liquefied gases, 78–81
Liquids, 40–42
Liquid Splash Suits for Hazardous Chemical Emergencies (NFPA 1992), 236, 245

Local effects, 159
Lockout/tag-out, 379–381
Logistics section, 304
Long-term effects, 160
Lower explosive limit (LEL), 71, 72
Low explosives, 68

Magnesium, 96, 97, 99
Malleable, 49
Material safety data sheet (MSDS), 6, 7–11, 36, 64, 72, 73, 84, 85, 87, 93, 102, 103, 115, 125, 181, 190, 239, 277
Matter
   combined, 38–39
   defined, 37
   states of, 40–43
Medical examinations, 20
Medical group leader, 314
Mendeleev, Dimitri, 43
Metabolic acidosis, 264
Metabolism, 147–148
Methyl ethyl ketone, 102, 181
Meyer, Lothar, 43
Mine Safety and Health Administration (MSHA), 220
Mitigation, 40
Mixtures, 51–52
   heterogeneous, 52
   homogeneous, 51–52
Monitoring programs, 21.
   See also Air monitoring
MSDS (material safety data sheet), 6, 7–11, 36, 64, 72, 73, 84, 85, 87, 93, 102, 103, 115, 125, 181, 190, 239, 277
MSHA (Mine Safety and Health Administration), 220
Mutual aid, 325

National Fire Protection Association (NFPA), 172
   identification system, 180–185

National Institute for Occupational Safety and Health (NIOSH), 161, 220
Neutralization, 129–130
Neutron, 37, 38
New technology programs, 23
NFPA (National Fire Protection Association), 172
NFPA 704 identification system, 180–185
   disadvantages of, 181, 183
   flammability, 184
   health hazards, 183
   reactivity hazards, 184–185
NIOSH (National Institute for Occupational Safety and Health), 161, 220
Noise hazards, 378
Nonentry rescue, 373
North American Emergency Response Guidebook (DOT), 176, 178, 190–197
   procedures for using, 195–197
North Chemical Resistance Guide, 261
Nucleus, 37, 38

Occasional site worker, 18
Occupational Safety and Health Act of 1970, 3, 5, 6
Occupational Safety and Health Administration (OSHA), 2, 3–5, 6, 12, 14, 161, 294, 325, 364
   respiratory protection regulations, 208–209
Odor threshold, 152
Olfactory glands, 367
Open head fiber drum, 201
Open head steel drum, 198, 200
Operational goals, 302
Operations section, 303
Organic peroxides, 102
Organophosphate pesticides, 108–110
Orthostatic hypotension, 263
OSHA (Occupational Safety and Health Administration), 2, 3–5, 6, 12, 14, 161, 294, 325, 364
   respiratory protection regulations, 208–209

Overbooties, 244
Overpressure, 66
Oxidation, 99–100
Oxidizers, 49, 99–105
   emergency procedures, 103–105
   health hazards of, 103
   organic peroxides, 102
   safe handling and storage of peroxides, 103
Oxygen, low levels of, 207, 365–366
   absorption, 366
   bacterial action, 366
   combustion, 366
   displacement, 365
   oxidation, 366
   painting, 366
Oxygen meters, 349–350

Pathogen, 157
PCB (polychlorinated biphenyl), 110
PEL (permissible exposure limits), 17, 18, 162–163, 213, 366
Penetration, of chemical, 258
Periodic table, 39, 43–50
   atomic number, 48–50
   atomic symbols, 47–48, 49
   atomic weight, 43, 45–47
Permanent gases, 78
Permeation, 255–258
Permissible exposure limits (PEL), 17, 18, 162–163, 213, 366
Permit required confined space (PRCS), 364–365
Personal protective equipment (PPE), 16, 20, 235–272
   chemical compatibility charts, 260–261
   conditions for use, 253–255
     level A, 253–254
     level B, 254
     level C, 254–255
     level D, 255
   flash fire protection, 251, 253, 254
   identifying hazardous product, 237–239
   selecting appropriate protection, 239–240
   type and level of protection, 240–251
     chemical protective clothing, 241–242, 255–260, 261–265
     high-temperature protective clothing, 240–241
     level A, 245, 250–251
     level B, 244–245, 247–249
     level C, 242–244
     level D, 242
     structural firefighting clothing, 240
Pesticides, 105
pH, 125
Photoionization detectors (PIDs), 352
Photon, 119
Photophobia, 367
Physical change, 53–54
Physical hazards, 363–386
   confined space, 364–378
   control of hazardous energy (lockout/tag-out), 378–381
   noise hazards, 378
Placards, 176–180
Planning, 323–342
   emergency plans, 325–327
     OSHA emergency action plans, 326
     OSHA fire prevention, 326–327
   emergency response plan, 333–334
   hazard communication, 327
   for hazardous substances release emergencies, 324–325
     coordinating planning efforts, 325
   process safety management, 327–332
   site safety plan, 331–332, 334–338
Planning section, 303
Pocket Guide to Chemical Hazards (NIOSH), 239
Pocket Guide to Hazardous Materials (NIOSH), 352

Index

Poisons, 105–116
  defined, 105
  emergency procedures, 115–116
  health hazards of, 108–112
  safe handling, use, and storage, 114–115
  toxic chemicals, 111
  toxic gases, 112
Polar substance, 124
Polyethylene, 199
Positive pressure, 224
Potential energy, 379
Potentiation interactions, 146–147
PPE (personal protective equipment), 16, 20
PRCS (permit required confined space), 364–365
Pressure, 56
  and gases, 56–59
Process safety management (PSM), 327–332
Propane, 70, 72, 78–81
Proton, 37–38
Proximity suits, 240
Public information officer (PIO), 304–305
Pulmonary edema, 160
Pyrophoric substances, 96

Radiation, 116–122, 123
  alpha, beta, and gamma radiation, 117–120
  emergency procedures, 121
  sources of, 120–121
Radiation Safety Officer, 121
Radiation sickness, 120
Radioactive isotopes, 117
Radioactives, 116–122
  alpha, beta, and gamma radiation, 117–120
  emergency procedures, 121
  radiation, 116–117
  sources of radiation, 120–121
  time-distance-shielding, 122, 123
RCRA (Resource Conservation and Recovery Act), 2, 3, 13, 24, 176, 329
Recommended Practice for Responding to Hazardous Materials Incidents (NFPA 471), 241–242
Refresher training, emergency responders, 31
Regulations, 2–3
  HAZWOPER, 12–32
Regulatory driver, 2
Relative humidity, 359
Release reporting and response, 329–330
Reportable quantity (RQ), 176
Resource Conservation and Recovery Act (RCRA), 2, 3, 13, 24, 176, 329
Respiratory protection, 205–234
  equipment, 210–215
    air purifying systems, 210–222
    air supplying respirators, 222–230
    maintenance, storage, and recordkeeping of equipment, 230–231
    protection fundamentals, 208–209
    respiratory hazards, 206–207
Retention factors, 147–148
Right to know, 6
Route of entry, 148–158

Safety and health program, 15–16
Safety officer, 304
Sanitation for temporary workplace, 23
Saponification, 128
SARA (Superfund Amendments and Reauthorization Act), 3, 5, 328
SCBA. *See also* Self-contained breathing apparatus
Self-contained breathing apparatus (SCBA), 77, 219, 221, 222, 223, 224–229, 282, 283
  advantages of, 226
  closed circuit type, 225
  disadvantages of, 226
  doffing procedures, 229
  donning procedures, 226–228
  maintenance, storage, and recordkeeping, 230–231
  open circuit type, 225

Sensitizers, 107
Shipping papers, 186–190
Shore breezes, 359
Short-Term Exposure Limit (STEL), 162
Site
    characterization and analysis, 16
    control, 16
    health and safety plan, 18
    Superfund, 3, 12
Site access control officer, 313, 315–316
Site safety plan, 331–332, 334–338
Site security officer, 313
Size-up, 308
Skasol, 201
Sodium chloride, 39
Solids, 40–42
Solvents, 91–93
Span of control, 294, 302
Specific gravity, 238
Spill Prevention Control and Countermeasure Plan (SPCC), 330
Stable electron, 37
Stage/staging, 308
Standard temperature and pressure (STP), 345, 346
State plan states, 4–5
States of matter, 40–43
Strategy, 296, 297
Stratification of gases, 369
Strong force, 38
Structural firefighting clothing, 240
Superfund Amendments and Reauthorization Act of 1986 (SARA), 3, 5, 328
Superfund site, 3, 12
Supervisor training, 18–19
Support Function Protective Garments for Hazardous Chemical Operations (NFPA 1993), 236, 243
Support zone, 242
Symbols, DOT, 179

Syncope, 368
Synergistic chemical interactions, 146
Systemic effects, 159

Tactics, 296, 297
Technical specialist, 316–317
Teratogen, 164
Threshold Limit Value—Ceiling (TLV/C), 162
Threshold Limit Values and Biological Exposure Indices, 162
Threshold Limit Value-Time Weighted Average (TLV-TWA), 162
Time-distance-shielding and radiation exposure, 122, 123
TNT (trinitrotoluene), 67–68
Tool drop, 287
Topography, 359
Toxic atmosphere in confined space, 366
Toxic materials, OSHA categories, 2
    highly toxic, 107
    toxic, 107
    toxic chemicals, 111
    toxic gases, 112
Toxicology, 141–169
    branches of, 143–144
    defined, 143
    effects of chemical exposure, 158–160
    exposure limits, 161–163
    exposure mechanisms, 144–158
    terminology, 163–165
Training requirements, 16–20
    general site worker, 17–18
    occasional site worker, 18
    supervisor, 18–19
Transportation of hazardous materials and wastes, 329
Treatment, storage, and disposal facilities (TSDFs), 24
Trinitrotoluene (TNT), 67–68

TSDF (treatment, storage, and disposal facility), 24
Tyvek, 261, 276

Umbilical type air system, 222, 223, 229–230
Uncontrolled energy sources, 378–381
Unity of command, 301–302
Upper explosive limit (UEL), 71, 72
Uptake rate, 145
U.S. Coast Guard National Response Center (NRC), 176

Vapor density, 45, 46, 47, 85, 86, 221
Vapor pressure, 49, 90–91

Vapor Protective Suits for Hazardous Chemical Emergencies (NFPA 1991), 236, 245, 250
Vascular, 155
Ventilators, 374–375
Viscosity, 42
Vital signs, 261

Warm zone, 274, 312
Warning property, 212–213
Water, 39, 91
    atomic formation, 39
Work practices, 20

Zeroing, 350